MATERIALS SCIENCE AND TECHNOLOGIES

RECENT ADVANCES IN CERAMIC MATERIALS RESEARCH

MATERIALS SCIENCE AND TECHNOLOGIES

Additional books in this series can be found on Nova's website
under the Series tab.

Additional E-books in this series can be found on Nova's website
under the E-book tab.

MATERIALS SCIENCE AND TECHNOLOGIES

RECENT ADVANCES IN CERAMIC MATERIALS RESEARCH

JOAN JOSEP ROA ROVIRA
AND
MERCE SEGARRA RUBI
EDITORS

nova publishers
New York

For permission to use material from this book please contact us:
Telephone 631-231-7269; Fax 631-231-8175
Web Site: http://www.novapublishers.com

Library of Congress Cataloging-in-Publication Data

ISBN: 978-1-62417-729-3

Published by Nova Science Publishers, Inc. † New York

CONTENTS

Preface		**vii**
Chapter 1	Hybrid Organoceramics Deriving from Dendritic Polymers, Methods of Preparation, Optimization Techniques and Prospected Applications *M. Arkas*	**1**
Chapter 2	Modified-(K,Na)NbO_3 Piezoelectric Ceramics for 'Hard' Lead-free Materials *J. B. Lim*	**31**
Chapter 3	How to Assess the Ceramic Suitability of Raw Clays? *M. Hajjaji and A. Khalfaoui*	**51**
Chapter 4	Aging of Dental Zirconia *C. A. M. Volpato, R. F. Kikko, D. K. Vasconcellos, I. M. Zani and M. C. Fredel*	**71**
Chapter 5	Peculiarities of the Structure and Phonon Spectrum in Nano- and Microstructure Ceramics *E. N. Khazanov and A. V. Taranov*	**87**
Chapter 6	Optical Ceramics for Fast Scintillator Materials *M. Nikl, T. Yanagida, H. Yagi, T. Yanagitani, E. Mihokova and A. Yoshikawa*	**127**
Chapter 7	Microstructure Evolution and Mechanical Properties of Silicon Nitride Based Ceramics Sintered by Spark Plasma Sintering (SPS) *Jow-Lay Huang and Pramoda K. Nayak*	**177**
Chapter 8	Friction, Hardness and Elastic Modulus Determined by AFM-FS and Nanoindentation Techniques for Advanced Ceramics Materials *G. Oncins, J. J. Roa, E. Rayón, J. Díaz, M. Morales, M. Segarra and F. Sanz*	**215**

Chapter 9 Rapid Fabrication of Bulk Ceramics and Glasses by High-Gravity Combustion Synthesis **251**
Guanghua Liu, Jiangtao Li and Kexin Chen

Index **293**

PREFACE

This book presents current areas of research in the field of Recent Advances in Ceramic Materials Research, which is a "catch-all" description of the sintering, texture and characterization of advanced ceramic materials that have applications in the real world.

In Chapter 1, hybrid devices have been developed by the integration of dendritic polymers with ceramics. This has been achieved by impregnating the ceramics with appropriate organic solutions, resulting in covalent attachment of organic compounds on the surface of ceramics or their non-covalent absorption inside their pores. Hybrid organoceramics have also been obtained through sol-gel synthesis of hybrid materials or through biomimetic processes, in which dendritic molecules of specific shape and properties are acting as templates, and induce the formation of hybrid nanospheres. Dendritic polymers can also grow on the surface of ceramic membranes. In this review, optimization techniques will also be described as well as applications of devices coupled with future prospects.

As explained in Chapter 2, lead-free piezoelectric ceramics, $(K_{1-X}Na_X)NbO_3$ modified by $K_4CuNb_8O_{23}$ [KNN-KCN$_x$] with 0.0 mol% $\leq x \leq$ 1.0 mol%, were synthesized using conventional ceramic processing. KNN-KCN$_x$ were observed in the perovskite structure with orthorhombic symmetry and revealed that the addition of KCN was effective in enhancing the density and decreasing phase transition temperatures ($T_{o\text{-}t}$, T_c). The asymmetrical polarization hysteresis loops exhibition should be related to the development of an internal bias field (E_i), resulting in high mechanical quality factor (Q_m) of >1500, with the dielectric loss of 0.6%. The formation of E_i is believed to be the result of defect dipoles of acceptor ions and oxygen vacancies, leading to piezoelectric "hardening" effect, by stabilizing and pinning of the domain wall motion. Furthermore, the piezoelectric properties were found to decrease, and the coercive field (E_c) increased when compared with the undoped material, exhibiting typical characteristics of "hard" behavior. The temperature usage range was limited by the polymorphic phase transition temperature. The full set of material constants was determined for the KNN-KCN$_x$ materials. Compared with "hard" PZT ceramics, the KNN-KCN$_x$ presents lower dielectric and piezoelectric properties; however, comparable values of Q_m, dielectric loss, and E_c were obtained. To further compositionally optimize KNN-KCN$_x$ near MPB, the ratio of K/Na in $(K_{1-x}N_x)$N-0.5KCN was varied from $0.45 \leq x \leq 0.60$. Results revealed that increase in Na was effective in changing the microstructure and relative density of $(K_{1-x}N_x)$N-0.5KCN. Further, varying the Na resulted in a slight shift of the phase transition temperatures ($T_{o\text{-}t}$ and T_c) toward low values. A high Q_m of 1850 was found at x = 0.54, which might be due to the build-up of E_i within $(K_{1-x}N_x)$N-0.5KCN. Thermal hysteresis was confirmed with Na content, resulting from structural changes. Thus, $(K_{1-x}N_x)$N-0.5KCN with x = 0.54

exhibited excellent piezoelectric properties with d_{33} (97 pC/N), k_p (36%), and Q_m (1850). Therefore, acceptor modified-KNN lead-free materials are promising candidates for application in high-power piezoelectric devices.

For investigating its shear mode property, optimized KNN-0.5KCN was characterized as a function of temperature. Shear-mode KNN-0.5KCN was found to possess comparable electric resistivity, piezoelectric and electromechanical properties to pure KNN. The dielectric permittivity K^T_{11} and piezoelectric coefficients d_{15} were found to increase against temperature, while the electromechanical coupling factors k_{15} maintained similar values till phase transition temperature ($T_{o-t} \sim 200^oC$), above which the properties dropped significantly due to depolarization. Good temperature-dependent properties, together with its high-Q_m, are demonstrating practical potential for high power shear mode applications.

Furthermore, clays are still considered as the main feeding materials for conventional ceramic products (bricks, tiles, faience, etc). As a matter of fact, the suitability of these materials depends on their inherent characteristics (chemical composition, mineralogical constitution, grain size and plasticity) as well as on extrinsic factors such as firing temperature, soaking time, heating rate and atmosphere, and shaping methods. Given the numerous influencing factors, the ceramic suitability of clays is commonly approached by trial. In general, changes of the microstructure of ceramic samples as a function of operating factors, e.g. firing cycle parameters, composition of ceramic bodies are investigated and the technical properties are evaluated and compared to standard data. Besides these investigations, some empirical diagrams involving grain size distribution and chemical or mineralogical composition are used. However, neither the firing conditions nor the technological magnitudes of the ceramic products can be predicted by the use of such diagrams. The ceramic suitability of a clay could be evaluated by adopting a new approach based on the response surface methodology (RSM). In Chapter 3, the RSM results are compared to those obtained by the use of empirical diagrams, and discussed in relation to the clay firing transformations and some important properties (strength, water absorption and firing shrinkage).

In Chapter 4, research focused on the microstructure and processing of ceramic materials has considerably increased in dentistry due to biological, mechanical and aesthetic demands. Currently, zirconia can be considered one of the most promising materials for biomedical applications because it combines excellent mechanical properties with biocompatibility. In dentistry, this ceramic has been widely used to make prosthesis substructures and implant abutments. To be applied for structural purposes, it must be stabilized in the tetragonal phase at room temperature. However, when the stabilized zirconia contacts with water or steam vapor, it may undergo a considerable degradation of its mechanical properties. The presence of moisture interferes with its metastability, creating an unfavorable condition for the longevity of the material, characterized by macro- and microcracks, or even fractures. This phenomenon is known as “low temperature degradation” or “aging”. The present study evaluates the degree of fit between ceramic abutment/implant and its resistance to fracture before and after thermal aging. The results presented no statistical difference in the degree of fit abutment/implant; however, numerically, the best results were found in the control group. The results were statistically significant when evaluating resistance, suggesting that aging decreased the fracture resistance of zirconia abutments.

The experimental results on transport characteristics of diffusion propagation of slightly non-equilibrium thermal phonons (NP) in a number of dense dielectric nano- and microstructured ceramics and composites on their base at helium temperatures are presented in Chapter 5. The connection between the diffusion coefficient of NP and grain boundaries (GB), grain size and structure, process-dependent parameters of synthesis was analyzed. The model of GB as a finite-depth flat layer and the NP transport in the polycrystalline structure was suggested under the assumption $qR >> 1$ (where, q is the phonon wave vector and R is the mean grain size). Under these conditions, the temperature dependence of the diffusion coefficient $D(T)$ is determined by spectral properties of GB. It allowed us to estimate the mean values of thickness and acoustic impedance of GB. Considering the thermophysical properties optimization, the restrictions on grain size due to the phonon scattering on grains (the case of $qR \sim 1$) leading to the arising of the gap in the phonon spectrum were investigated. This type of gap has been revealed in the composites based on polycrystalline yttrium-stabilized zirconium (YSZ) matrix with nanocrystalline fraction of corundum (r = 20-30nm). The gap characteristics are controlled by both commensurability of the phonon wavelength with fragment size and GB structure. The presence of inclusions or secondary phases and defects with characteristic sizes smaller than grains of the main ceramic material shifts the top gap edge to higher frequencies. The inclusion of nano-sized fragments of α-Fe (<5%) to the polycrystalline corundum matrix, substantially (over two orders of magnitude) suppress the NP diffusion. In this case, the phonon kinetic is determined by the size of metal fragments as well as by the surrounding spinel layer. It was shown that transport properties of NP in the $Ce_{1-x}Gd_xO_{2-y}$ ceramic solid solutions may depend on the amount of impurities in the grain as well as on the formation energy of defect (vacancy) in the anion sub-lattice. The value of relaxation energy can be experimentally determined when it is comparable to the energy of NP. It was shown that the better quality of GB and grain structure (and hence the better acoustical, optical and thermophysical properties) of the optically transparent ceramics, on the base of cubic oxides and fluorides, can be achieved when the material structure is formed by the twinning process as the main mechanism of plastic deformation. The peculiarities of phonon transport in the ferroelectric ceramics having both sharp and diffused phase transition, corresponding to the crystal and glassy state, have been investigated. The impact of domain structure to the kinetics of non-equilibrium phonons in $BaTiO_3$ single crystal was examined. The value of reflection coefficient of sub-terahertz NP on domain boundaries was estimated.

Optical ceramics (OC) are transparent or translucent materials constituted by tight aggregating crystallite micro- or nanograins, each randomly oriented with respect to their neighbours. They have been under development as an alternative to single crystal materials to provide bulk optical elements in the case that single crystals cannot form or when ceramic materials show superior optical and luminescence properties. Development of OC for the scintillator applications was triggered by the needs of Computer Tomography (CT) medical imaging in the 90's of last century. The manufacturing and characterization of $(Y,Gd)_2O_3$:Eu, Gd_2O_2S:Pr,Ce,F and $Gd_3Ga_5O_{12}$:Cr,Ce, and later on also the Eu^{3+} or Tb^{3+}-doped Lu_2O_3 slow ceramic scintillators were reported. More recently, fast optical ceramics with scintillation response in the tens-hundreds ns time scale, and based on the Ce^{3+}-doped $Y_3Al_5O_{12}$ (YAG), was reported in late 1990's and its higher density analog based on LuAG host doped with Ce^{3+} or Pr^{3+} was reported only several years ago for the first time. In Chapter 6, the latest studies in the field of fast scintillation optical ceramics in the last decade will be reviewed, focusing on the interconnection among the manufacturing technology, scintillation performance, and defect occurrence and nature in the discussed material systems. The most studied group of materials are the cubic structure aluminum or multicomponent garnets, but interesting results have been achieved also in sesquioxide, silicate, hafnate, complex perovskite or rare earth halide compounds. Both basic physics and application aspects will be considered in the description of the above mentioned systems.

Chapter 7 presents silicon nitride as one of the major structural ceramics that possess excellent mechanical properties such as high flexural strength, good facture resistance and high hardness. In general, microstructural changes of Si_3N_4 in the matrix greatly influence the mechanical properties. In order to control the final microstructure and properties, the powder synthesis and sintering process play important roles in fabricating Si_3N_4-based ceramics. Recently developed Spark Plasma Sintering (SPS) is regarded as an energy saving technology due to short process time and fewer processing steps. The powder preparation and densification of this nano-powder using SPS technique have been reviewed. In the present study, a commercially available nanosized β-Si_3N_4 doped with sintering additives (6 wt% Al_2O_3 and 8 wt% Y_2O_3) has been used as initial power and consolidated by an SPS technique at different heating rates (50, 100 and 200 °C/min) by varying temperature from (1550 °C to 1700 °C). At slower heating rates (50 and 100 °C/min), the nanosized grains are maintained after sintering at 1600° C for 5 min, while anisotropic grain growth is accelerated above 1500 °C by applying a rapid heating cycle (200 °C/min). In addition to the dynamic Ostwald ripening that occurs during the sintering process, the presence of Morie′ fringes and dislocations is attributed to grain rotation and misfit strain between the subgrains and elongated, large grains. Grain coalescence is found to occur in the rapid heating process. The microstructures of the developed Si_3N_4 based ceramics were achieved, and the grain length, grain width, and aspect ratio for these sintering bulks were found to increase gradually when elevating sintering temperature. The fracture toughness for these β-Si_3N_4 based nanoceramics also has been evaluated for specimens sintered at different sintering temperatures. Furthermore, Si_3N_4 based nanocomposites incorporating electroconductive nanoparticles are regarded as promising materials for cutting tools and making electrical discharge machining possible. In this regards, conductive TiC nanopowder has been incorporated into nanosized β-Si_3N_4-based powder and consolidated by a SPS technique with a rapid heating process. The influence of this conductive phase on the microstructure development of the Si_3N_4 matrix has been demonstrated. Furthermore, this chapter will be interesting for scientists and engineers concerned with the novel processing of ceramics using SPS and use of these ceramics for engineering applications.

In the last years, nanoindentation by means of Atomic Force Microscopy-Force Spectroscopy (AFM-FS) or Nanoindenters has become a powerful tool to study the nanomechanics of all type of materials at micro-, nano- and also picometric scale, from soft metals, like copper, to brittle materials, as ceramics. The experimental basis of these techniques is the evaluation of the response of a material to an applied vertical load (in order to obtain the hardness, H, and the elastic modulus, E) or to a shear force (Lateral Force Microscopy, LFM) so as to obtain the friction coefficient, μ. In Chapter 8, the different methods to analyze friction, hardness and elastic modulus by means of AFM are explained for several examples as diamond single crystals, SiC single crystals, titanium dioxide coatings, mica and silicon oxide. Besides, examples of elastic deformation using nanoindentation are included. Moreover, several examples of results obtained by means of a Nanoindenter are also described in detail. A particular emphasis on ceramic coatings and advanced ceramic materials, such as YBaCuO superconductors, yttria-stabilized zirconia, doped ceria electrolytes for fuel cells and yttria stabilized polycrystalline tetragonal zirconia, are reported and commented. The interest of these techniques is evidenced by the increasing quantity of nanomechanics-related papers published in the last decades, near a thousand of which appeared during the last five years. Unfortunately, a lot of practical information about nanomechanics of hard materials is still scarce in the literature (only one percent of the above mentioned publications are related to ceramic materials). This chapter aims to present the basic principles and methods applied to extract the different mechanical properties and also to review and comment real examples related to the cited techniques.

Chapter 9 presents a rapid method to fabricate bulk ceramics and glasses by high-gravity combustion synthesis. This chapter briefly reviews the recent progress on bulk ceramic and glass materials by a new method called high-gravity combustion synthesis. By this new method, both the processing time and energy consumption can be reduced in comparison with conventional powder sintering. Furthermore, in this review, several ceramic and glass materials prepared by high-gravity combustion synthesis will be presented as examples at first. Then, a systematic investigation on the reaction kinetics in high-gravity combustion synthesis will be made, with a special emphasis on phase separation and microstructure evolution.

In: Recent Advances in Ceramic Materials Research
Editors: Joan J. Roa Rovira and Merce Segarra Rubi
ISBN: 978-1-62417-729-3

Chapter 1

HYBRID ORGANOCERAMICS DERIVING FROM DENDRITIC POLYMERS, METHODS OF PREPARATION, OPTIMIZATION TECHNIQUES AND PROSPECTED APPLICATIONS

*M. Arkas**
Institute of Physical Chemistry, NCSR 'Demokritos',
Attiki, Greece

ABSTRACT

Hybrid devices have been developed by the integration of dendritic polymers with ceramics. This has been achieved by impregnating the ceramics by appropriate organic solutions resulting in covalent attachment of organic compounds on the surface of ceramics or their non-covalent absorption inside their pores.

Hybrid organoceramics have also been obtained through sol-gel synthesis of hybrid materials or through biomimetic processes in which dendritic molecules of specific shape and properties are acting as templates and induce the formation of hybrid nanospheres. Dendritic polymers can also grow on the surface of ceramic membranes. In this review optimization techniques will also be described as well as applications of devices coupled with future prospects.

1. INTRODUCTION

Ceramic materials in various forms such as ceramic membranes, nanoceramics and bioceramics are widely used for a variety of purposes. Ceramic membranes are applied in separation technology, in particular for the preparation of micro-filtration (MF) or ultra-filtration (UF) membranes [1] and in gas separation [2-4].

* Corresponding Author's e-mail address: markas@chem.demokritos.gr (M. Arkas).

They are also used as supports for catalysts [5-7], for water purification [8-9] and wastewater treatment [10-11]. The basic characteristics that make ceramic membranes suitable for these applications are their high thermal stability, good mechanical properties [12], chemical inertia, resistance at corrosive environments, stability of operating characteristics, low energy consumption, long working life, compactness of the installation and possibility to effect full automation of the process. However their use in highly specialized applications is rather limited since they lack for example the ability to remove organic contaminants at ppb levels because the ceramic materials do not show selective activity.

In addition, silica, alumina, zirconia and titania, are used for the formulation of ceramic nanoparticles. Nanoceramics derived from these materials have several advantages; their preparation is simple and usually requires ambient temperature conditions. They can be adjusted to the desirable size, shape, porosity, and are extremely inert [13]. Because of their inert nature they are highly compatible with biological systems and nowadays these nanoparticles are important drug-delivery vehicles. Their ultra-low size (less than 50 nm) renders them effective in evading the uptake by the reticuloendothelial system [14]. They do not swell or change in porosity according to pH and thus can protect biomacromolecules against denaturation. They are not susceptible to microbial attack. In addition their surfaces can be easily functionalized for conjugation to target-specific ligands such as monoclonal antibodies.

Bioceramics are building blocks for the construction of biomaterials. These materials can bond directly to living tissues such as bone or tooth and can be used as replacement. For example, when biologically mineralized apatite crystals form with precise regulation, structure, crystal shape, size, composition and surface area, they can template the structure of bone tissue [15].

Dendritic polymers consist of symmetric dendrimers (i.e. polymers with the shape of a tree) and their non-symmetric hyperbranched analogues. Some examples of symmetric compounds include diaminobutanepoly(propylene imine) DAB (Figure 1a) and poly(amido amine) PAMAM dendrimers (Figure 1c). Examples of non-symmetric hyperbranched polymers are poly(ethylene imine) PEI (Figure 1b) and polyglycerol PG (Figure 1d).Their development receives very high scientific interest since this type of polymers has applications in the fields of drug delivery as controlled release systems [16], solvent extraction [17], liquid crystalline materials [18-22], dyes [23], and antioxidants [24]. As discussed above for ceramic membranes, many different applications in separation science techniques involve dendrimers, for example, size exclusion chromatography [25-30] ion exchange chromatography [31-32], high performance liquid chromatography [33-35], inverse gas chromatography [36], electrokinetic capillary chromatography [37-45], capillary [46-48] and gel [49-51] electrophoresis.

The combination of functionalized dendritic polymers, with suitably developed ceramic filters leads to novel hybrid devices that combine the encapsulation properties of the polymers and the mechanical properties of ceramics which can act as a stable substrate rendering these specific organoceramics promising candidates for a variety of applications such as water treatment [52-54], catalysts [55]and magnetoceramics [56]. The scope of this review is to give an overview on the methods currently used for the preparation of these hybrid materials; their current applications; the problems encountered together with the implemented solutions and their future prospects.

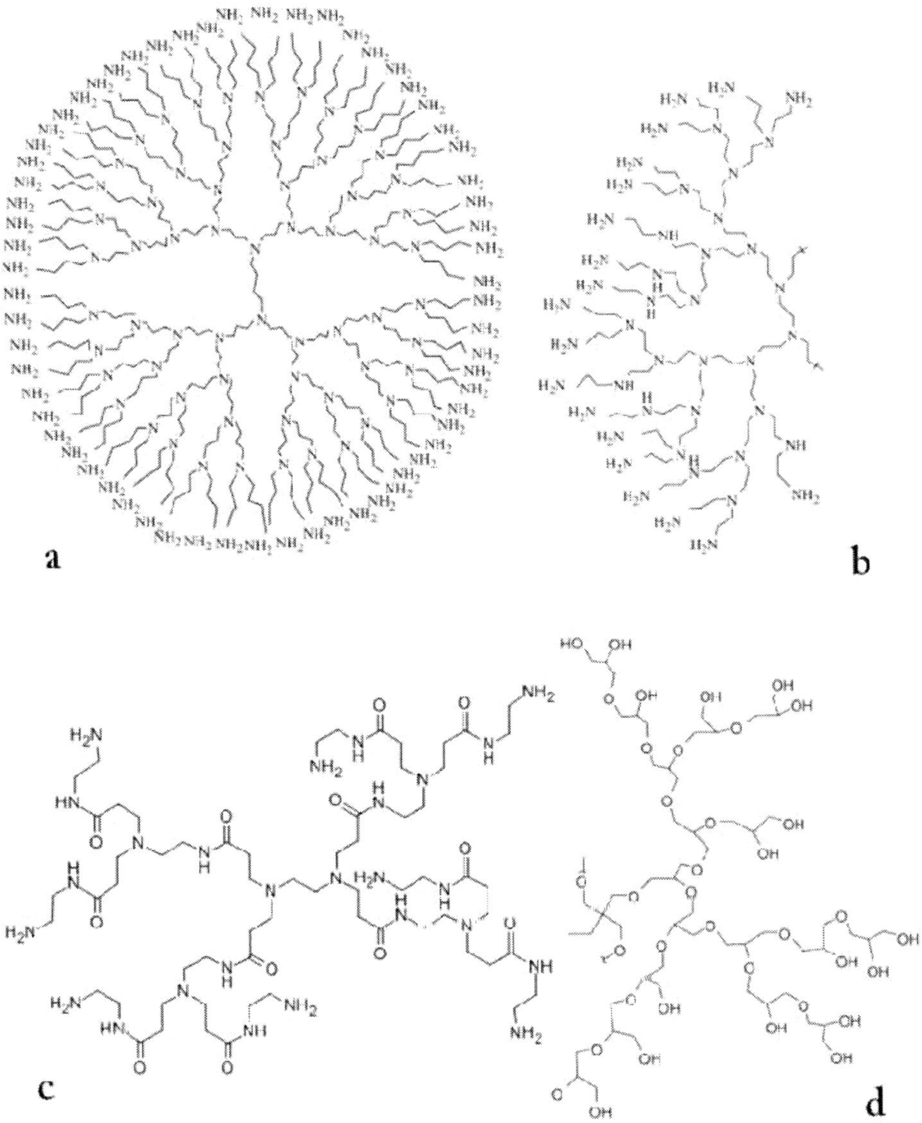

Figure 1. Chemical Structures of a) Diaminobutanepoly(propylene imine) DAB, b) Poly(ethylene imine) PEI, c) Poly(amido amine) PAMAM, and d) polyglycerol PG.

2. METHODS OF PREPARATION

2.1. Impregnation

2.1.1. Absorption of Dendritic Compounds into the Pores of the Ceramic Support

The most obvious and simplest technique to prepare ceramic/dendritic composite materials is the direct immersion of porous ceramic filters into a solution of dendritic polymer. This method provides significant impregnation with dendritic content reaching up to

22% [57]. The absorption of the dendritic polymer often proceeds under inert atmosphere to avoid oxidation and at higher than ambient temperatures in order to enhance the solubility of the polymer by lowering at the same time its viscosity. For the same ceramic material, impregnation is directly proportional to the concentration of the dendritic polymer as can be seen for example for an alumina ceramic filter, in a wide range of octyl functionalized polyethylene imine PEI (Figure 1b) hyperbranched polymer concentrations (0.2–30%, w/w) (Figure 2a). The loading capacity of the ceramic membrane depends on the filter pore structure (Figure 2a). Furthermore, the slope of these straight lines (the impregnation percentage/polymer concentration ratio) is clearly proportional to the specific surface area of the ceramic material (Figure 2b).The latter suggests a complete coverage of the pore surface by the polymer.

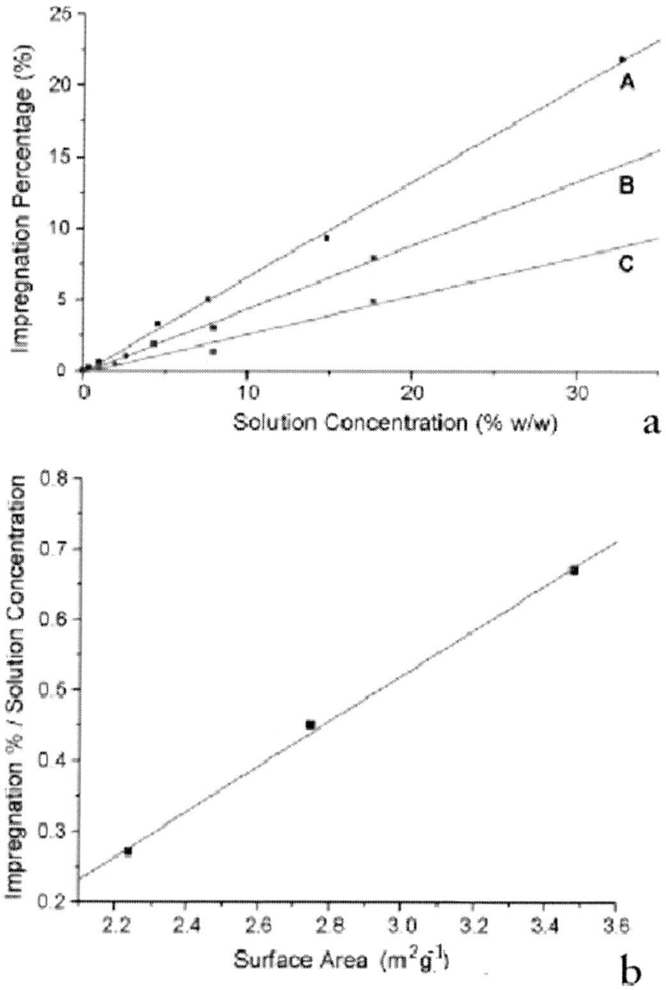

Figure 2. a) Octyl functionalized polyethylene imine PEI impregnation percentage in alumina tube filters A, B, and C of various surface areas as a function of polymeric solution concentration; and b) the effect of pore surface area on the ratio of impregnation percentage/concentration of polymer solution. (Reprinted with permission from Ref 57).

Ceramic membranes manufactured from different materials exhibit different impregnation degree when immersed into dendritic polymer solutions of the same concentration. This can clearly be deduced from the results presented in Figure 3 with ceramics made from Al_2O_3, TiO_2 and SiC immersed into a 30% solution of ethoxysilylated dendritic polymers such as Diaminobutane poly(propylene imine) DAB, Poly(ethylene imine) PEI or Polyglycerol PG. Impregnation degree is dependent on the porosity of the ceramic filters and thus, they are higher in Al_2O_3 and SiC and lower in TiO_2 membranes. For ceramic membranes with the same porosity (Al_2O_3 and SiC) the sample with smaller pores has a larger pore surface and therefore Al_2O_3 filters exhibit higher impregnation degree. Furthermore, ethoxysilylated polyglycerol impregnation degree was higher for all ceramic membranes. This is attributed to solution viscosity.

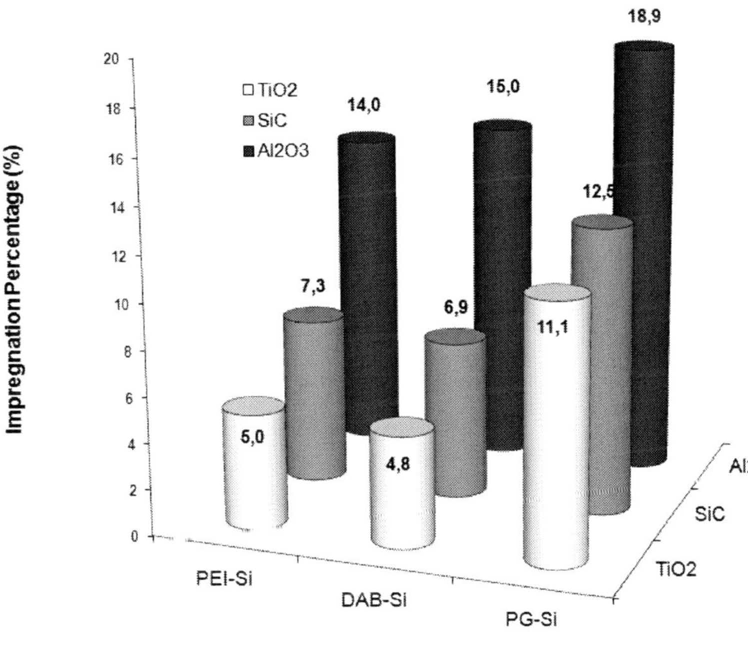

Figure 3. Impregnation degree for three ceramic membranes (TiO_2, SiC, Al_2O_3) after immersion into 30% solutions of the triethoxysilylated dendritic polymers. (Reprinted with permission from Ref 54).

2.1.2. Chemical Attachment of Dendritic Compounds on the Ceramic Surface

Nanocomposites comprising of dendritic polymer moieties chemically attached onto a ceramic framework are promising materials for solubility-based separations catalysis and drug delivery. This approach allows to obtain hybrid materials that exhibit the desired chemical properties designed by an easy and versatile process. In addition, dendritic polymers covalently bound on the ceramic surface, are expected to be susceptible to repeated use with subsequent optional regeneration without loss of the organic material. Chemical bond formation of dendritic polymers with different ceramic substrates can be achieved via sol–gel reactions, by employing for example triethoxysilyl derivatives of dendritic polymers diaminobutane poly(propylene imine) and polyglycerol (DAB, PEI, PG Figure 1). These triethoxysilyl derivatives are susceptible to facile hydrolysis yielding silanols which in turn condensate, by loss of water, with other neighboring hydroxy groups, either intramolecularly or intermolecularly. These dendritic silanols can also condensate with hydroxyl groups, which are present on the surface of activated ceramics(TiO_2, SiC and Al_2O_3), (Figure 4), forming a chemically bound film [52]. This activation of the ceramics can be performed by various methods. For example by etching through immersion into 1 M NaOH solution for several hours, repeated washing with water and methanol and drying under vacuum as described by Ottenbrite [58]. After curing, a nanostructured dendrimeric network covalently bound on the ceramic filter is developed.

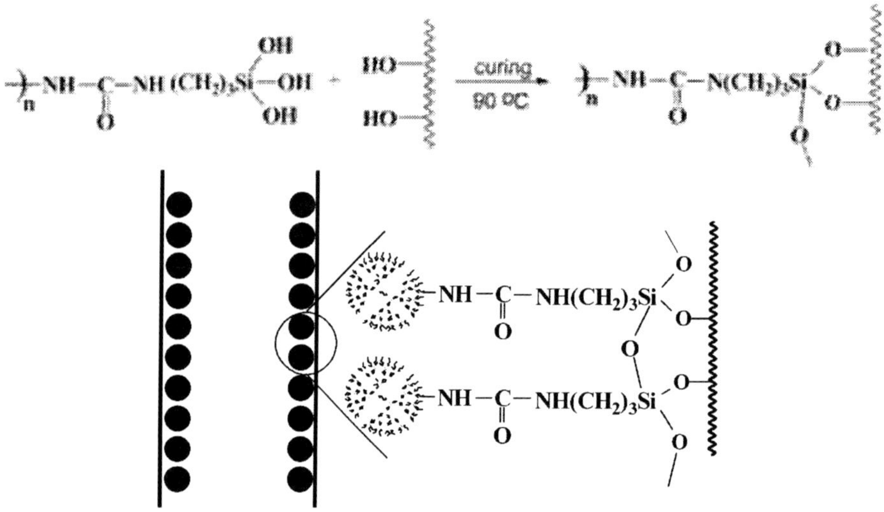

Figure 4. Chemical attachment of functionalized dendritic polymers on the surface of silica ceramic filters. (Reprinted with permission from Ref 52).

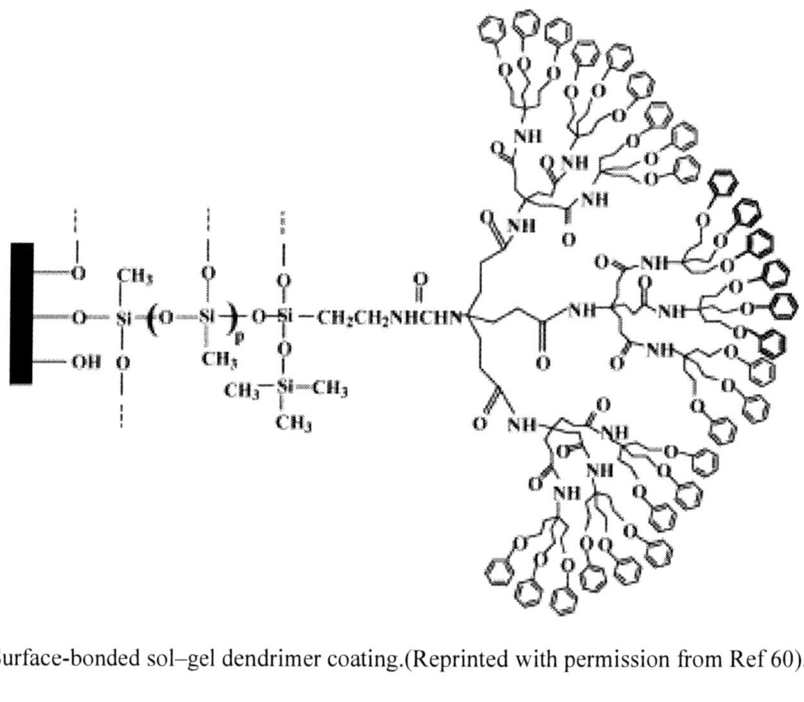

Figure 5. Surface-bonded sol–gel dendrimer coating.(Reprinted with permission from Ref 60).

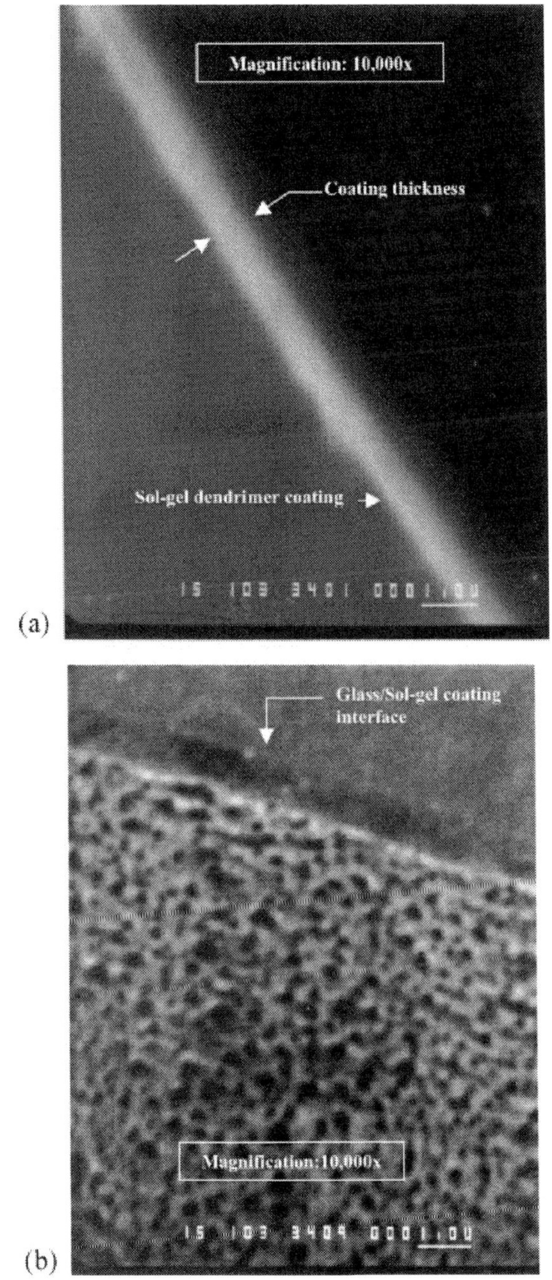

Figure 6. Scanning electron microscopic (SEM) images of a 250 μmi.d. sol–gel dendrimer-coated microextraction capillary (a) illustrating the coating thickness, andb) illustrating the typical roughened surface obtained by sol–gel coating process.(Reprinted with permission from Ref 60).

Sol–gel dendrimer coating on the inner walls of fused silica capillaries has been applied in solventless microextraction and preconcentration. [59-60]. Benzyl-terminated dendrimers were treated by 3-(triethoxysilyl)propylamine and formed sol-gel active dendritic reagents. These were used together with methyltrimethoxysilane in a coating solution.

Hydrolysis and polycondensations in a variety of ways as described above creates a sol–gel network. (Figure 5). The sol–gel dendritic network developed in the vicinity of the fused silica capillary inner surface chemically anchors to the column walls forming a surface-bonded stationary phase film with remarkable uniformity (Figure 6a) and a roughened, porous texture (Figure 6b). This introduction of silylated dendritic polymer on silica provides another easy method for manufacturing composite materials for separation purposes.

In another experiment, polyethylenimine (PEI) is grafted onto the surface of magnetic ceramic porous adsorbent i.e. a porous powder composed of SiO_2 and Fe_3O_4 prepared by a dispersion polymerization method via the coupling effect of γ-chloropropyltrimethoxysilane. The magnetic porous composite was firstly activated by 2% methane sulfonic acid to introduce silanol bonds, followed by reacting with coupling agent of γ-chloropropyltrimethoxysilane at 80 °C. The treated magnetic porous composite was added into a 10% PEI solution and reacted at 90 °C for 6 h to obtain the expectant adsorbent (Figure 7) [61].

Alternatively, instead of γ-chloropropyltrimethoxysilane, 3-glycidoxypropyl-trimethoxysilane has been applied in order to develop a silica supported chiral catalyst based on PAMAM dendrimer (Figure 8) for the enantioselective addition of diethylzinc to benzaldehyde.

The reaction for the grafting of the *nth* generation of the dendrimer proceeds in methanol under reflux. The latter has been further modified with the introduction of a chiral functionality at the periphery by reaction with (1*R*, 2*R*)-(+)-1-phenylpropylene oxide with the optional presence of a long alkyl chain spacer [62].

Titania can also be covalently bonded to dendritic polymers. For example, PAMAM dendrons (generation 1-3) which have a siloxy focal point and long alkyl (hexyl, C6) chain spacers were attached to TiO_2 nanoparticles synthesized by the hydrolysis of tetraisopropylorthotitanate.

Both reactions proceeded simultaneously in the same pot by simply mixing the modified PAMAM and the titania precursor at various ratios at 0°C in 1-propanol and then pouring the mixture in water. This functionalization of TiO_2 nanoparticles enhanced their photocatalytic activity towards 2,4-dichlorophenoxyacetic acid [63].

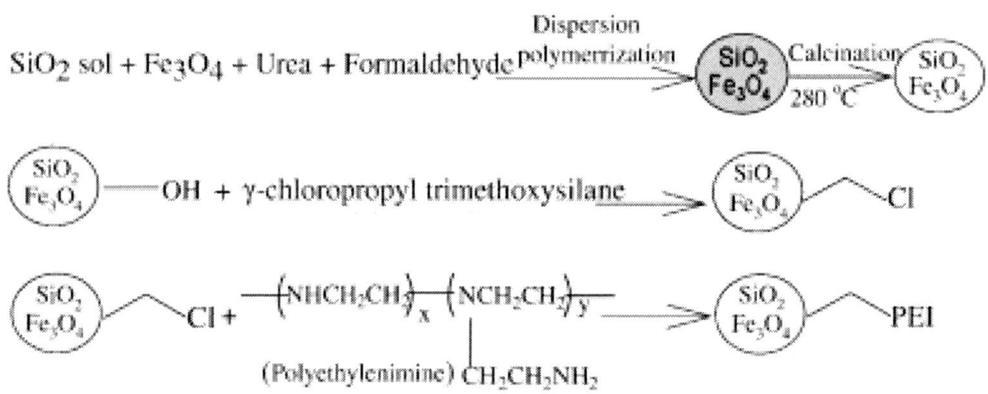

Figure 7. Schematic depiction of formation route for PEI-grafted magnetic porous adsorbent.(Reprinted with permission from Ref 60).

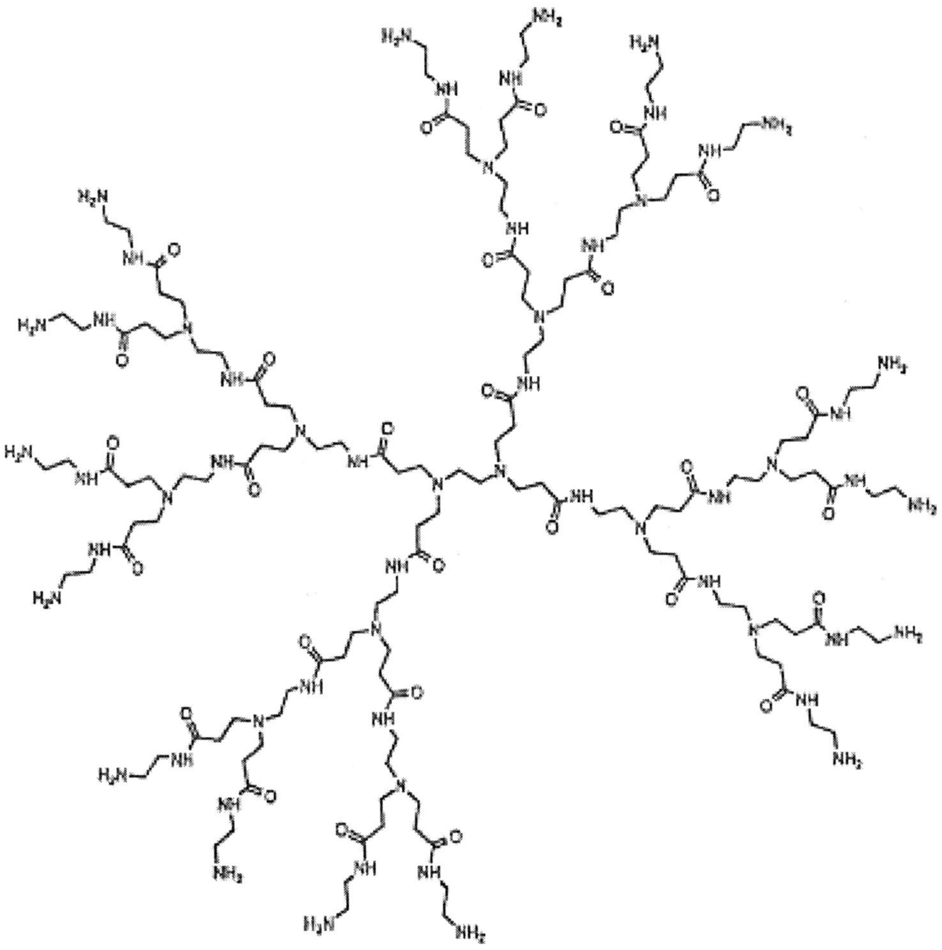

Figure 8. Generation 2 PAMAM Dendrimer.

2.2. Sol Gel Synthesis

Silicon-containing dendrimers (PAMAMOS) are part of a limited variety of commercially available dendritic polymers [64]. Among them alkoxysilyl-functionalized dendrimers present the potential for dendrimer cross-linking via conventional sol-gel reactions. These sol-gel reactions are exactly the same as those described for the chemical attachment of dendritic compounds on the surface of ceramics i.e. hydrolysis alkoxysilyl (Si-O-R) end-groups into the corresponding silanols, Si-OH, and subsequent condensation of the silanol intermediates into siloxane (Si-O-Si) interdendrimer bridges. Curing of the solvent-free products with the optional use of a catalyst such as bis(2-ethylhexanoate)tin, at a selected temperature for a desired period of time in an oven open to atmospheric moisture results in a wide variety of different organic-inorganic materials with completely different properties depending of the selection of these parameters and the generation of the dendrimer. Furthermore multifunctional cross-linking agents such as tetramethoxysilane or tetraethoxysilane can be employed (Figure 9) in order to enhance the cross-link density of the dendrimers and content of the resulting materials in silica [65]. Furthermore controlled thermal degradation of PAMAM molecules permits the development to nano-porous structures in which the diameters of pores correspond to the diameters of the PAMAM cores used (Figure 13) [66].

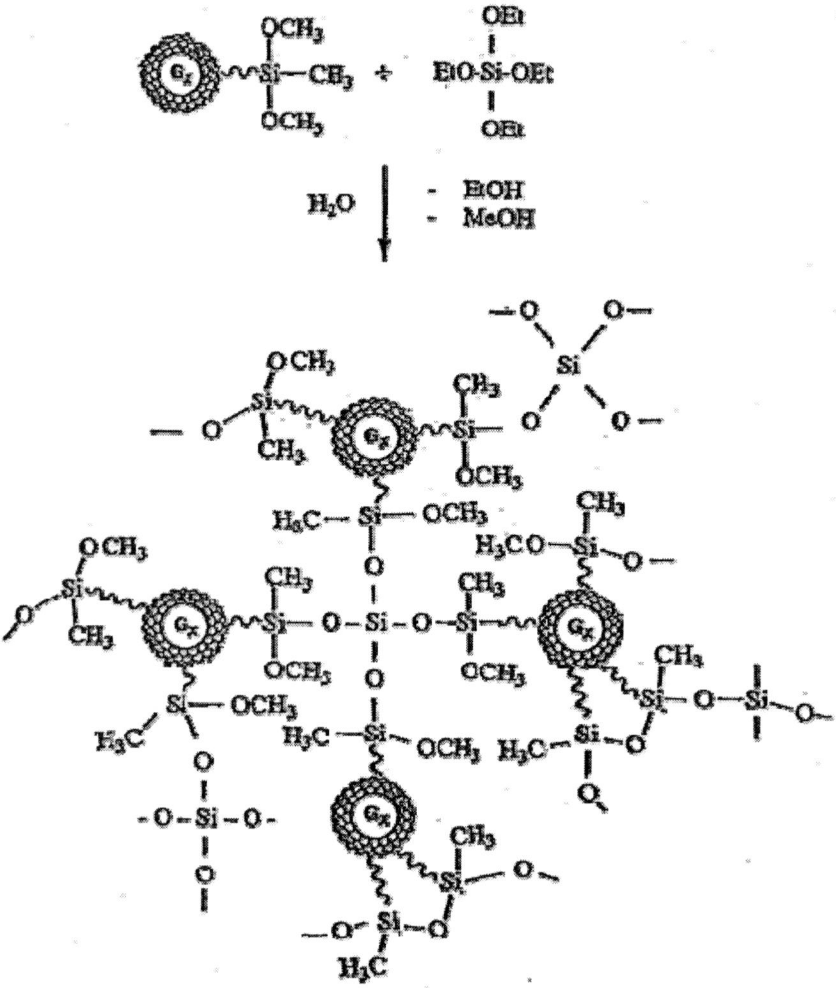

Figure 9. Preparation of honeycomb-like PAMAMOS networks from alkoxysilyl dendrimer precursors. (Reprinted with permission from Ref 65).

Production of hybrid silica can also be achieved using hyperbranched polyethylenimine polymers (PEI) of two different molecular weights (5000 and 25000) functionalized by 3-(triethoxysilyl)-propylisocyanate as silica precursor in N,N-Dimethylformamide (DMF) or chloroform. Cross-linking of hyperbranched macromolecules having surface groups able to react with silica precursors along with hydrolysis and poly-condensation reactions lead to the formation of a novel silica network architecture. This hybrid material can be used in its native state, impregnate or covalently bind to a ceramic substrate. In addition by firing the dendritic template is removed and a silica powder of nanoporous structure and advanced textural properties is obtained. Nanostructures obtained exhibit specific surface area values varying between 300 and 500 $m^2 \cdot g^{-1}$depending on synthesis conditions and mean pore diameter around 2 nm.

2.3. Biomimetic Processes

It has been recently established that biogenic silica is formed rapidly in water under mild conditions through a process known as biosilicification. This process proceeds at ambient temperature, pressure and neutral pH. By this method, an amazing diversity of nanostructure frameworks is produced [67-69]. Examples of such biosilicification are the diverse array of nanopatterned silica in many diatoms or the protective glass spicules of some sponges. These forms of biogenic silica are species-specific indicating that the organisms control silica condensation over a wide length scale. The phenomenon is attributed to the involvement of biopolymers and has attracted the attention of many researchers. Various attempts have been performed to reproduce the process in vitro and to prove this involvement of the biopolymers. In this context it has been shown that polycationic peptides called silafins [67], and the R5 peptide [70] which was isolated from diatom cell walls, generate networks of silica nanospheres within seconds when added to a solution of silicic acid. Apart from the use of biopolymers further work conducted during the current decade has shown that conventional synthetic polymers can perform the same task. For example a variety of homopeptides such as poly(arginine) and poly(lysine) [71], a combination of triethanolamine and cetyltrimethylammonium chloride [72] can also act as biomimetic templates forming silica nanoparticles.

Amine terminated dendrimers of various generations such as poly(amidoamine) PAMAM and poly(propylene imine) PPI have highly localized concentrations of amino moieties and therefore can also act as well-defined biomimetic templates for silica nanosphere production[73]. The sol–gel reactions proceed at room temperature by the addition of metastable silicic acid into the dendrimer solution in phosphate buffer, pH 7.5. The polymeric templates are encapsulated by the growing nanospheres and precipitated from water in a manner similar to that previously described for bioactive peptides. Silica production depends on the amine concentration (Figure 10). In addition, silica is produced exclusively in the form of nanospheres with a distinct size distribution. It has been proven that the formation of silica is attributed to the presence of amino groups since analogous hydroxyl-terminated PAMAM dendrimers did not yield silica [74]. Such nanomaterials could be suitable for various applications including water treatment and catalysis provided that dendritic polymers, although incorporated in the nanospheres, retain their properties.

Organic-inorganic hybrid ceramic oxides powders (ZrO_2, CeO_2, and Y_2O_3) have also been prepared in the presence of poly(amidoamine), PAMAM (G 5.5) anionic starburst dendrimers which are functionalized with surface carboxylate groups[75]. The latter were obtained when the methylester-terminated dendrimers were hydrolyzed with stoichiometric amounts of NaOH. In this case, sol–gel reactions proceed by the addition of aqueous solutions of $ZrOCl_2$, $CeCl_3$, and YCl_3 into an aqueous dendrimer solution, with the addition of NH_4OH. The as-synthesized nanoparticles were amorphous whereas by firing at 600 ^{o}C for 2 hours nanocrystalline powders are obtained.

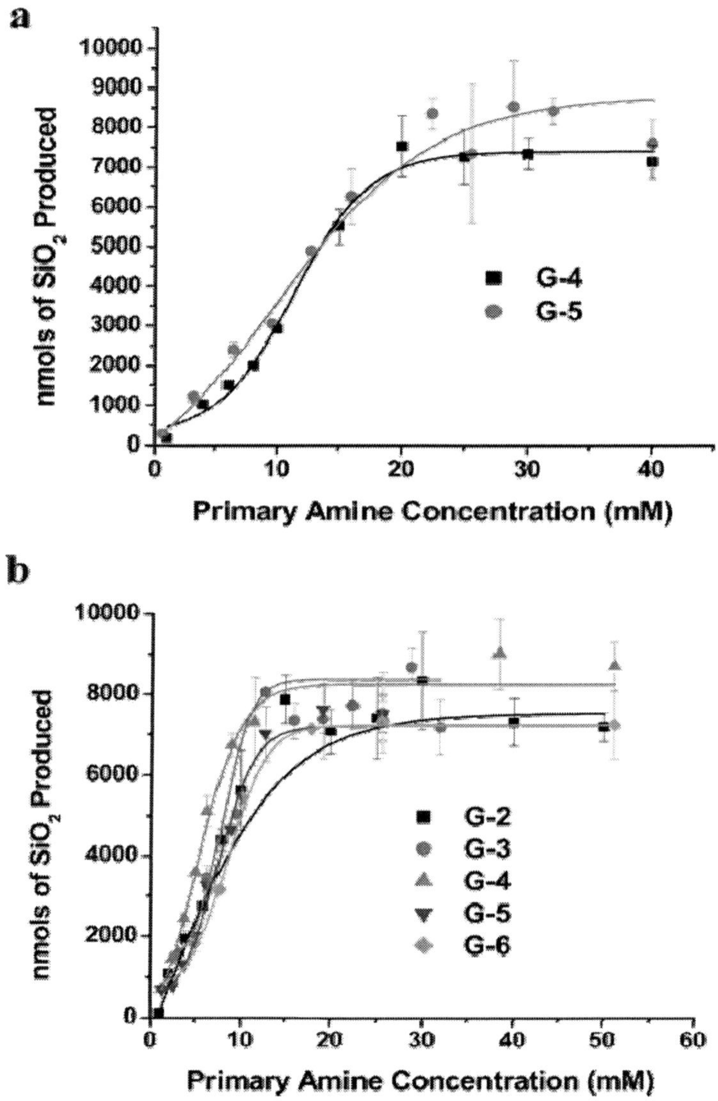

Figure 10. Biomimetic silica condensation as a function of primary amine concentration for (a) PPI dendrimers G4 and G5, and (b) PAMAM dendrimers G2−G6.(Reprinted with permission from Ref 73).

This method was also employed in order to obtain organic-inorganic hybrid ceramic nanoparticles for pollutant absorption and water purification and to avoid the conventionally employed technique, i.e. functionalization by attaching hydrophobic groups or by performing cross-linking reactions, and finally impregnation into ceramic membranes. Ceramic nanospheres were produced by employing hyperbranched poly(ethylene-imine) polymers and silicic acid prepared from tetramethylorthosilicate [76]. In this case, silica precipitation even occurs in the absence of phosphate buffer but to a much lesser yield, while no silica precipitation is observed in phosphate buffer in the absence of dendritic polymer. Scanning electron microscopy confirmed the formation of spherical nanoparticles (Figure 11) i.e. nanospheres.

Figure 11. SEM micrographs of spherical composite a) PEI–silica nanoparticles , and b) thermally treated polymer-free silica nanoparticles.(Reprinted with permission from Ref 76).

In the formation of polymer/silica nanocomposites, the dendritic polymers likely serve several roles in the silica condensation reaction. At pH 7 the polymers bearing amino groups are partially protonated (in the form of NH_3^+) [67,70,77], and assume the traditional role of charge balance between the dendritic template and the surface of growing silica nanoparticles. This leads to the formation of an extensive network between NH_3^+, Si–O− groups and water molecules. This interaction causes the encapsulation of the dendrimers within the silica nanospheres. Overall the nanospheres show a negative surface potential at low pH that decreases to about −80 mV at pH 11 as indicated by zeta-potential measurements in the case of the nanocomposites containing hyperbranched PEI (Figure 12). The isoelectric point (IEP) of heat treated silica nanospheres (i.e. without dendtiric polymer) is about 2, in accordance with what is typically observed for amorphous silica. The presence of dendritic polymers in the nanospheres shifts the IEP towards higher pH. Indeed, the hybrid organic-inorganic nanospheres have an IEP of about 8 resulting from the presence of the protonated amino groups of the polymer. They have a positive surface potential of about +50 mV at pH 2 that decreases to about −70 mV at pH 11.

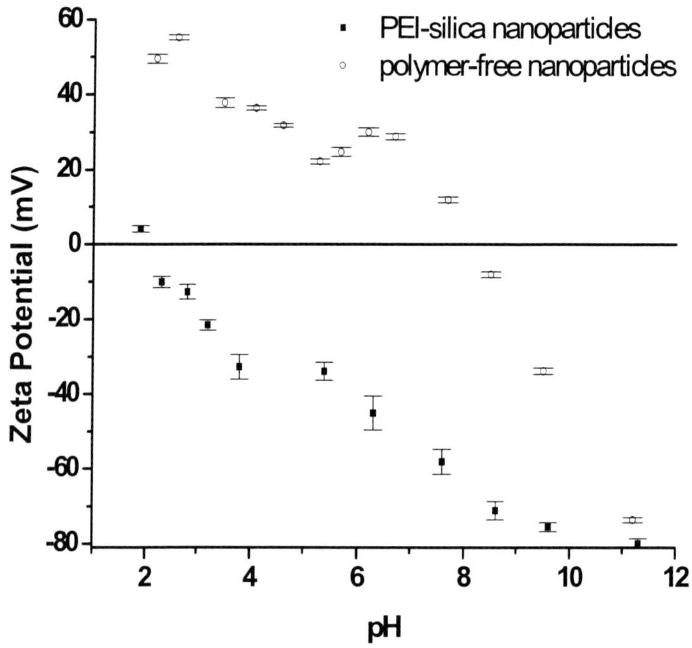

Figure 12. Zeta-potential as a function of solution pH for hybrid PEI–silica nanoparticles and polymer-free nanoparticles. (Reprinted with permission from Ref 76). (The length of the figure has been extended).

Having in mind the above mentioned electrostatic interaction it is not surprising that the concentration of phosphate buffer in solution has been identified as the key factor for the control of the size of the nanosheres produced by PAMAM, i.e. particle size increases with increasing phosphate concentrations Figure 13. At pH 7.5, when the phosphate buffer was replaced with individual salt solutions, SEM analysis of the particle size revealed a linear increase in particle diameter as salt concentrations increased to 100 mM (Figure 14).

Dendritic polymers can act as templates for the biomimetically controlled formation of biopolymers. Carboxylic acid terminated poly(amidoamine) (PAMAM) dendrimers (4,5 generation i.e carboxylic acids that have not reacted with amines to form the 5th amide generation of the poly amide) are used as templates for the preparation of spherical hydroxyapatite nanoparticles (Figure 14) [78] while hydrothermal treatment with PAMAM (generation 5,5) produces highly crystalline hydroxyapatite nanorods (Figure 15,16,17) [79]. Conventional PAMAM dendrimers with amido groups at the external surface afforded hydroxyapatite nanoparticle with size and morphology depending on their generation [80]. The shapes of crystals were rod-like in the presence of G1.0 or G2.0 PAMAM dendrimers, while their morphology became ellipsoid-like when PAMAM molecules of higher generations were used.

As the generation number of the PAMAM dendrimer increased from G1.0 to G4.0, the lengths of these particles were decreased from 82 to 38 nm. The same effect was observed with increased concentrations of PAMAM G4.0. Carboxylic terminated G2.5 PAMAM form nano-rod nanoparticles having an average length ca. 65 nm and diameter ca. 26 nm. in majority, and elliptical particles in minority while polyhydroxy analogs consist of elliptical particles in majority with an average grain size ca. 20 nm and short nano-rods in minority exhibiting length ca. 30 nm and width ca. 25 nm[81].

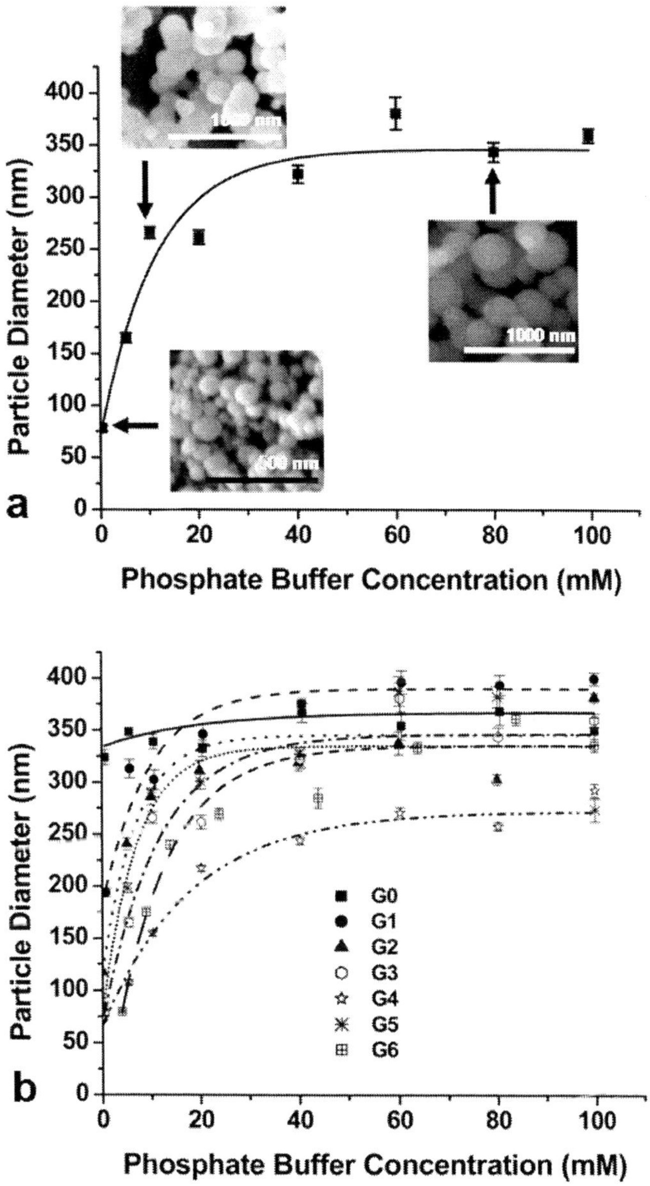

Figure 13. a)Representative size distributions and micrographs of various nanospheres produced by the G3 PAMAM dendrimer at various phosphate concentrations. Note: when the phosphate concentration was zero, no silica precipitated. b)Representative size distributions for nanospheres produced from PAMAM dendrimers at phosphate buffer concentrations greater than 0.1 mM. Error bars represent the standard error associated with the sample size analyzed for the particle size distributions.(Reprinted with permission from Ref 74).

Diaminobutanepoly(propylene imine) dendrimers (DAB) direct the formation of hydroxyapatite (HAP) in preference to all other calcium phosphate phases [82] even at room temperature. Elongated rod-like nanoparticles are developed, the dimensions of which depend greatly on the conditions used. The samples kept at room temperature present a very fine nanostructure with the mean length and width reaching down to 11 and 5 nm, respectively. By increasing the temperature of hydrothermal treatment from 80° to 130°C, a considerable rise of mean particle size is observed. Poly (propylene imine) dendrimer functionalized with long alkyl chains when mixed with single-chain surfactans such as octadecylamine haxedecyltrimethylammonium bromide and sodium dodecyl sulfate afforded a dendrimer/HAP composite that, “in contrast to other synthetic composite materials, shows plastic deformation (as in bone) while retaining a high stiffness”[83]. Furthermore they control nanoparticle size distribution and morphology.

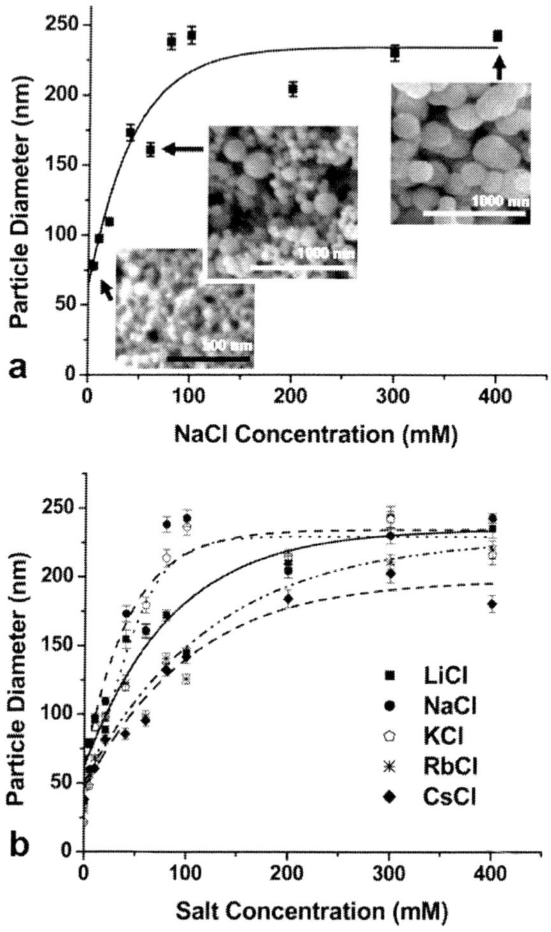

Figure 14. a) Silica production activity (20 mM primary amine). Representative size distributions and micrographs of various nanospheres produced by G4 PAMAM dendrimer at various NaCl concentrations. b) Representative size distributions for nanospheres produced from PAMAM dendrimers at salt concentrations greater than 0.1 mM. Error bars represent the standard error associated with the sample size analyzed for the particle size distributions.(Reprinted with permission from Ref 74).

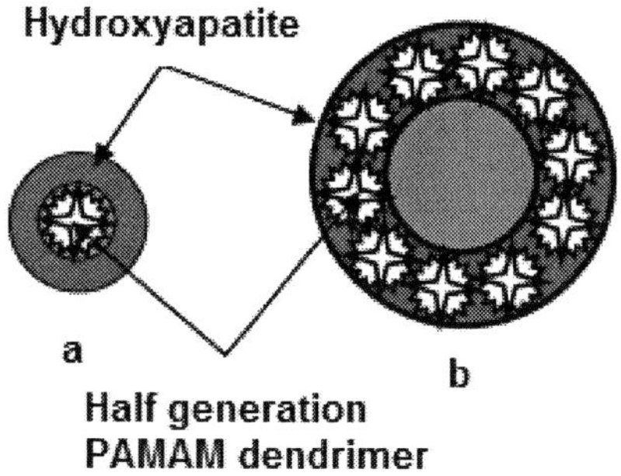

Figure 15. Possible mechanisms of slow hydroxyapatite growth.a) Dendrimeris a nucleus for hydroxyapatite precipitation and growth. b) Dendrimer supramolecular aggregate is a nucleus for hydroxyapatite precipitation and growth. (Reprinted with permission from Ref 78).

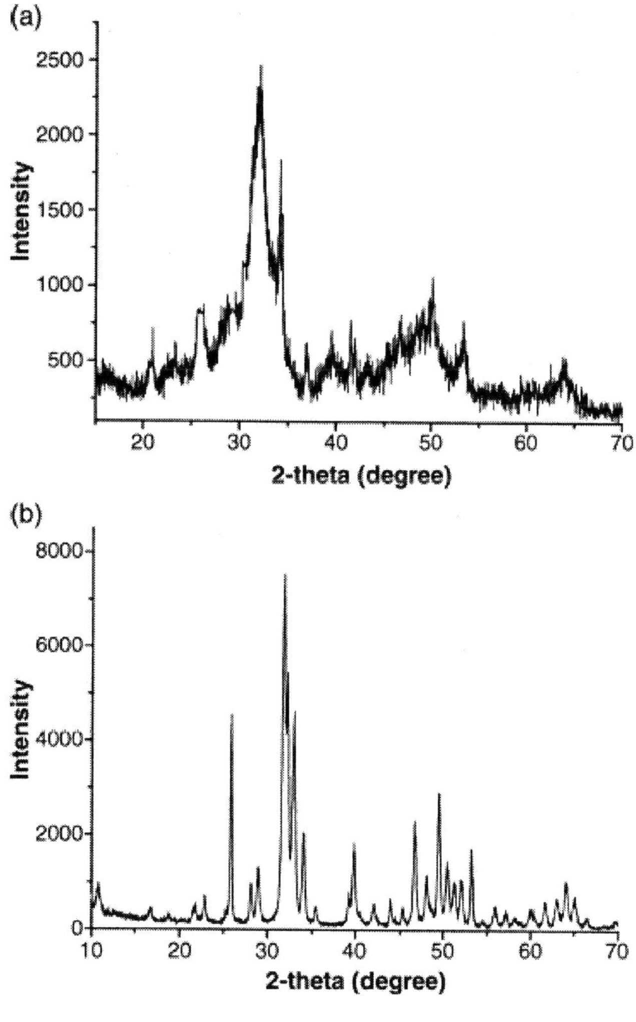

Figure 16. The X-ray diffraction patterns for hydroxyapatite prepared at room temperature a) and hydrothermally treated at 150 °C b) in the presence of anionic starburst dendrimer. (Reprinted with permission from Ref 79).

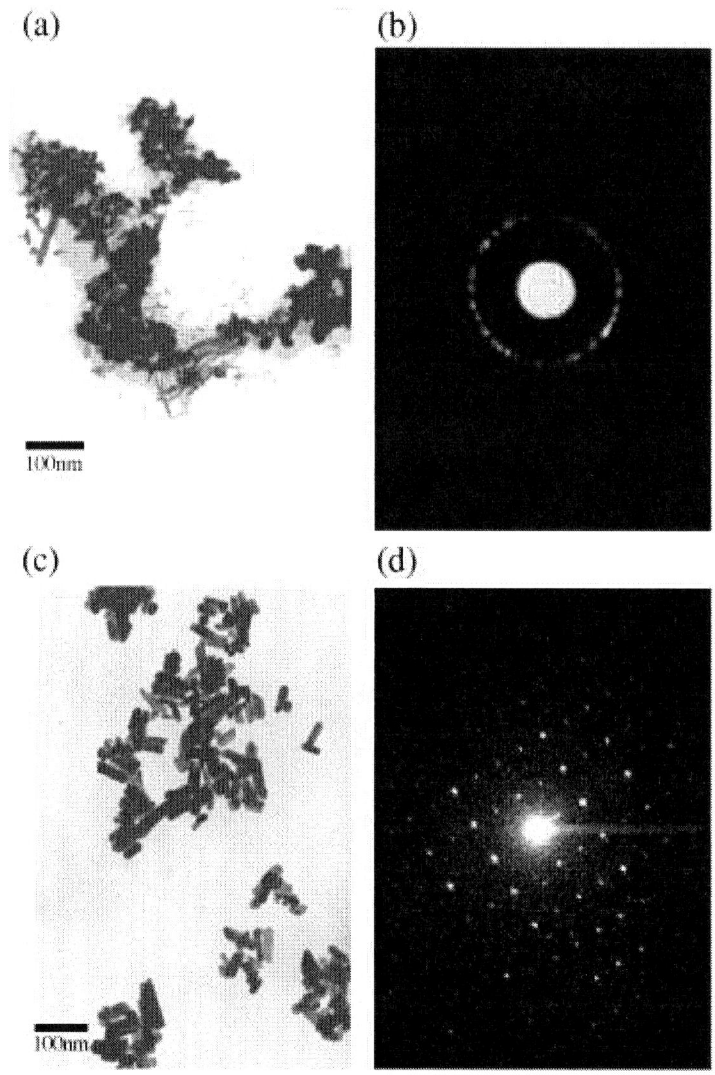

Figure 17. The transmission electron micrography (a, c) and electron diffraction patterns (c, d) of hydroxyapatite obtained at room temperature (a, c) and by hydrothermal treatment at 150 °C (b, d). (Reprinted with permission from Ref 79).

2.4. Direct Growth of Dendritic Polymerson the Surface of Ceramics

Another approach for the modification of the surface of ceramics by dendritic polymers is the direct propagation of dendrons on their surface through conventional divergent pathway. This solid-phase synthesis technique is a quick and simple alternative for the preparation of hybrid dendritic polymer/ceramic materials. The majority of the examples of this method have been employed on silica gel. Nevertheless the reactions for the activation of the surface, the binding of the dendron core and the propagation of the dendritic branches for silica gel and silica ceramics are exactly the same.

The propagation of PAMAM (polyamidoamine) dendrons on silica can be accomplished by a series of alternate Michael addition and amidation reactions [84,85]. Firstly introduction of amino groups onto the activated (as described in section 2.1.2) silica surface can be performed by reacting the surface silanol groups with 3-aminopropyltriethoxylsilane APES. This reaction is carried out at 110 °C in toluene for 48 h. PAMAM dendrons are then constructed by the repetitive additions of a branching unit. Michael addition of methyl acrylate (MA) to the terminal amine generates amino propionate esters. Subsequent amidation of the ester moieties with ethylene diamine (EDA) eventually completes the first generation. Repetitive Michael addition and amidation reactions yield the higher generations [86] (Figure 18). PAMAM modified silica was further functionalized with an n-octadecylisocyanate to introduce long aliphatic chain to the periphery of the dendron (Figure 18). In this manner amphiphilic silica bearing hydrophilic dendritic core and hydrophobic surface is produced for applications in pollutant encapsulation. PAPAM dendritic polymers were also phosphonated in order to provide an attractive coordination site for rhodium. The phosphonated dendrimers were then readily prepared by reaction with chloro(dicarbonyl)rhodium(I) dimer. The resulting complexes were used for the catalytic hydroformylation of olefins [87]. Another modification of the periphery of PAMAM dendrons by salicylaldehyde gives a Schiff base. The reaction of this Schiff base with $Mn(OAc)_2 \cdot 4H_2O$ yields a Mn(II) complex, immobilizing Mn(II) onto PAMAM–SiO2 dendrimers for the preparation of catalysts for olefin epoxidation [88].

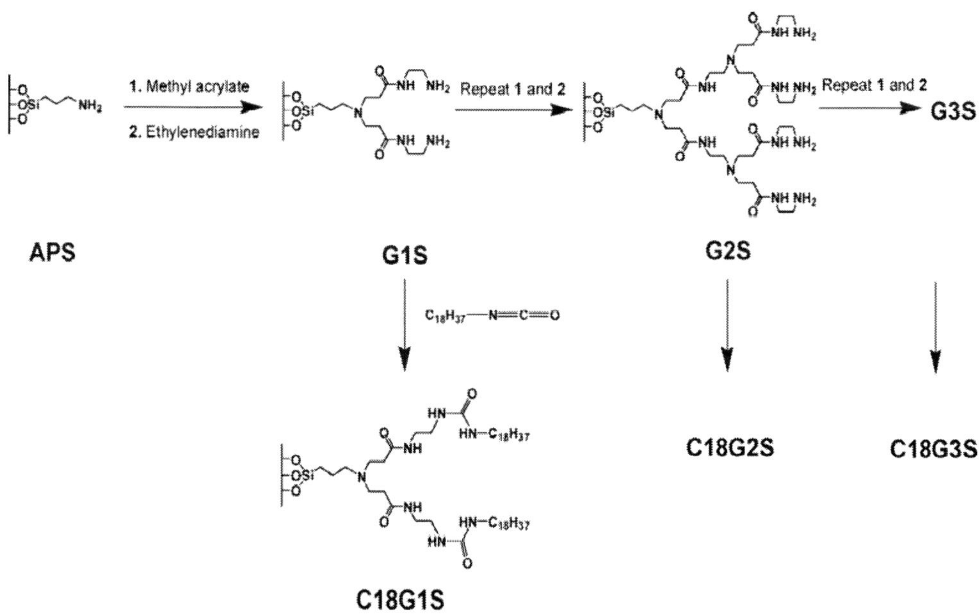

Figure 18. Construction of PAMAM dendrons on silica surface via conventional divergent pathway and modification of amino-end-groups by octadecylisocyanate to yield amphiphilic dendron-surface-modified silica.(Reprinted with permission from Ref 86).

In another example the synthesis of a series of melamine-based dendrimers of various generations grown on the surface of mesoporous silica is reported. The synthesis of these composites commences with the functionalization of the surface with 3-

aminopropyltriethoxylsilane followed by iterative reactions with triazinetrichloride and 4-aminomethylpiperidin [89]. The derivatized with amino groups silica is alternatively incubated into solutions of these reagents in 4°C for 24h.

In a similar method γ-alumina membranes consisting of mesoporous γ-alumina layer deposited on a macroporous alumina support tube were subjected in a special activation treatment [90]. Then 3-aminopropyldimethylethoxysilane (APDMES) was attached on this support and dendrimers were grown directly off the surface by replacing 4-aminomethylpiperidin by another diamine such as diisopropylethylamine or piperazine and following the same reaction as above. Finally this modified membrane was further functionalized by dodecylamine (Figure 19) to provide applications in the field of reverse-selective membranes.

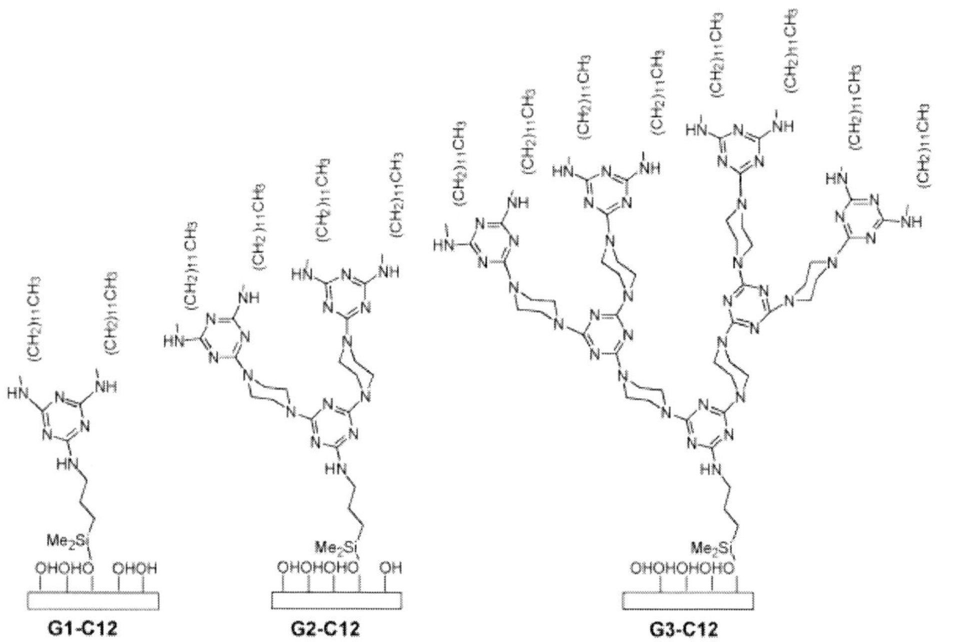

Figure 19. Structures of dodecylamine-capped dendrimers of different generations. (Reprinted with permission from Ref 90).

Although the solid-phase strategy is slightly difficult for generating dendrons bearing perfect symmetrical branch structure, it however provides a convenient approach to dendritic graft polymers onto various ceramic structures such as silica, alumina and titanium dioxide, through the solid–liquid reaction.

3. Optimization Techniques

3.1. Impregnation

Functionalized dendritic polymer coating of the ceramic surface at high impregnation degrees followed by scanning electron microscopy (SEM) does not cover the surface as a continuous layer but has cracks attributed to the stress arising during the drying of the layer (Figure 20). It is also clear, that by increasing polymer impregnation the ceramic surface is gradually becoming smoother (Figure 21) thus suggesting a decrease in pore surface area and the blocking of small sized pores.

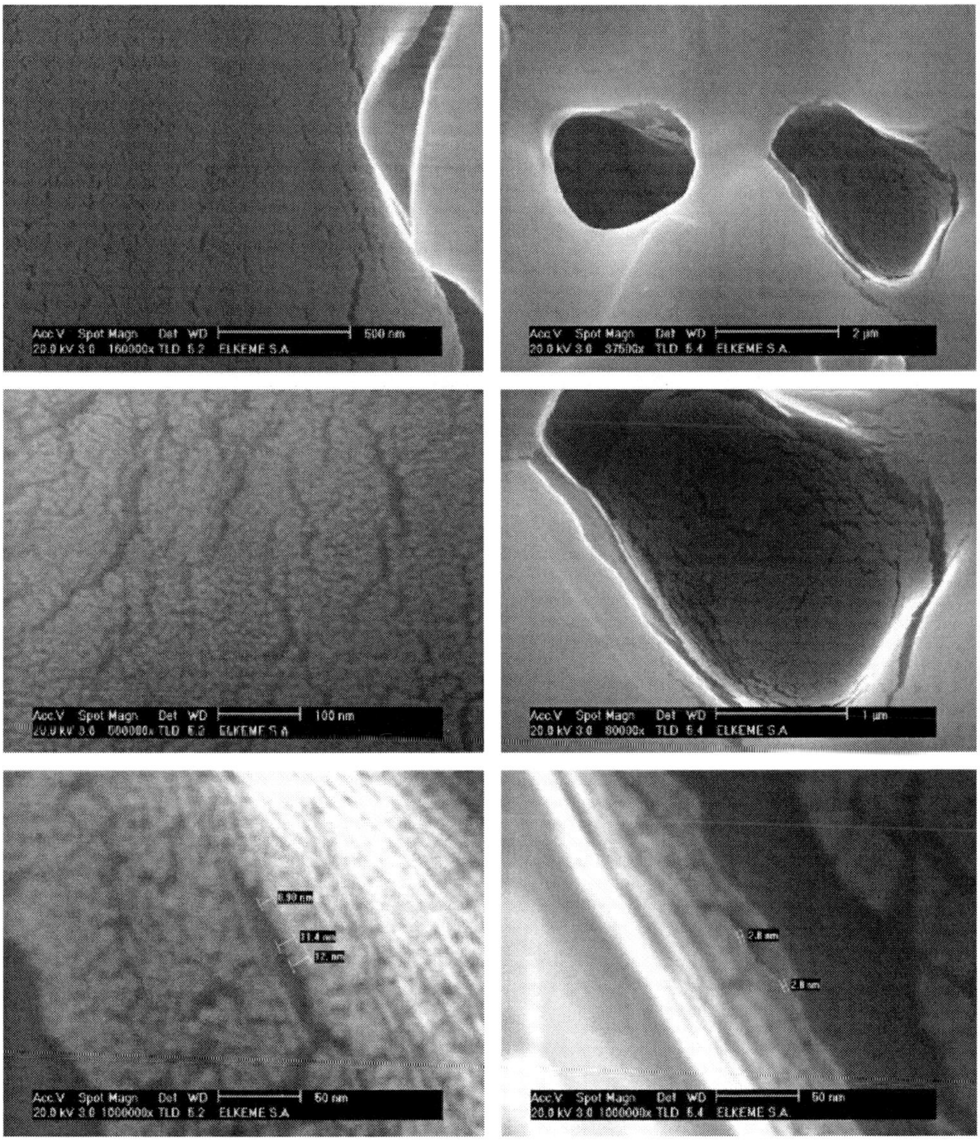

Figure 20. SEM micrographs of a ceramic surface impregnated with 22% octyl substituted PEI5 (left column) and alternative with octyl substituted 0.2/4% PEI5 employing a two step immersion (right column).(Reprinted with permission from Ref 57).

Pore blocking is confirmed by mercury porosimetry studies performed in ceramic membranes, both empty and impregnated, with various polymer loadings. As can be seen in Figure 22, as the polymer loading increases, the porosity systematically decreases and the mean pore size follows the same trend, although it is affected in a much lesser degree. This behaviour is more pronounced in ceramics with lower porosities and smaller mean pore size. In addition, the fast blocking of the small pores prevents further access to some of the bigger ones leading to further loss of the accessible pore volume and finally shifting the pore size distribution to smaller values.

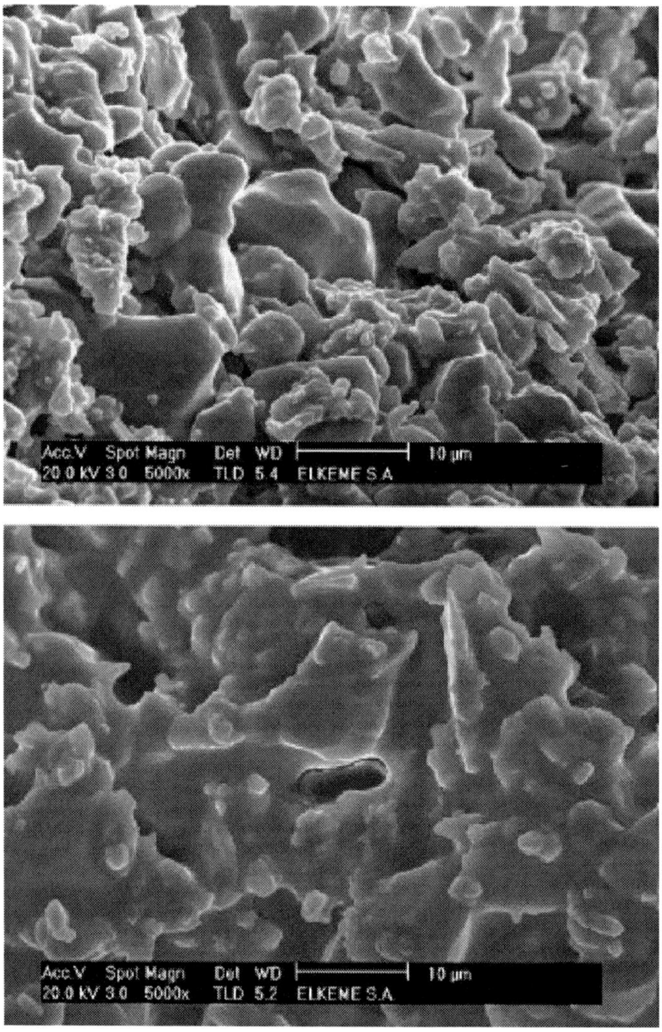

Figure 21. SEM micrographs of a) filter B impregnated with 7.5% octyl substituted PEI5 and b) of filter A impregnated with 22% octyl substituted PEI5. (Reprinted with permission from Ref 57).

In order to avoid the problems of cracks and pore blocking, a two step procedure is proposed using a very dilute solution in the first step [57].The porosity decrease observed is much less than the respective values measured at the same polymer loading degree obtained by the one step method (Figure 22). This two step immersion process also affords ceramic membranes with polymer considerably lower (ca. 2–3 nm) layer thickness (Figure 20). In conclusion in order to attain optimum characteristics every polymer impregnation protocol must be experimentally defined according to the pore structure characteristics of the ceramic membrane.

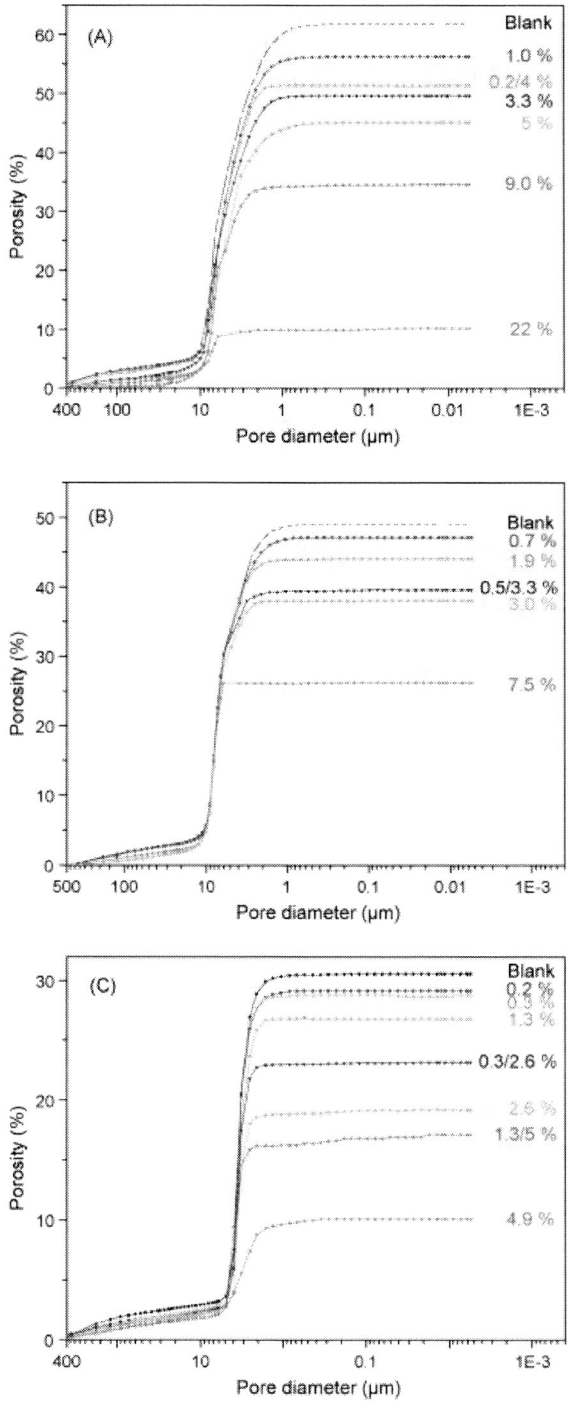

Figure 22. Pore size distribution of filters (A porosity 62%, mean pore size 5.3 BET surface area 3.48, B porosity 49%, mean pore size 7.1 BET surface area 2.75 C porosity 31%, mean pore size 3.5 BET surface area 2.24) impregnated with octyl substituted PEI5 at different percentages.(Reprinted with permission from Ref 57).

3.2. Directed Growth of (Dendritic) Polymers on the Surface of Ceramics

During the formation steps for the propagation of PAMAM dendrons on silica surface side reactions occur causing structural diversifications such as dimer (a), and intramolecular cyclization (b) (Figure 23).

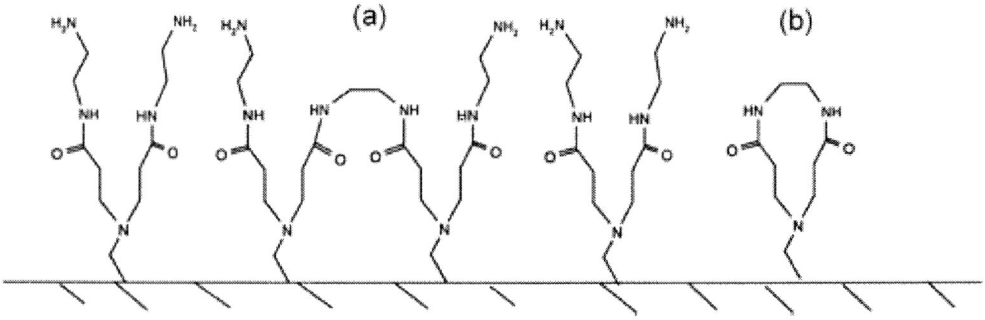

Figure 23. Structural Defects Caused by the Cross-Linking Reaction. (Reprinted with permission from Ref 85).

Initial amino group content of 3-aminopropyltriethoxylsilane attached to the silica surface is found to play a crucial role to the formation of these defects. A low initial concentration of amino group decreases steric hindrance and additionally increases the distance between the amino groups, reducing subsequently the formation of cross-linked structures [85] in higher generations of dendrons (Figure 24). For the remediation of this situation the immobilization of 3-(triethoxysilyl) propionitrile on silica is proposed. This act as a "spacer" to reduce the initial density of amino groups (Figure 25), thereby improving the synthesis efficiency.

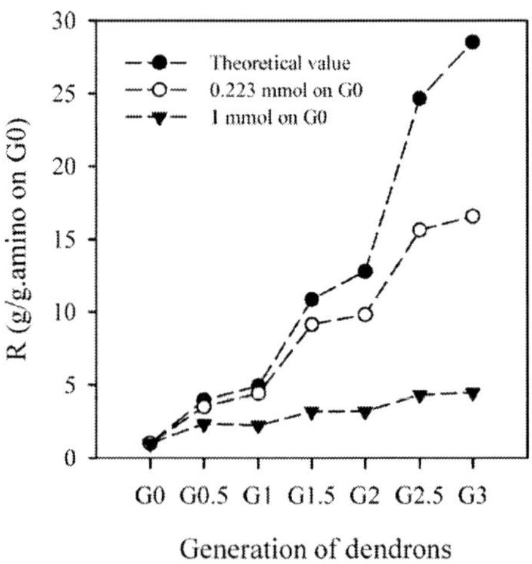

Figure 24. Influence of initial amino concentration. R is the ratio of the grafted amino groups to the initial amino group content (Ao).(Reprinted with permission from Ref 85).

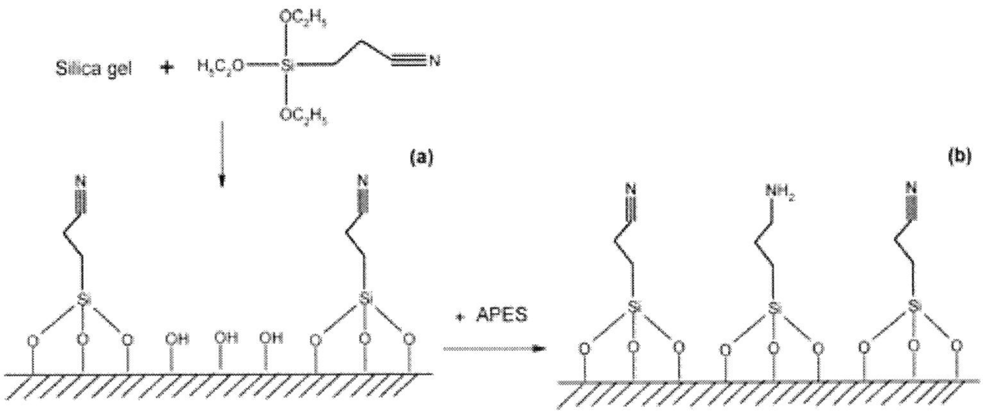

Figure 25. Nitrile Grafting on Silica for the elimination of structural defects.(Reprinted with permission from Ref 85).

Another parameter responsible for the steric crowding, is pore size and the irregular nature of the pore structure of amorphous silica. A proposed solution includes the application of silica with ordered, for example hexagonally packed, channels with a narrow size distribution such as MCM-41[86] In the same approach although an enlargement of the pores of this same material via post synthesis treatment caused a disorganization of the pore system permitted the preparation of PAMAM up to the fourth generation with a very limited percentage of defects.

4. Prospected Applications

PAMAM polymers have been used to stabilize semiconductor quantum dots formed by nanometer-scale CdS clusters [91]. These systems could be further stabilized by precipitation within silica nanospheres. Dendritic polymers also stabilize colloidal suspensions of gold by in situ reduction of $HAuCl_4$ encapsulated into PAMAM by aqueous solution of $NaBH_4$ [92]. Colloidal silver nanoparticles are obtained by the reduction Ag^+ ions encapsulated into PEI even in the absence of another reductant [93]. The visible absorption spectra, macroscopic optical image and size of these nanoparticles depend on the $AgNO_3$/PEI ratio (Figure 26). Biogenic nanocomposites, i.e. ceramic/dendritic polymer/noble metal materials, have unique potential for applications such as catalysis. The ceramic apart from stabilizer also acts as solid substrate. The application of solid catalysts of this type permits easy retrieval regeneration and reuse.

Dendritic polymers produce a vast variety of differently colored metal complexes. This capability is preserved to the hybrid materials. Possible applications for the detection of explosives chemical warfare agents and other substances even in small concentrations can be investigated.

Silica nanoparticles modified by aminopropyltriethoxysilane (APTES) and methyl triethoxysilane (MTES) were used to form ternary complexes with DNA-dendrimer [94]. Silica nanoparticles are used because their dense nature can concentrate DNA at the surface of cells grown in culture. This allows efficient uptake of DNA by an endosomal-lysomal route. Efficiency of the DNA-transfection reagent-silica nanoparticle complex was increased by a factor of 10 due to this mechanism. The use of biogenic silica produced by a complex of DNA with a dendrimer bearing amino groups, provided that the ternary complex is not toxic, could simplify the formation of the silica nanoparticles and the overall transfection process.

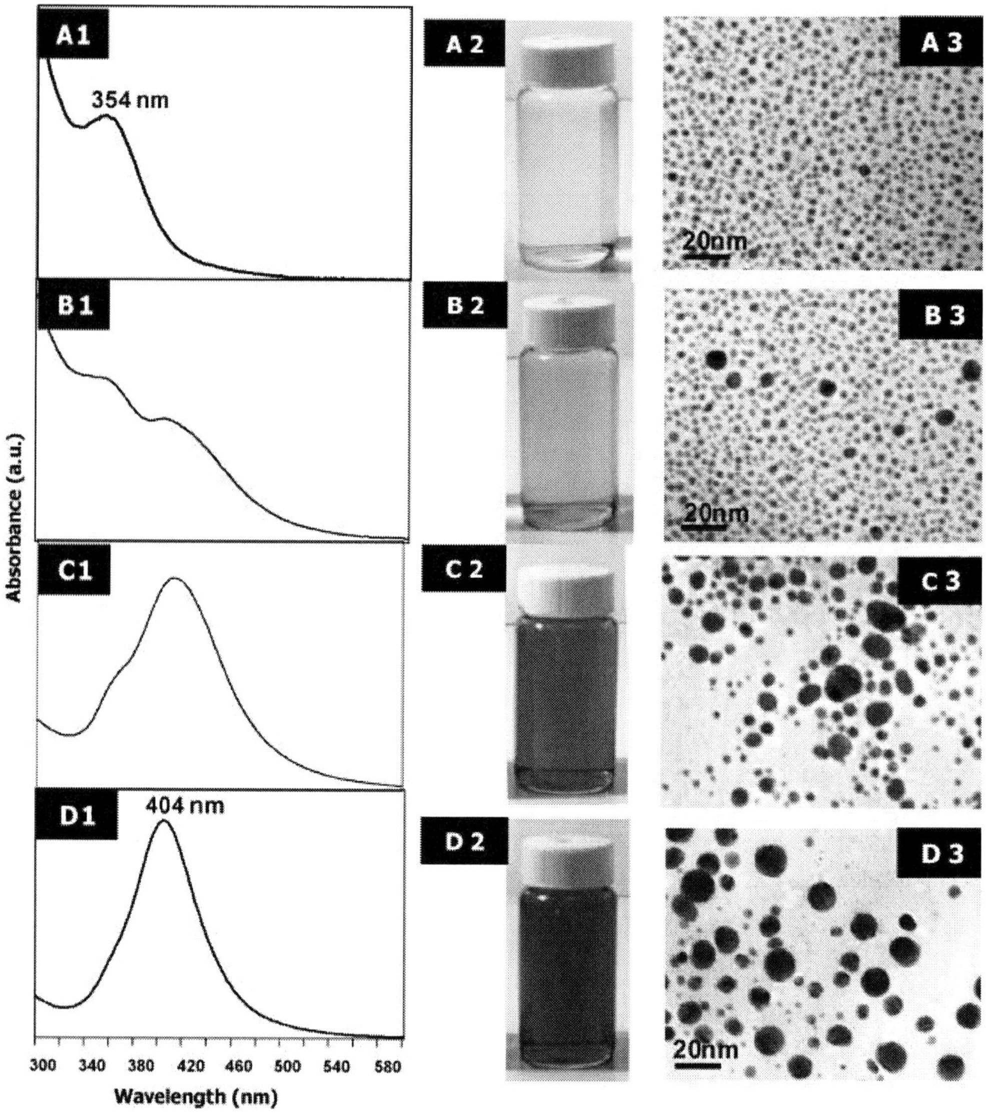

Figure 26. (A1–A3) UV–vis absorption spectra,photograph of the colloidal solution, and corresponding TEM image of AgNO3/PEI ratio (X) = 1;(B1–B3) AgNO3/PEI ratio (X) = 3;(C1–C3) AgNO3/PEI ratio (X) = 5;(D1–D3) AgNO3/PEI ratio (X) = 8.(Reprinted with permission from Ref 93).

Conclusion

The development of hybrid organic-inorganic materials by the integration of dendrimeric derivatives with ceramic substrates can be achieved by employing several convenient methods and techniques including immersion, and chemical reactions such as sol-gel synthesis of a sillylated dendritic compound, biomimetic synthesis, covalent attachment of a dendritic compound to the inorganic support or direct propagation of a dendron from the inorganic matrix. Incorporation percentages are polymer specific and depend on several parameters such as the concentration of the dendritic SPACE polymer, organic/inorganic ratio and the geometric parameters of the ceramics such as the porosity and the pore size. Thus in order to attain uniform and optimum results each experimental protocol must be defined for each dendritic polymer/ceramic combination. The resulting organoceramics proved promising devices for a variety of applications and exhibit improved properties compared to the unprocessed ceramic analogs. The diversity of both dendritic polymers and ceramic membranes will permit, by careful selection of the suitable materials and processing parameters, the formation of composite devices which will have extended applications and will be adopted for specific cases.

ACKNOWLEDGMENTS

I thank Dr C. M. Paleos and Dr. D. Tsiourvas for their helpful comments.

REFERENCES

[1] R. S. A. de Lange, K. Keizer, A. J. Burggraaf. *J. Membr. Sci.*, 104(1-2), 81 (1995).
[2] M. Acharya, H. C. Foley. *J. Membr. Sci.*, 161(1-2), 1 (1999).
[3] A. Conesa, A. Fernández Roura, J. A. Pitarch, I. Vicente-Mingarro, M. A. Rodríguez. *J. Membr. Sci.*, 155(1), 123 (1999).
[4] L. Montanaro, G. Saracco. *Ceram. Int.*, 21(1), 43 (1995).
[5] P. Nair, F. Mizukami, T. Okubo, J. Nair, K. Keizer, A. J. Burggraaf. *AIChE J.*, 43(11 A), 2710 (1997).
[6] A. I. Rat'ko, V. E. Romanenkov, E. V. Bolotnikova, Zh. V. Krupen'kina. *Kinet. Katal.*, vol. 45(2), 279 (2004).
[7] M. Bodzek, K. Konieczny. *Desalination*, 119 (1-3), 191 (1998).
[8] C. J. Russo. *Membr. Sci.*, 288(1-2), 239 (2007).
[9] P. Wang, N. Xu, J. Shi. *J. Membr. Sci.*, 173(2), 159 (2000).
[10] C. Yang, G. Zhang, N. Xu, J. Shi. *J. Membr. Sci.*, 142 (2), 235 (1998).
[11] S. Benfer, P. Árki, G. Tomandl. *Adv. Eng. Mater.* 6 (7), 495 (2004).
[12] Yu. I. Komolikov, L. A. Blaginina. *Refract. Ind. Ceram.*, 43(5-6), 181 (2002).
[13] O. M. Koo, I. Rubinstein, H. Onyuksel. *Nanomed-Nanotechnol.*, vol. 1(3), 193 (2005).
[14] S. Parveen, R. Misra, S. K. Sahoo. *Nanomed-Nanotechnol.*, 8(2), 147 (2012).
[15] D. Tsiourvas, A. Tsetsekou, M. Arkas, S. Diplas, E. Mastrogianni. *J. Mater. Sci.-Mater. M.*, 22(1), 85 (2011).
[16] P. Kolhe, E. Misra, R. M. Kannan, S. Kannan, M. Lieh-Lai. *Int. J. Pharm.*, vol. 259(1-2), 143 (2003).
[17] M. Seiler, D. Köhler, W. Arlt. *Sep. Purif. Technol.*, 30(2), 179 (2003).
[18] J. H. Cameron, A. Facher, G. Lattermann, S. Diele. *Adv. Mater.*, vol. 9(5), 398 (1997).

[19] R. M. Richardson, S. A. Ponomarenko, N. I. Boiko, V. P. Shibaev. *Liq. Cryst.* 26(1), 101 (1999).
[20] M. W. P. L. Baars, S. H. M. Söntjens, H. M. Fischer, H. W. I. Peerlings, E. W. Meijer. *Chem. Eur. J.*, 4(12), 2456 (1998).
[21] S. Bauer, H. Fischer, H. Ringsdorf. *Angew. Chem. Int. Ed.,* 32(11), 1589 (1993).
[22] D. Tsiourvas, T. Felekis, Z. Sideratou, C. M. Paleos. *Macromolecules*, 35(16), 6466 (2002).
[23] S. M. Burkinshaw, P. E. Froehling, M. Mignanelli. *Dyes Pigm.*, 53(3), 229 (2002).
[24] H. Bergenudd, P. Eriksson, C. De Armitt, B. Stenberg, E. Malmström Jonsson. *Polym. Degrad. Stab.*, 76(3), 503 (2002).
[25] A. Morikawa, M. Kakimoto, Y. Imai. *Macromolecules*, 24(12), 3469 (1991).
[26] T. H. Mourey, S. R. Turner, M. Rubinstein, J. M. J. Fréchet, C. J. Hawker, K. L. Wooley. *Macromolecules*, 25(9), 2401 (1992).
[27] G. Shah, P. L. Dubin, J. I. Kaplan, G. R. Newkome, C. N. Moorefield, G. R. J. Baker. *Colloid Interface Sci.*, 183(2), 397 (1996).
[28] A. M. Striegel, R. D. Plattner, J. L. Willett. *Anal. Chem.*, 71(5), 978 (1999).
[29] L. J. Bu, W. K. Nonidez, J. W. Mays, N. B. Tan. *Macromolecules*, 33(12), 4445 (2000).
[30] S. M. Ahmed, P. M. Budd, N. B. McKeown, K. P. Evans, G. L. Beaumont, C. Donaldson, C. M. Brennan. *Polymer*, 42(3), 889 (2001).
[31] B. Jendrusch-Borkowski, J. Awad, F. Wasgestian. *J. Inc. Phenom. Macromol. Chem.*, 35(1-2), 355 (1999).
[32] Y. J. Pan, W. T. Ford. *Macromolecules*, 33(10), 3731 (2000).
[33] S. van der Wal, Y. Mengerink, J. C. Brackman, E. M. M. de Brabander, C. M. Jeronimus-Stratingh, A. P. Bruins. *J. Chromatogr. A*, 825(2), 135 (1998).
[34] Y. J. Pan, W. T. Ford. *J. Org. Chem.*, 64(23), 8588 (1999).
[35] F. Tajarobi, M. El-Sayed, B. D. Rege, J. E. Polli, H. Ghandehari. *Int. J. Pharm.*, 215(1-2), 263 (2001).
[36] A. Polese, C. Mio, A. Bertucco. *J. Chem. Eng. Data*, 44(4), 839 (1999).
[37] S. A. Kuzdzal, C. A. Monnig, G. R. Newkome, C. N. Moorefield. *J. Chem. Soc., Chem. Commun.*, (18), 2139 (1994).
[38] N. Tanaka, T. Fukutome, K. Hosoya, K. Kimata, T. Araki. *J. Chromatogr. A*, 716(1-2), 57 (1995).
[39] N. Tanaka, T. Fukutome, T. Tanigawa, K. Hosoya, K. Kimata, T. Araki, K. K. Unger. *J. Chromatogr. A*, 699(1-2), 331 (1995).
[40] M. Castagnola, L. Cassiano, A. Lupi, I. Messana, M. Patamia, R. Rabino, D. V. Rossetti, B. Giardina. *J. Chromatogr. A,* 694(2), 463 (1995).
[41] N. Tanaka, H. Iwasaki, T. Fukutome, K. Hosoya, T. Araki. *J. High Res. Chromatogr.,* 20(10), 529 (1997).
[42] S. A. Kuzdzal, C. A. Monnig, G. R. Newkome, C. N. Moorefield. *J. Am. Chem. Soc.,* 119(9), 2255 (1997).
[43] A. L. Gray, J. T. Hsu. *J. Chromatogr. A*, 824(1), 119 (1998).
[44] C. Stathakis, E. A. Arriaga, N. J. Dovichi. *J. Chromatogr. A*, 817(1-2), 233 (1998).
[45] J. L. Haynes, S. A. Shamsi, J. Dey, I. M. Warner. *J. Liq. Chromatogr. Relat. Technol.*, 21(5), 611 (1998).
[46] D. Stöckigt, G. Lohmer, D. Beider. *Rapid Commun. Mass Spec.*, 10(5), 521 (1996).
[47] D. J. Pesak, J. S. Moore, T. E. Wheat. *Macromolecules*, 30(21), 6467 (1997).

[48] L. Balogh, A. de Leuze-Jallouli, P. Dvornic, Y. T. Kunugi, A. Blumstein, D. A. Tomalia. *Macromolecules*, 32(4), 1036 (1999).
[49] J. S. Choi, E. J. Lee, Y. H. Choi, Y. J. Jeong, J. S. Park. *Bioconjugate Chem.*, 10(1), 62 (1999).
[50] D. S. Shah, T. Sakthivel, I. Toth, A. T. Florence, A. F. Wilderspin. *Int. J. Pharm.*, 208(1-2), 41 (2000).
[51] H. Arima, F. Kihara, F. Hirayama, K. Uekama. *Bioconjugate Chem.*, 12(4), 476 (2001).
[52] M. Arkas, D. Tsiourvas, C. M. Paleos. *Chem. Mater.*, 7(13), 3439 (2005).
[53] M. Arkas, R. Allabashi, D. Tsiourvas, E. –M. Mattausch, R. Perfler. *Environ. Sci. Technol.*, 40(8), 2771 (2006).
[54] R. Allabashi, M. Arkas, G. Hörmann, D. Tsiourvas. *Water Res.*, 41(2), 476 (2007).
[55] A. V. Gaikwad, V. Boffa, J. E. Ten Elshof, G. Rothenberg. *Angew Chem - Int Edit*, 47(29), 5407 (2008).
[56] Q. Sun, K. Xu, W. Y. J. Lam, J. A. K. Cha, X. Zhang, B. Z. Tang. *Mater. Sci. Eng. C*, 16(1-2), 107 (2001).
[57] A. Tsetsekou, M. Arkas, A. Kritikaki, S. Simonetis, D. Tsiourvas. *J. Membr. Sci.*, 311(1-2), 128 (2008).
[58] R. M. Ottenbrite, H. Zeengin, J. Siddiqui. *Polym. Prepr.*, 39(2), 585 (1998).
[59] G. R. Newkome, K. S. Yoo, A. Kabir, A. Malik. *Tetrahedron Lett.*, 42(43), 7537 (2001).
[60] A. Kabir, C. Hamlet, K. S. Yoo, G. R. Newkome, A. Malik. *J. Chromatogr. A*, 1034(1–2), 1 (2004).
[61] Y. Pang, G. M. Zeng, L. Tang, Y. Zhang, Y. Y. Liu, X. X. Lei, Z. Li, J. C. Zhang. *Desalination*, 281(1), 278 (2011).
[62] Y. –M. Chung, H. –K. Rhee. *Catal. Lett.*, 82(3-4), 249 (2002).
[63] Y. Nakanishi, I. Toyoko. *J. Colloid Interf. Sci.*, 297(1) 122 (2006).
[64] P. R. Dvornic. *J. Polym. Sci. Part A*, 44(9), 2755 (2006).
[65] P. R. Dvornic, J. Li, A. M. de Leuze-Jallouli, S. D. Reeves, M. J. Owen. *Macromolecules*, 35(25), 9323 (2002).
[66] Y. Chujo, H. Matsuki, S. Kure, T. Saegusa, T. Yazawa. *J. Chem. Soc. Chem. Commun.*, 5, 635 (1994).
[67] N. Kröger, R. Deutzman, C. Bergsdorf, M. Sumper. *Proc. Natl. Acad. Sci. U. S. A.*, 97 (26), 14133 (2000).
[68] N. Kröger, R. Deutzman, M. Sumper. *Science*, 286(5442), 1129 (1999).
[69] M. Sumper, E. Brunner. *Adv. Funct. Mater.*, 16(1), 17 (2006).
[70] M. R. Knecht, D. W. Wright. *Chem. Commun.*, 9(24), 3038 (2003).
[71] T. Coradin, O. Durupthy, J. Livage. *Langmuir*, 18(6), 2331 (2002).
[72] J. Kobler, K. Möller, T. Bein. *ACS Nano*, 2(4), 791 (2008).
[73] M. R. Knecht, D. W. Wright. *Langmuir*, 20(11), 4728 (2004).
[74] M. R. Knecht, S. L. Sewell. *Langmuir*, 21(5), 2058 (2005).
[75] F. Zhang, S. –P. Yang, W. –M. Wang, H. –M. Chen, Z. -H. Wang, X. –B. Yu. *Mater. Lett.*, 58(26), 3285 (2004).
[76] M. Arkas, D. Tsiourvas. *J. Hazar. Mater.*, 170(1), 35 (2009).
[77] T. Mizutani, H. Nagase, N. Fujiwara, H. Ogoshi. *Bull. Chem. Soc. Jpn.*, 71(8), 2017 (1998).
[78] A. J. Khopade, S. Khopade, N. K. Jain. *Int. J. Pharm.*, 241(1), 145 (2002).

[79] F. Zhang, Z. H. Zhou, S. P. Yang, L. H. Mao, H. M. Chen, X. B. Yu. *Mater. Lett.*, 59(11), 1422 (2005).
[80] S. J. Yan, Z. H. Zhou, F. Zhang, S. P. Yang, L. Z. Yang, X. B. Yu. *Mater. Chem. Phys.*, 99(1), 164 (2006).
[81] Z. H. Zhou, P. L. Zhou, S. P. Yang, X. B. Yu, L. Z. Yang. *Mater. Res. Bull.*, 42(9), 1611 (2007).
[82] D. Tsiourvas, A. Tsetsekou, M. –I. Kammenou, N. Boukos. *J. Am. Ceram. Soc.*, 94(7), 2023 (2011).
[83] J. J. J. M. Donners, R. J. M. Nolte, N. A. J. M. Sommerdijk. *Adv. Mater.*, 15(4), 313 (2003).
[84] N. Tsubokawa, H. Ichioka, T. Satoh, S. Hayashi, K. Fujiki. *React. Funct. Polym.,* 37(1-3), 75 (1998).
[85] J. Bu, R. Li, C. W. Quah, K. Carpenter. *J. Macromolecules*, 37(18), 6687 (2004).
[86] C. –C. Chu, N. Ueno, T. Imae. *Chem. Mater.*, 20(8), 2669 (2008).
[87] S. C. Bourque, H. Alper, L. E. Manzer, P. Arya. *J. Am. Chem. Soc.*, 122(5), 956 (2000).
[88] J. Bu, Z. M. A. Judeh, C. B. Ching, S. Kawi. *Catal. Lett.*, 85(3-4), 183 (2003).
[89] E. J. Acosta, C. S. Carr, E. E. Simanek, D. F. Shantz. *Adv. Mater.*, 16(12), 985 (2004).
[90] S. Yoo, S. Yeu, R. L. Sherman, E. E. Simanek, D. F. Shantz, D. M. Ford. *J. Membr. Sci.*, 334(1-2), 16 (2009).
[91] K. Sooklal, L. H. Hanus, H. J. Ploehn, C. J. Murphy. *Adv. Mater.*, 10(14), 1083 (1998).
[92] M. E. Garcia, L. A. Baker, R. M. Crooks. *Anal. Chem.*, 71(1), 256 (1999).
[93] M. Manoth, K. Manzoor, M. K. Patra, P. Pandey, S. R. Vadera, N. Kumar. *Mater. Res. Bull.*, 44(3), 714 (2009).
[94] R. A. Gemeinhart, D. Luo, W. M. Saltzman. *Biotechnol. Prog.*, 21(2), 532 (2005).

In: Recent Advances in Ceramic Materials Research
Editors: Joan J. Roa Rovira and Merce Segarra Rubi
ISBN: 978-1-62417-729-3

Chapter 2

MODIFIED-(K,NA)NBO$_3$ PIEZOELECTRIC CERAMICS FOR 'HARD' LEAD-FREE MATERIALS

J. B. Lim[*]
MLCC R&D, LCR Division, Samsung Electro-Mechanics,
Suwon, Republic of Korea

ABSTRACT

Lead-free piezoelectric ceramics $(K_{1-X}Na_X)NbO_3$ modified by $K_4CuNb_8O_{23}$ [KNN-KCNx] with 0.0 mol% $\leq x \leq$ 1.0 mol% were synthesized using conventional ceramic processing. KNN-KCNx were observed to exist in the perovskite structure with orthorhombic symmetry. It is found that the addition of KCN was effective in enhancing the density and decreasing phase transition temperatures (T_{o-t}, T_c). The asymmetrical polarization hysteresis loops observed in present case is due to development of an internal bias field (E_i), resulting in high mechanical quality factor (Q_m) of >1500 with the dielectric loss of 0.6%. The formation of E_i is believed to be the result of defect dipoles of acceptor ions and oxygen vacancies, leading to piezoelectric "hardening" effect, by stabilizing and pinning of the domain wall motion. Furthermore, the piezoelectric properties were found to decrease and the coercive field (E_c) increased when compared with the undoped material, exhibiting typical characteristics of "Hard" behavior. The temperature usage range was limited by the polymorphic phase transition temperature. The full set of material constants was determined for the KNN-KCNx materials. Compared with "Hard" $Pb(Zr,Ti)O_3$(PZT) ceramics, the KNN-KCNx possessed lower dielectric and piezoelectric properties; however, comparable values of Q_m, dielectric loss, and E_c were obtained. To further compositionally optimize KNN-KCNx near morphotrophic phase boundary(MPB), the ratio of K/Na in $(K_{1-x}N_x)$N-0.5KCN was varied from $0.45 \leq x \leq 0.60$. Results revealed that increase in Na was effective in changing the microstructure and relative density of $(K_{1-x}N_x)$N-0.5KCN. Further, varying the Na resulted in a slight shift of the phase transition temperatures (T_{o-t} and T_c) toward low values. A high-Q_m of 1850 was found at x = 0.54, which might be due to the build-up of E_i within $(K_{1-x}N_x)$N-0.5KCN. Thermal hysteresis was confirmed with Na content, resulting from structural changes. Thus, $(K_{1-x}N_x)$N-0.5KCN with x = 0.54 exhibited

[*] Corresponding Author's e-mail address: limjongbong@gmail.com (J. B. Lim).

excellent piezoelectric properties with d_{33} (97 pC·N^{-1}), k_p (36%), and Q_m (1850). Therefore, acceptor modified-KNN lead-free materials are promising candidates for application in high-power piezoelectric devices.

For investigating its shear mode property, optimized KNN-0.5KCN was characterized as a function of temperature. Shear-mode KNN-0.5KCN was found to possess comparable electric resistivity, piezoelectric and electromechanical properties to pure KNN. The dielectric permittivity K^T_{11} and piezoelectric coefficients d_{15} were found to increase against temperature, while the electromechanical coupling factors k_{15} maintained similar value till phase transition temperature ($T_{o\text{-}t}$ ~ 200°C), above which, the properties dropped significantly due to depolarization. Good temperature-dependent properties, together with its high-Q_m, are demonstrating practical potential for high power shear mode applications.

1. INTRODUCTION

High power applications, such as ultrasonic motors, transformer and high intensity focused ultrasound (HIFU) driven with piezoelectric materials require "Hard" piezoelectrics with high mechanical quality factor (low mechanical loss) and low dielectric loss to deliver high acoustic power without excessive heat generation and/or thermal runaway [1,2].

Currently, "Hard" Pb(Zr,Ti)O_3 (PZT) piezoelectric ceramics, including PZT4 and PZT8 [3] have been the mainstay for high power applications. However, it has recently been desirable to develop lead-free piezoelectric ceramics with comparable performance to replace lead-based materials because of the toxicity of lead oxide in PZT ceramics [4]. For the increase in applications of high power, the PZT ceramics should be replaced with lead-free piezoelectric ceramics. Hence, the development of lead-free piezoelectric materials, such as tungsten bronze-type, Bi-layer structured, and perovskite materials, has been intensively studied [5-7]. Among the lead-free materials, alkali-based perovskite compounds in $K_xNa_{1-x}NbO_3$ are environmentally friendly materials and are promising candidates for alternative lead-based system [8-11]. The most importance factor in $K_xNa_{1-x}NbO_3$ system is the ratio of K to Na in $K_xNa_{1-x}NbO_3$. Even though most studies were carried out on pure $K_{0.5}Na_{0.5}NbO_3$ or $K_{0.5}Na_{0.5}NbO_3$-based systems [12-15], the ideal composition in $Na_xK_{1-x}NbO_3$ system is still not clear. To date, the highest piezoelectric properties in $K_xNa_{1-x}NbO_3$ system were observed at the optimum Na/K ratio of 0.535/0.480 [16]. In preliminary works, the highest properties in $0.95(K_xNa_{1-x})NbO_3$-$0.05LiTaO_3$ system was observed at the K/Na ratio of 0.45/0.55. According to previous studies on the Na/K ratio in $K_xNa_{1-x}NbO_3$, the Na-rich region exhibited higher piezoelectric properties than the K-rich region. For high power applications, optimized dielectric and piezoelectric properties can be obtained with the addition of $K_4CuNb_8O_{23}$ to Na-rich $(K_{0.45}Na_{0.55})NbO_3$[17,18]. To further study the theoretical and practical applications of hard lead-free piezoelectric materials, it is desirable to have a full set of materials constants, including elastic constant, electromechanical coupling factors, piezoelectric coefficients, and dielectric permittivity, and also understand the temperature-dependent piezoelectric behavior.

In the case of piezoelectric constant (d_{ij}), it has been well-known that the piezoelectric coefficient is usually large in the order of $d_{15} > d_{33} > d_{31}$.[19] A large value of the piezoelectric strain coefficient is useful for generating large deformation by low voltage in actuators and high power applications [20]. However, development of novel piezoelectric actuators and motors with systems would make use of the piezoelectric shear response of piezoelectric ceramics where the actuators are often subject to a high driving electric field. Understanding of how the shear response of the material varies with the driving field amplitude is crucial for the proper design and use of these materials. In the case of $K_xNa_{1-x}NbO_3$ system, however, much less is known about the piezoelectric coefficient d_{15} related to the shear strain S_5 which is attributed to the field induced displacements of (ferroelastic) non-180° walls when the electric field is applied perpendicular to the polar axis.

Therefore, (K,Na)NbO$_3$- $K_4CuNb_8O_{23}$ (KNN-KCN) in this study was systematically investigated for hard lead-free ceramics in relation to high field ferroelectric behavior and temperature dependence of the dielectric and piezoelectric properties. In addition, the shear-mode piezoelectric behavior for KNN-KCN was studied as a function of temperature and characterized with the full set of material constants determined according to IEEE piezoelectric standard [21.22].

2. Experimental Procedure

Conventional mixed oxide processing was used to prepare the KNN and KCN compositions, respectively. KNN and KCN were fabricated with high purity oxide powders (99%, Alfa Aesar, Ward Hill, MA), K_2CO_3, Na_2CO_3, Nb_2O_5, and CuO. Powders for KNN and KCN were stoichiometrically weighed and mixed using ball-milling in an anhydrous ethanol solution for 24 h. The mixtures were dried and calcined at 850°C for 3h. The calcined powders of KNN and KCN were vibratory-milled with an acrylic binder prior to pressing into pellets. After binder burn-out at 550°C, samples were sintered at 1100°C for 2h. The sintered samples were polished to 1 mm thickness, and then painted with silver paste on either sides followed by annealing at 800°C for 15 min. All samples sintered were first dried in the oven for about 30 min to minimize the effect of moisture on piezoelectric properties.

The density was measured by the Archimedes method. The phase purity and microstructure were examined by X-ray powder diffraction (XRD; D/MAX 2200, Rigaku, Japan) and scanning electron microscopy (FE-SEM; MIRA-II LMH, Tescan, USA). High field measurement of the *P-E* (polarization-field) hysteresis behavior was determined using a modified Sawyer-Tower circuit, driven by a high voltage power supply (TREK Model 610). After poling with 60 kV/cm electric field at 150°C, piezoelectric coefficient (d_{33}) was measured by a Berlincourt-d_{33} meter. For the measurement of shear-mode property, the electrodes were removed and redeposited by sputtering on the surface of a sample which is perpendicular to the poling direction to allow determination of the shear mode properties. The dielectric permittivity was measured as a function of temperature at 1 kHz using a multifrequency LCR meter (HP4284A, Hewlett-Packard, Palo Alto, CA). The shear mode impedance curves were measured by HP4194A impedance-gain phase analyzer as function of temperature, the electromechanical coupling coefficients (k_{15}) were calculated from the resonance frequencies (f_r) and anti-resonance frequencies (f_a) using the following equation:

$$k_{15}^2 = \frac{\pi}{2}\frac{f_r}{f_a}\cot\left(\frac{\pi}{2}\frac{f_r}{f_a}\right) \tag{1}$$

The piezoelectric coefficients (d_{15}) (strain constant) were calculated according to

$$d_{15}^2 = k_{15}^2 K_{11}^T s_{44}^E \quad (2)$$

where s_{44}^E is the elastic compliance (stiffness). The resistance (R) and resistivity ($\rho=1/\sigma$) were calculated from the current measured at 100 V by a source meter (Keithley 2410C) as a function of temperature.

3. Results

3.1. Effect of $K_4CuNb_8O_{23}$ in $(K,Na)NbO_3$ "Hard" Lead-free Materials

Fig. 1 shows X-ray diffraction patterns of the $(K_{0.45}Na_{0.55})NbO_3$-$xK_4CuNb_8O_{23}$[KNN-KCNx] with 0.0 mol % $\leq x \leq$ 1.0 mol% sintered at 1100°C for 2h. All of the prepared KNN were orthorhombic regardless of the KCN content, as shown in Fig. 1(a). When 0.7 mol% KCN was added, small peaks for KCN were detected and their intensity increased with increasing the amount of KCN up to 1.0 mol% doped level. It can be considered that below 0.5 mol% KCN, it reacted and formed a liquid phase which leads to the dense microstructures in KNN. Above 0.5 mol% KCN, on the other hand, it is possible that a large amount of KCN liquid phase was formed and crystallized during cooling and observed in KNN. Fig. 1(b) shows an enlarged diffraction patterns from 28 ° to 33 °. The relative intensity of (020)/(200) to (111) reflection shows differences between pure KNN and KCN-doped KNN as well as the shape of (220) peak also changes remarkably. It might be related to a phase transition between two orthorhombic phases. The composition in this study is very close to the boundary between L and M phases, according to Ahtees' work [23]. Therefore, the addition of a small amount of KCN may cause KNN to shift the phase composition in the vicinity of MPB.

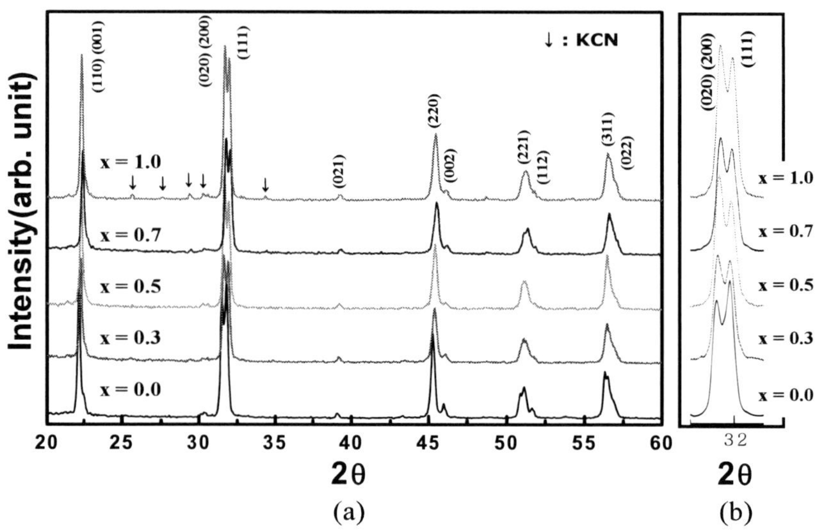

Figure 1. X-ray diffraction patterns of KNN-KCNx with 0.0 mol% $\leq x \leq$ 1.0 mol%.

Figure 2(a) shows the relative density of KNN–KCNx sintered at various temperatures for 2 h, respectively. When the sintering temperature was increased, the relative densities of KNN were increased and slightly saturated. In addition, the relative densities of KNN were found to be increased with KCN contents, giving a maximum value of 96.4% for KNN-KCN1.0. It can be inferred that the liquid phase related KCN was formed inside samples during the sintering and assisted the densification of KNN. Furthermore, the variation in relative density is closely related to the microstructure, as shown in Figure 2(b). Therefore, it can be concluded that KCN is effective in improving the sinterability of KNN, and the optimum sintering temperature of KNN was found to be 1100°C.

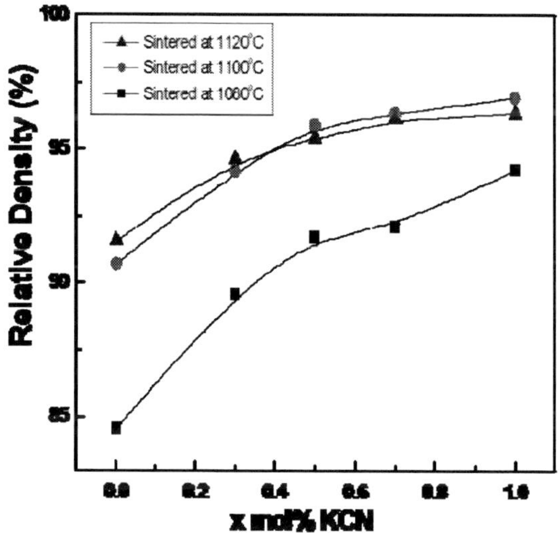

(a)

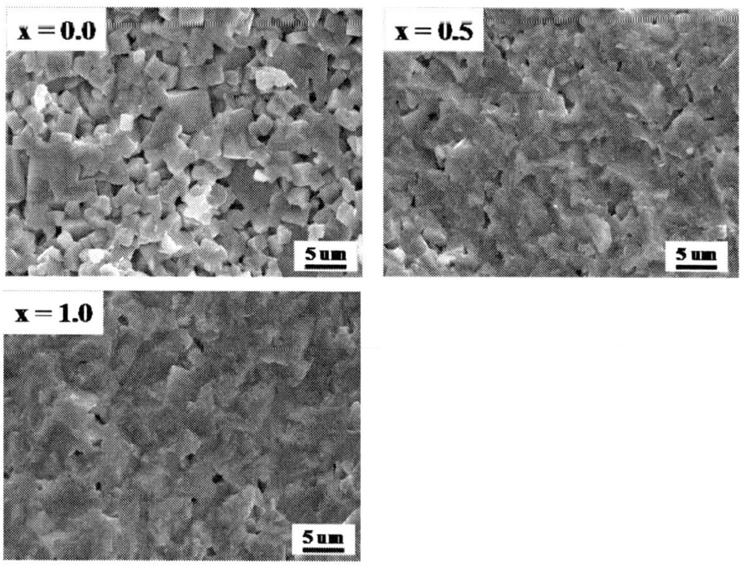

(b)

Figure 2. (a) Relative density and (b) fractured surface sintered at 1100°C for KNN-KCNx.

At room temperature, the dielectric permittivity and *tan δ* measured at 1 kHz were found to be on the order of 330 and 0.2% for the KNN-KCN0.5, respectively. Figure 3 shows the temperature dependence of the dielectric permittivity of KNN-KCNx. The inset in Figure 3 shows the phase transition temperatures of $T_{o\text{-}t}$ and T_c, respectively, which is the same as the sequence of phase transition of $KNbO_3$ [24]. For pure KNN, the phase transition temperatures were observed at 211°C of $T_{o\text{-}t}$ and 419°C of T_c, respectively. With increasing KCN content, both $T_{o\text{-}t}$ and T_c were slightly shifted downward and then saturated up to KNN-KCN1.0. Therefore, the addition of KCN can cause the phase transition temperature to decrease in the KNN.

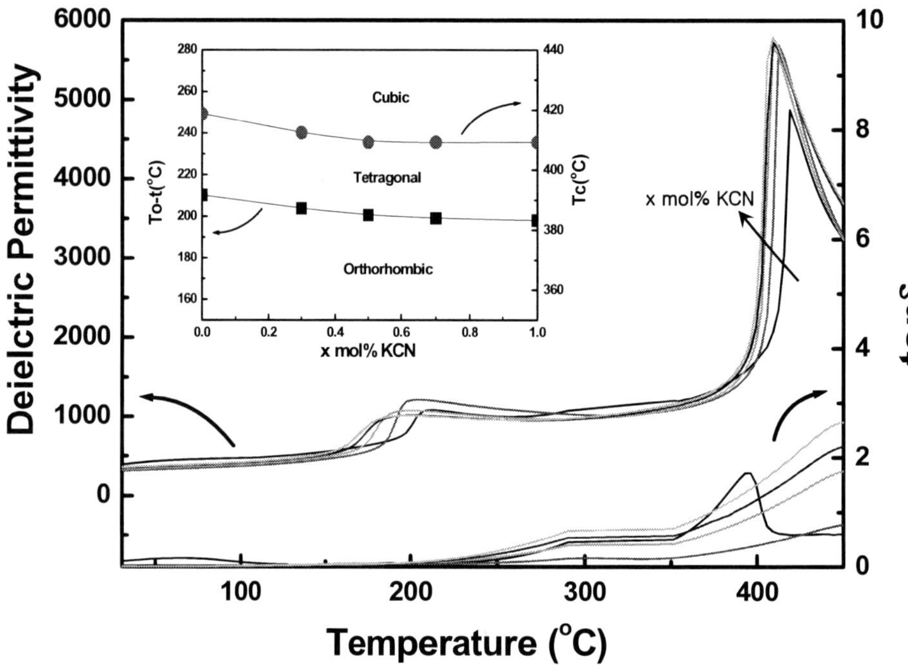

Figure 3. Temperature dependence of the dielectric permittivity of KNN-KCNx with 0.0 mol% ≤ x ≤ 1.0 mol% . The inset shows the variation of $T_{o\text{-}t}$ and T_c.

Piezoelectric properties of KNN were also measured as a function of KCN content. Figure 4 shows piezoelectric constant (d_{33}), electromechanical coupling factor (k_p), and mechanical quality factor (Q_m) for KNN-KCNx. The value of d_{33} for pure KNN was 64 pC/N (Figure 4). With increasing the KCN content, however, the value of d_{33} of KNN increased and then slightly decreased, giving a maximum value of 100 pC/N at KNN-KCN0.5. The k_p also exhibited a tendency similar to that of d_{33}, which reached a maximum value of 33% for KNN-KCN0.5. The small decrease of d_{33} and k_p up to 1.0 mol% might be due to the coexisting KCN observed in Figure 1 (XRD). The Q_m for pure KNN in Figure 4(b) was low, approximately 100. When the small amount of KCN was added, Q_m was significantly increased up to 1500 for KNN-KCN0.5, which is 4 times higher than that of the hot-pressed ceramics [25]. Takahashi *et al.* reported in PZT that the Q_m is affected by the internal bias field (E_i), which is due to the formation of complexes of acceptor and oxygen vacancies [26]. Moreover, the E_i should reduce hysteresis loss, especially through prohibiting domain wall motions [26]. Therefore, it can be explained that the combination of acceptor Cu^{2+} and oxygen vacancy has to be along one of the pseudocubic [001] direction, while the spontaneous polarization in orthorhombic KNN is known to develop along one of the pseudocubic [011] direction. The angle between them might be around 45°, 135°, or 90° degrees. Preferable alignment of these defect associates may lead to internal bias field and thus result in properties similar to hard PZT.

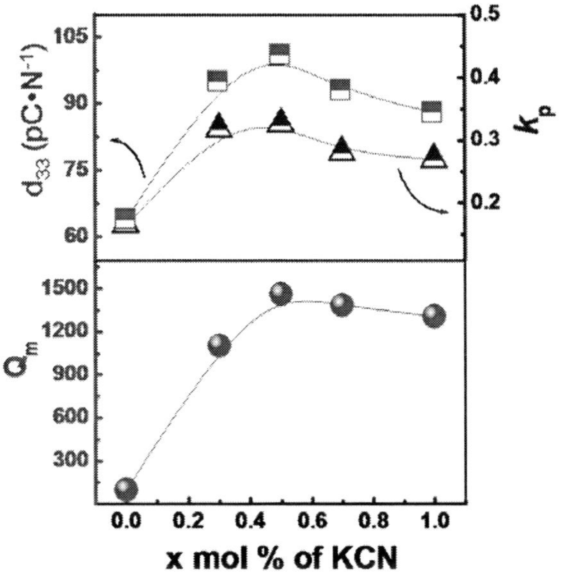

Figure 4. Piezoelectric properties of KNN-KCNx with 0.0 mol% ≤ x ≤ 1.0 mol%.

In order to characterize the ferroelectricity, *P-E* loops of KNN were evaluated as a function of applied field. Figure 5 shows the *P-E* loops of KNN-KCNx with 0.0 mol% ≤ x ≤ 1.0 mol%. P_r and E_c for pure KNN were 21 μC·cm^{-2} and 13 kV·cm^{-1}, respectively, which were found to maintain similar values with increasing KCN content (Figure 5). Furthermore, it was found that internal bias field (E_i) defined by the shift of ferroelectric hysteresis loops along the field axis was formed, as shown in the inset of Figure 5. Based on *P-E* loops in Figure 5, the variation of E_c and E_i are illustrated as a function of KCN contents in Figure 6. As can be seen in Figure 6, E_c shows small variation with the KCN content, which indicates that E_c is less dependent to the KCN content. In contrast, the E_i sharply increased to 4.1 kV/cm for 0.5 mol% KCN and then saturated up to 1.0 mol% KCN, which is similar to the behavior of Q_m value. According to these results, it can be suggested that E_i could be developed and increased for KNN-KCN0.5 because the number of defect dipoles, oxygen vacancy-acceptor increased with KCN content. A similar behavior was found by K. Carl *et. al.*[27].

In order to further confirm a relation between E_i and Q_m, a *P-E* loop and Q_m-value for the KNN-KCN0.5 were measured as a function of temperature, ranging from -100°C to 25°C in Figure 7. As can be seen, E_c is linearly increased up to 20 kV·cm^{-1} against a decrease in temperature because the domain wall was hard to be switchable. However, E_i was slightly increased until -25°C and saturated even with further decreasing temperatures up to 4.7 Kv·cm^{-1} at -100°C. With a decrease of temperature, Q_m was abruptly decreased until -25°C and, then, seemed to be gradually saturated with further decreasing temperatures. From these results, it could be considered that E_i is not much affected by the temperature, and the behavior of E_i is similar to that of Q_m-value. A study on the relation between internal bias E_i and Q_m needs further experimental works to better understand for hard materials.

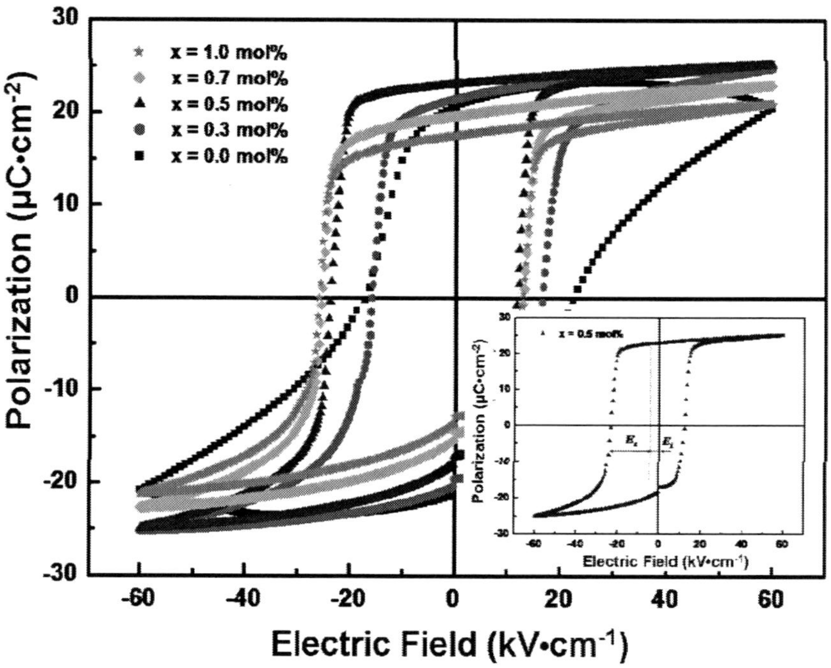

Figure 5. P-E hysteresis loops for KNN with various KCN contents.

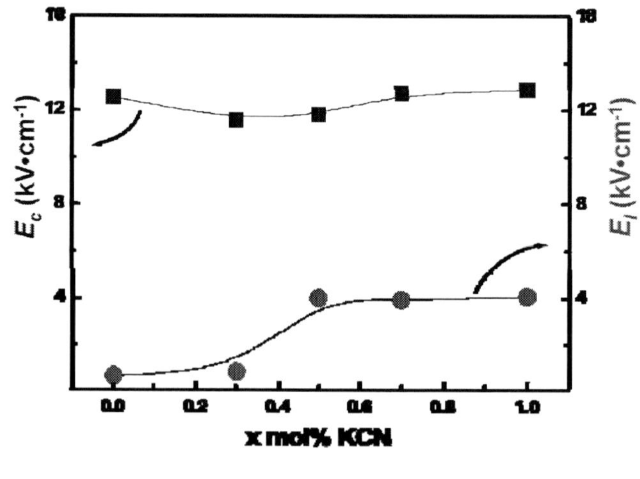

Figure 6. Dependence of E_c and E_i for KNN as a fuction of KCN content.

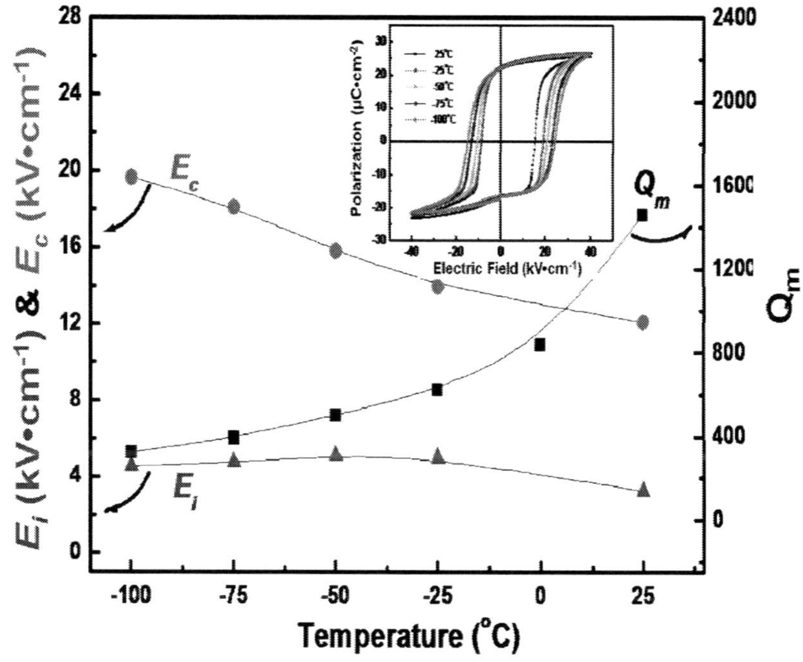

Figure 7. Temperature dependence of E_c, E_i and Q_m for KNN-KCN0.5.

Table 1. Elastic compliance s_{ij} (10^{-12} $m^2 \cdot N^{-1}$) and elastic stiffness c_{ij} (10^{10} $N \cdot m^{-2}$)

Material	s_{11}^E	s_{33}^E	s_{44}^E	s_{11}^D	c_{11}^E	c_{33}^E	c_{44}^E	c_{11}^D
KNN-KCN0.5	8.6	9.3	22.0	8.2	14.6	12.9	4.6	14.8
PZT4	12.3	15.5	39.0	10.9	13.9	11.5	2.6	14.5
PZT8	11.5	13.5	31.9	10.4	13.7	12.3	3.1	14.0

Table 2. Piezoelectric Coefficients, d_{ij} ($pC \cdot N^{-1}$), e_{ij} ($C \cdot m^{-2}$), g_{ij} ($10^{-3}Vm \cdot N^{-1}$), h_{ij} ($10^8 V \cdot m^{-1}$), Electromechanical coupling factors k_{ij} and mechanical quality Q_m

Material	d_{33}	d_{31}	d_h	e_{33}	e_{31}	g_{33}	g_{31}	h_{33}	h_{31}
KNN-KCN0.5	100	-32	26	8.4	-2.2	34.9	-12.3	50.1	-13.2
PZT4	289	-123	37	15.1	-5.2	25.1	-10.7	26.87	-9.25
PZT8	225	-97	31	13.2	-4.0	25.4	-10.9	25.71	-7.79
Material	k_{33}	k_{31}	k_t	k_p	Q_m				
KNN-KCN0.5	0.60	-0.21	0.50	0.36	1500				
PZT4	0.70	-0.33	0.51	0.58	500				
PZT8	0.64	-0.30	0.48	0.51	1000				

Table 3. Dielectric constants, ε_{ij}, and dielectric impermeability constants, $\beta(10^{-4}\cdot e_0^{-1})$, for commercial PZT

Material	ε^T_{33}	ε^T_{11}	ε^S_{33}	ε^S_{11}	β^T_{33}	β^T_{11}	β^S_{33}	β^S_{11}
KNN-KCN0.5	292	395	190	310	34.2	25.3	52.6	32.3
PZT4	1300	1475	635	730	7.69	6.78	15.75	13.70
PZT8	1000	1290	580	900	10.00	7.75	17.24	11.11

The full-set of material constants of lead-free materials, including elastic constants, piezoelectric coefficients, electromechanical coupling factors, and dielectric permittivity, are determined and listed in Tables 1, 2, and 3, and compared to the values of hard-PZT materials. The elastic compliance constants of KNN-KCN0.5 were found to be lower than the values of PZT-based hard materials, with higher elastic stiffness constants. The piezoelectric strain coefficients d_{ij} and piezoelectric stress coefficients e_{ij} were found to decrease when compared to the hard-PZT materials, while the piezoelectric voltage coefficients, g_{ij} , and piezoelectric stiffness coefficients, h_{ij} , increase due to its low dielectric permittivity, being on the order of 292.

3.2. Effect of K/Na Ratio in Modified-(K,Na)NbO_3 "Hard" Lead-free Materials

X-ray diffraction patterns of 0.55 mol% $K_4CuNb_8O_{23}$-added $(K_{1-x}Na_x)NbO_3$ [KCN-KNNx] with $0.45 \leq x \leq 0.60$ are shown in Figure 8(a). All the prepared samples exhibited similar diffraction patterns, representing a single orthorhombic perovskite structure. No second phase was seen over the wide composition range of x = 0.45 - 0.60 (Figure 8(a)). Figure 8(b) shows enlarged diffraction patterns, from 30.5° to 32.8°. The main diffraction peaks shifted slightly to higher diffraction angles with increasing x. This is readily explained by the smaller ionic radius of Na^+ (1.39 Å) as compared to that of K^+ (1.66 Å). In contrast to previous reports [28], the relative intensities of the main peaks for (002) and (111) remained unchanged with increasing x, indicating that no phase transition was identified by the change of K/Na ratio in this range. According to phase diagram of $KNbO_3$-$NaNbO_3$ [23], the morphotrophic phase boundary (MPB) region existing two orthorhombic phases (O_1 and O_2) is very narrow. In this study, however, the changing interval of K/Na ratio in KCN-KNNx was relatively large, being of the order of 1.0 mol% near the MPB region. Therefore, it can be considered that the variation of K/Na ratio in KCN-KNNx is too large to identify the phase transition of two orthorhombic phases.

Relative densities and fractured surface of KCN-KNNx are shown in Figure 9(a). For x = 0.45, the relative density was found to be low, with ~ 92.5%. When x was increased, the relative density increased up to 95.7% for x = 0.55, and then slightly decreased with a further increase in x. In the case of the fractured surface shown in Figure 9(b), the sample with x = 0.45 exhibited a relatively porous microstructure. However, with an increase in Na content(x), the density of the microstructure increased (Figure 9(a)). Despite the addition of KCN as a sintering agent, the relative densities of $x \leq 0.50$ were observed to be low with porous microstructures. According to our previous experimental data, K-rich KNNx without dopant usually show a higher relative density than Na-rich one. This suggests that K, rather than Na, is more effective in decreasing the sintering temperature of KNNx. Thus, we can infer that when x ≤ 0.50 in K-rich KCN-KNNx, both K and KCN formed a large amount of liquid phase during sintering, resulting in the formation of the porous microstructure, and they existed as an amorphous phase at room temperature, which caused no observation of second phase. In contrast, it can be suggested that for x ≥ 0.53 in Na-rich KCN-KNNx, the increase in x induced a decrease in the liquid phase during sintering. Thus, the Na content can be varied to enable the optimization of the amount of liquid phase and the densification of KCN-KNNx.

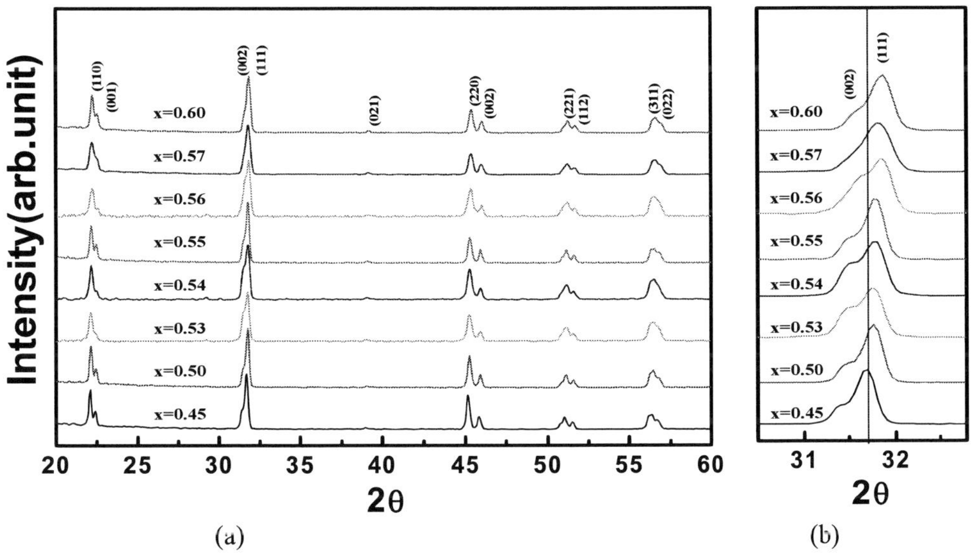

Figure 8. X-ray diffraction patterns of (a) KCN-KNNx with 0.45 ≤ x ≤ 0.60 and (b) enlarged diffraction patterns from 30.5° to 32.8°.

The temperature dependence of the dielectric permittivity of KCN-KNNx was measured at 100 kHz during the heating cycle, as shown in Figure 10. For x = 0.45, the orthorhombic–tetragonal transition temperature (T_{o-t}) was 197.7°C and the Curie temperature (T_c) was 399.5°C. With an increase in Na content(x), both the phase transition temperatures were found to slightly shift downward to 190.2°C(T_{o-t}) and 394°C(T_c) for x = 0.6, respectively. Thus, we can conclude that increasing x can cause the phase transition temperature to decrease. The inset of Figure 10 shows the variation in the phase transition temperatures (T_{o-t} and T_c) through thermal hysteresis during the heating and cooling cycles, respectively. With increase in x, both phase transition temperatures (T_{o-t} and T_c) were found to be in thermal hysteresis, but it was not significant. Usually, the different temperature behavior at the phase transition region related to thermal hysteresis is very sensitive to the temperature ramping rate during the heating and cooling cycles. In this study, to minimize the influence of thermal fluctuation on thermal hysteresis, we have tried to control exactly both the heating and cooling rate at $1^{o}C \cdot min^{-1}$, respectively, but we cannot also exclude the possibility of temperature ramping rate effect completely, affecting to thermal hysteresis. In addition, even though Na^+ (1.39 Å) was substituted for K^+ (1.66 Å) in KCN-KNNx, attributed to a structural distortion the variation in thermal hysteresis was not significant as a function of Na content. Therefore, it can be suggested that the thermal hysteresis revealed in this system might be related to the different temperature-dependent behavior of two phases at the phase transition regions (T_{o-t} and T_c), which is the first order phase transition during the heating and cooling cycles, respectively.

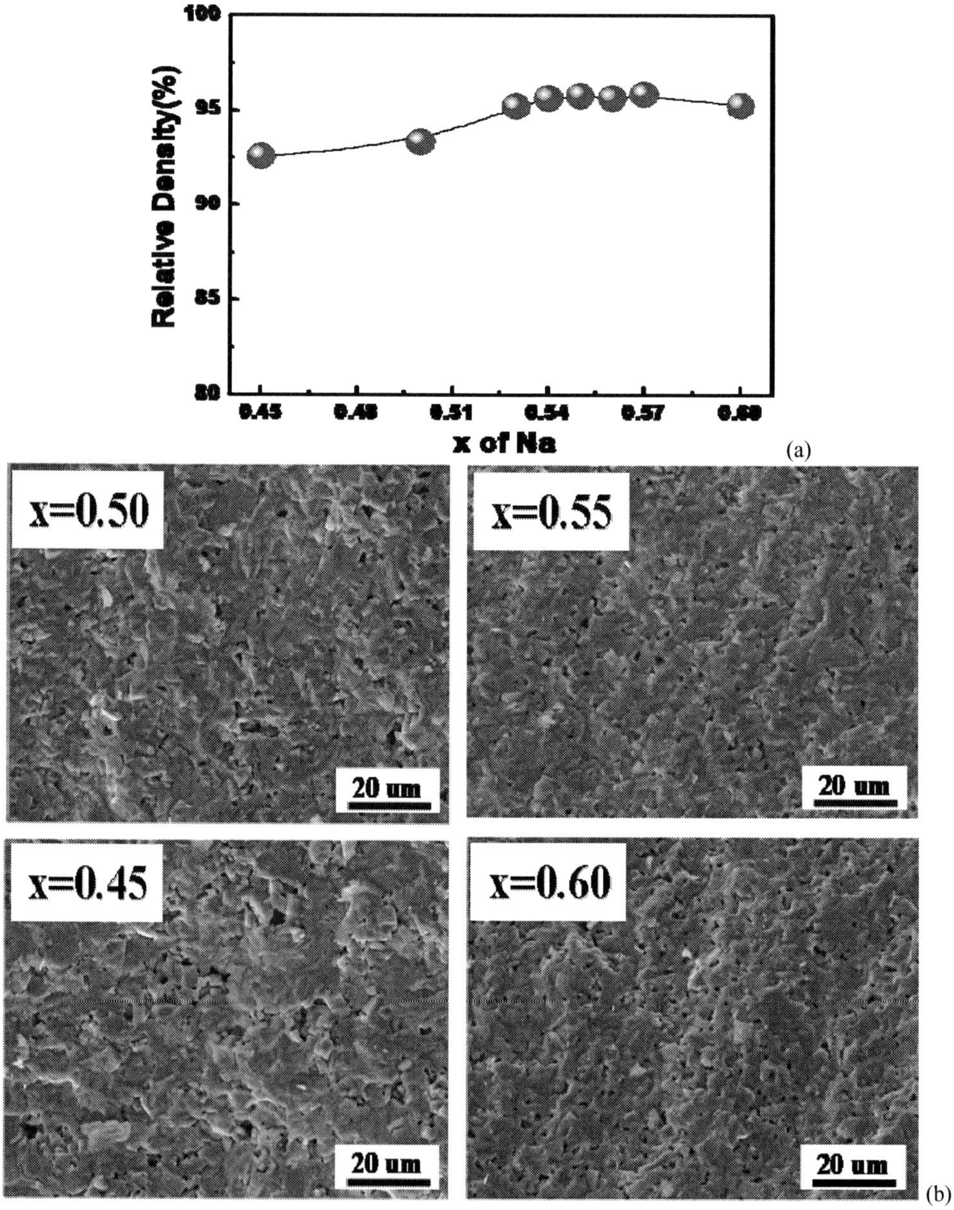

Figure 9. (a) Relative density and (b) fractured surface of KCN-KNNx as a function of x.

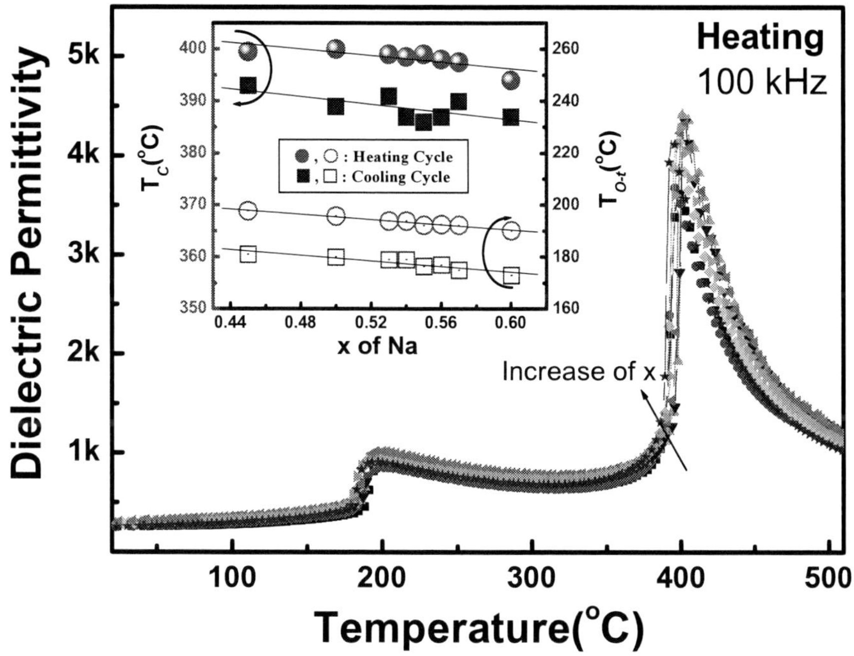

Figure 10. Temperature dependence of the dielectric permittivity of KCN-KNNx during the heating cycle. The inset in Figure 10 shows the variations of $T_{o\text{-}t}$ and T_c through the heating and cooling cycles.

Figure 11 shows the piezoelectric constant (d_{33}) and dielectric permittivity (ε_r) of KCN-KNNx as a function of Na content (x). As can be seen, with an increase in x, the d_{33} showed a small variation, with a maximum value of 98 pC·N^{-1} for x = 0.56, whereas the value of ε_r increased slightly up to 294 for x = 0.57. However, both d_{33} and ε_r exhibited similar tendency, i.e., small variations with changes in Na content (x).

Figure 12 shows *P-E* hysteresis loops of KCN-KNNx with $0.45 \leq x \leq 0.60$ as a function of the Na content and frequency. As shown in Figure 12(a), the *P-E* loops exhibited a typical square shape as "Hard" lead-free materials. The remnant polarization (P_r) in the inset of Figure 12 was plotted against x, and it was observed to increase slightly with x, which is not significat. Figure 12(b) shows the frequency dependence of the *P-E* hysteresis loops for x = 0.54 measured in the range 2–100 Hz. As the frequency was below 10 Hz, no frequency relaxation was observed, indicating nearly constant true remnant polarization and coercive field (E_c). However, increasing the frequency to 100 Hz dramatically reduced both P_r and E_c because of polarization relaxation resulting from the loss of switchable remnant polarization in the ferroelectric materials. Besides, according to the above results for frequency dependence, it can be considered that at low frequencies (≤ 10 Hz), the KCN-KNNx system is more stable than commercial "Hard" materials such as PZT-4H [29]. This finding shows that KCN-KNNx are promising candidates for high power applications.

The mechanical quality factor (Q_m) was measured as a function of Na content (x), as shown in Figure 13. The value of Q_m was found to be on the order of 820 for x = 0.45. With increase of Na content (x), the value of Q_m increased significantly up to 1850 for x = 0.54, which is seven times greater than that of hot-pressed ceramics [25]. With further increase to x = 0.60, the value of Q_m decreased to 1100.

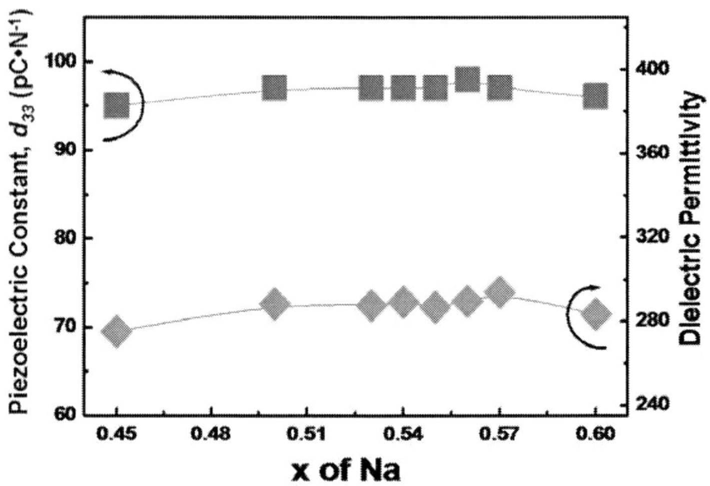

Figure 11. Piezoelectric constant (d_{33}) and dielectric permittivity (ε_r) of KCN-KNNx as a function of x.

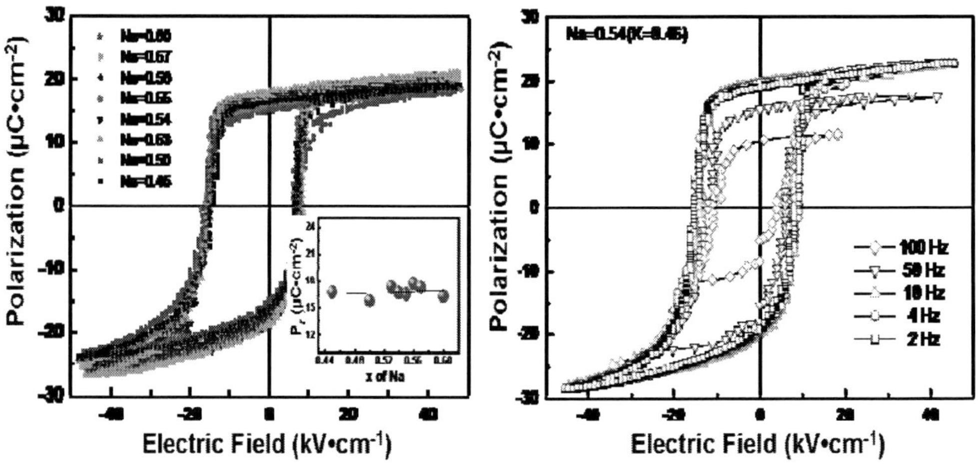

(a) (b)

Figure 12. P-E hysteresis loops of KCN-KNNx with $0.45 \leq x \leq 0.60$ as a function of (a) the Na content and (b) frequency.

The inset of Figure 13 shows the variations in E_c and internal bias (E_i) for the P-E loops shown in Figure 12, respectively. Both E_c and E_i in the inset of Figure 13 exhibited small variation, but it showed almost no dependence against x. In addition, high E_i by KCN was apparent in all KNNx samples, regardless of increase in Na content (x) (see the inset of Figure 13). According to previous reports [26, 28], high Q_m is affected by the formation of E_i through the acceptor and oxygen vacancy complexes. Moreover, E_i reduces hysteresis loss, especially by prohibiting domain wall motions [26]. Even though high E_i was seen for $x \leq 0.50$, the low Q_m in this study cannot be explained by the formation of E_i. In general, the Q_m is also sensitive to microstructural variations, such as those in density, liquid phase, grain size, etc. Based on the microstructural analysis results in Figure 9, it can be suggested that the formation of the large amount of liquid phase related to K^+ and KCN with a porous microstructure could be attributed to the reduction in the energy barrier of the defect migration and the development in a local depolarization field, which led to the observed decrease in the Q_m for KCN-KNNx with x ≤ 0.50. This is the first report of a systematic study on the variation of Q_m with changes in the K/Na ratio for "Hard" lead-free KNNx. The study on the effect of K/Na ratio needs further experimental work to gain a better understanding of the "Hard" lead-free materials. Detailed piezoelectric properties of KCN-KNN0.54 compared to PZT8 materials are summarized in Table 4.

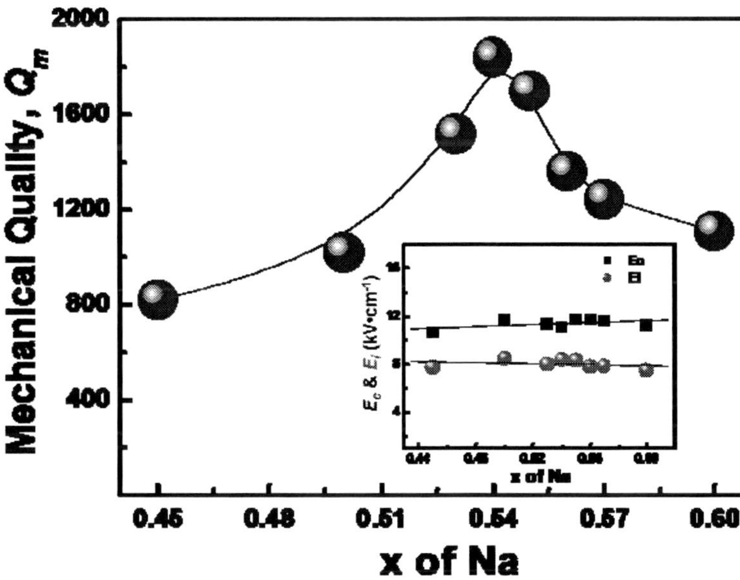

Figure 13. Variation of mechanical quality factors (Q_m) as a function of x. Inset describes the variations of coercive field (Ec) and internal bias (E_i).

Table 4. Detailed piezoelectric properties of KCN-KNN0.54 compared to PZT8 materials

Material	T_{ot} / T_c (°C)	d_{33} (pC·N^{-1})	ε^T_{33}	k_p	k_t	Q_m	$tan\delta$
KCN-KNN0.54	193 / 398	97	290	0.36	0.53	1850	0.003
KNN	215 / 420	95	205	0.40	0.46	230	0.030
PZT8	-/300	225	900	0.51	0.48	1000	0.004

3.3. Shear Mode Property in Modified-(K,Na)NbO$_3$ "Hard" Lead-free Materials

The measured resistivity (ρ_0) for (K$_{0.45}$Na$_{0.55}$)NbO$_3$[KNN] and (K$_{0.45}$Na$_{0.55}$)NbO$_3$-0.5 mol% K$_4$CuNb$_8$O$_{23}$[KNN-KCN0.5] was illustrated as a function of temperature(200°C - 500°C), shown in Figure 14. The resistivity of KNN and KNN-KCN0.5 were found to decrease linearly with increasing temperature. At high temperature, the resistivity of KNN-

KCN0.5 was higher when compared to the resistivity for KNN. The activation energy (E_a) was calculated from the slopes in Figure 14 according to Arrhenius equation:

$$\rho = \rho_0 \exp\left(\frac{-E_a}{\kappa T}\right) \tag{3}$$

where ρ_0 is pre-exponential factor, k is Boltzmann's constant, T is absolute temperature in degrees Kelvin [K]. The activation energy for KNN-KCN0.5 was found to be on the order of 1.06 eV, higher than KNN (0.88 eV), as shown in Figure 14. The activation energy is expected to reflect defect migration energy at high temperatures. Thus the high activation energy was thought to be related to the long range defects (acceptor–oxygen vacancy) ordering, which effectively traps the defects in deeper potential wells to overcome higher activation energy as they move through the lattice. Therefore, it can be considered that the higher resistivity for KNN-KCN0.5 makes the oxygen (vacancies) diffusion more difficult, thus higher activation energy is needed to overcome the thermal band gap at high temperature.

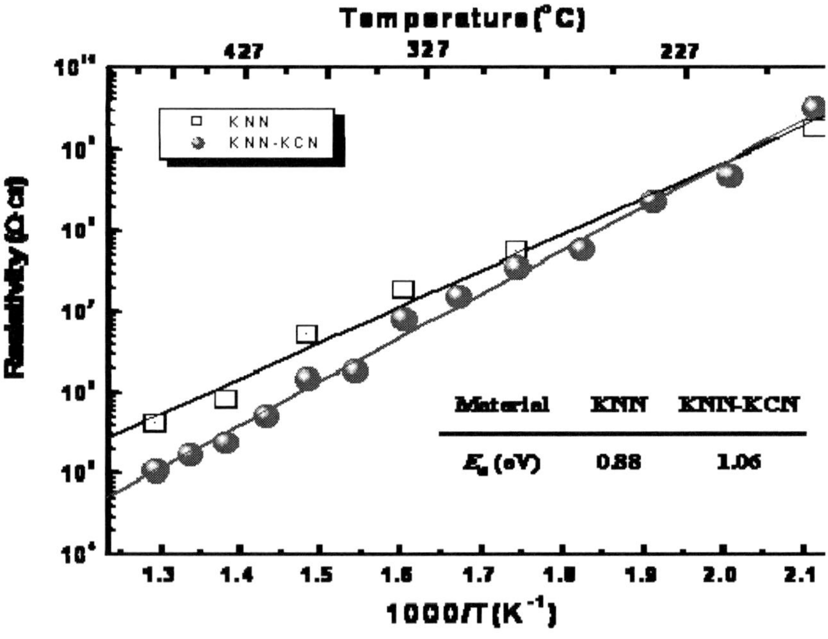

Figure 14. Variation of the resistivity for KNN-KCN and unmodified-KNN as a function of temperature.

Figure 15 shows dielectric permittivity K_{11}^T and dielectric loss as a function of temperature (20°C ~ 200°C) for KNN-KCN0.5. For comparison, unmodified KNN is also plotted in Figure 15. The dielectric permittivity and loss of KNN-KCN0.5 at room temperature were found to be on the order of 425 and 0.3%, respectively. The permittivity was found to be higher than the pure KNN, with lower dielectric loss (K_{11}^T=290 and $tan\delta$ =

3% for pure KNN). Dielectric permittivity for KNN-KCN0.5 gradually increased with temperature, till 180°C, approaching its phase transition temperature (orthorhombic to tetragonal phase (T_{o-t} ~ 188°C))[28]. The dielectric loss (*tan δ*) for KNN-KCN0.5 was found to follow the same trend as observed in dielectric permittivity, reaching peak value of 6% around 180°C.

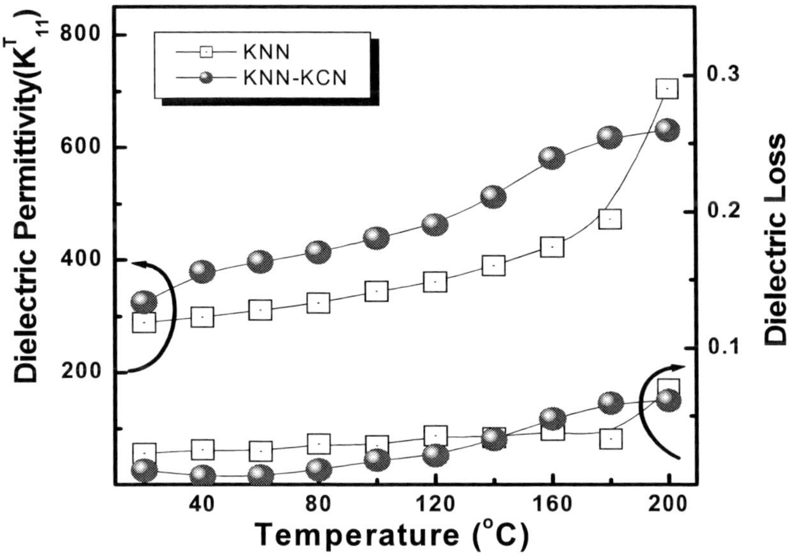

Figure 15. Temperature dependence of dielectric permittivity K_{11}^{T} and dielectric loss.

The temperature dependence of the shear-mode piezoelectric coefficient and electromechanical coupling factor for KNN and KNN-KCN0.5 were shown in Figure 16 over the temperature range from 20°C ~ 200°C. Both KNN and KNN-KCN0.5 exhibited an increase in piezoelectric coefficient at phase transition temperature, which is similar to the temperature dependent dielectric permittivity. The electromechanical coupling factors for KNN and KNN-KCN0.5, on the other hand, exhibited minimal temperature dependence, resulting in a stable coupling over the temperature range studied.

The elastic compliance also demonstrated the similar temperature dependent properties to dielectric and piezoelectric properties (d_{15}), as shown in Figure 17. For unmodified KNN, the elastic compliance at room temperature was found to be on the order of $38 \cdot 10^{-12}$ $m^2 \cdot N^{-1}$, while KNN-KCN0.5 exhibited to be lower than that of the unmodified KNN, being on the order of $33 \cdot 10^{-12}$ $m^2 \cdot N^{-1}$, owing to the "hardening" effect by the acceptor–oxygen vacancy defect dipoles.[17-18,27-28] The detailed dielectric and piezoelectric properties for KNN-KCN0.5 and unmodified KNN were summarized and compared to commercial PZT8 in Table 5. As listed in Table 5, piezoelectric properties of KNN-KCN0.5 as "Hard" lead-free materials exhibited comparable properties to commercial PZT8 ceramics.

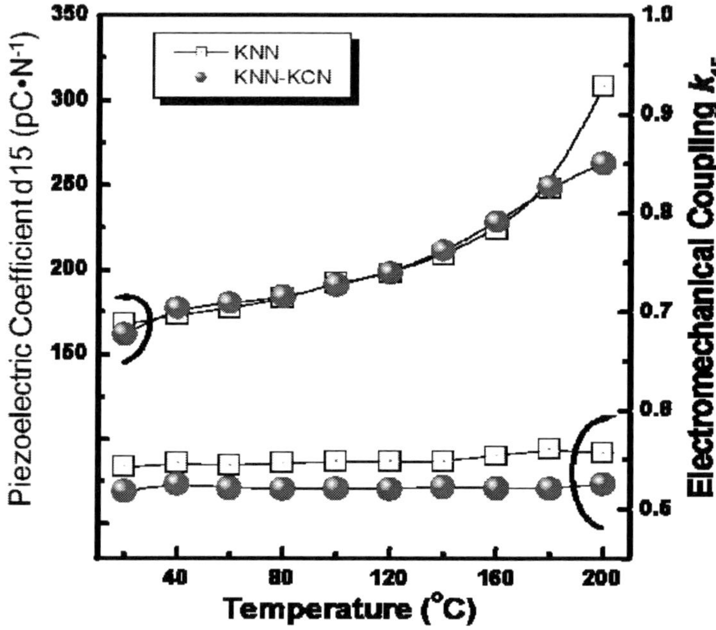

Figure 16. Variations in the piezoelectric coefficient d_{15} and electromechanical coupling factor k_{15} as a function of temperature.

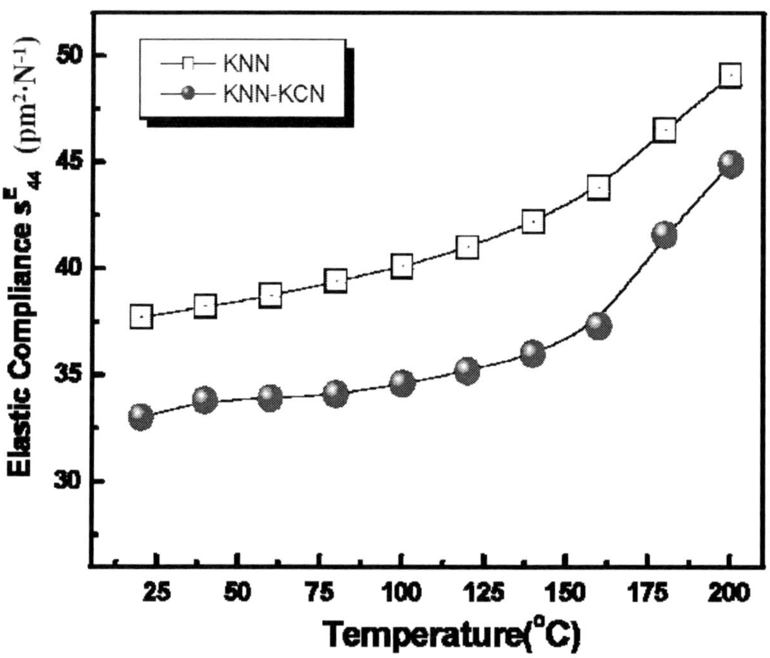

Figure 17. Elastic compliance constant s_{44}^{E} as a function of temperature.

Table 5. Detailed shear mode electrical properties of KNN-based lead-free materials compared to PZT8 materials

Material	T_{ot} / T_c (°C)	d_{15} (pC·N^{-1})	ε_{11}^T	ε_{11}^s	k_{15}	s_{44}^E (m^2·N^{-1})	c_{44}^E (N·m^{-1})	g_{15} (Vm·N^{-1})
KNN-KCN	188 / 410	162	425	310	0.52	$33 \cdot 10^{-12}$	$4.0 \cdot 10^{10}$	$43 \cdot 10^{-3}$
KNN	215 / 420	168	290	205	0.54	$38 \cdot 10^{-12}$	$2.7 \cdot 10^{10}$	$65 \cdot 10^{-3}$
PZT8	- / 326	1330	1290	900	0.55	$32 \cdot 10^{-12}$	$3.1 \cdot 10^{10}$	$29 \cdot 10^{-3}$

CONCLUSION

(KNa)NbO$_3$-K$_4$CuNb$_8$O$_{23}$ system was synthesized by conventional solid-state method and characterized as "Hard" lead-free piezoelectric ceramics.

Effect of K$_4$CuNb$_8$O$_{23}$ on phase structure and piezoelectric properties of (K$_{0.45}$Na$_{0.55}$)NbO$_3$ [KNN-KCNx] with 0.0 mol% $\leq$ x $\leq$ 1.0 mol% was investigated. The addition of KCN was effective in improving the sinterability and, as a consequence, piezoelectric properties.

The phase transition temperature of KNN modified by KCN was found to be shifted toward low temperatures.

The value of Q_m was found to be on the order of 1500 for KNN-KCN0.5, which is due to the stablization of domain wall movement by the build-up of E_i. Therefore, acceptor modified-KNN system could be expected to be excellent candidate of the hard lead-free piezoelectric materials for application in high-power piezoelectric devices.

Variation of the ratio of K/Na in 0.55 mol% K$_4$CuNb$_8$O$_{23}$-added (K$_{1-x}$Na$_x$)NbO$_3$ [KCN-KNNx] was characterized. Varying x with 0.45 $\leq$ x $\leq$ 0.60 in KCN-KNNx was effective in altering the microstructure, and thus the piezoelectric properties. Increase in x resulted in shifting of the phase transition temperature towards lower values, and inducing thermal hystersis.

The value of Q_m was found to be high, i.e., ~ 1850, for KCN-KNN0.54. This result was attributed to formation of E_i caused by defect dipole pinning domain wall movement. However, when x $\leq$ 0.50, the Q_m was relatively low because of the formation of a large liquid phase, resulting in a porous microstructure.

Shear-mode piezoelectric properties of pure (K$_{0.45}$Na$_{0.55}$)NbO$_3$ and (K$_{0.45}$Na$_{0.55}$)NbO$_3$-0.5 mol% K$_4$CuNb$_8$O$_{23}$ [KNN-KCN] were characterized as a function of temperature. The dielectric, piezoelectric, and electromechanical properties of KNN-KCN were found to be comparable to the commercial PZT8 ceramics. Compared to its pure counterpart, the KNN-KCN materials exhibited comparable electric resistivity and piezoelectric properties with improved values of Q_m. Both the pure and doped KNN materials show flat temperature dependence of the shear electromechanical coupling factors with moderate increasing of dielectric and piezoelectric properties. Therefore, for KNN-KCN system, further experimental studies are needed to better understand the theoretical and practical applications of "Hard" lead-free materials.

REFERENCES

[1] S. J. Zhang, R. Xia, L. Lebrun, D. Anderson, T. R. Shrout. *Materials Letters.*, 59, 3471(2005).

[2] K. Uchino. *Ferroelectric Devices.* New York: Marcel Dekker. (2000).

[3] R. C. Pohanka, P. L. Smith. *Recent advances in piezoelectric ceramics: Electronic Ceramcis-Properties, Devices and Applications.* New York: Marcel Dekker (1987).

[4] For example: the legislation will be enforced in the EU as the draft directives on waste from electrical and electronic equipment (WEEE), restriction of hazardous substances (RoHS) and end-of life vehicles (ELV).

[5] T. Wada, K. Toyoike, Y. Imanaka, Y. Matsuo. *Jpn. J. Appl. Phys.*, 40, 5703 (2001).

[6] H. Ishii, H. Nagata, T. Takenaka. *Jpn. J. Appl. Phys.*, 40, 5660 (2001).

[7] T. R. Shrout, S. J. Zhang. *J. Electroceram.*, 19, 111 (2007).

[8] R. Wang, R. Xie, T. Sekiya, Y. Shimoju, Y. Akimune, H. Hirosaki, M. Ito. *Jpn. J. Appl. Phys.*, 41, 7119 (2002).

[9] T. Wada, K. Tsuju, T. Saito, Y. Matsuo. *Jpn. J. Appl. Phys.*, 42, 6110 (2003).

[10] S. J. Zhang, R. Xia, T. R. Shrout, G. Zang, J. F. Wang. *Solid State Commun.*, 141, 675 (2007).

[11] S. J. Zhang, R. Xia, T. R. Shrout. *Appl. Phys. Lett.*, 91, 132913(1) (2007).

[12] E. Hollenstein, M. Davis, D. Damjanovic, N. Setter. *Appl. Phys. Lett.*, 87, 182905(1) (2005).

[13] Y. Saito, H. Takkao, T. Tani, T. Nonoyama, K. Takatori, T. Homma, T. Nagaya, M. Nakamura. *Nature [London]*, 432, 84 (2004).

[14] Y. P. Guo, K. Kakimoto, H. Ohsato. *Appl. Phys. Lett.*, 85, 4121 (2004).

[15] G. Z. Zang, J. F. Wang, H. C. Chen, W. B. Su, C. M. Wang, P. Qi, B. Q. Ming, J. Du, L. Zheng, S. J. Zhang, T. R. Shrout. *Appl. Phys. Lett.*, 88, 212908(1) (2006).

[16] P. Zhao, B. P. Zhang, J. F. Li. *Appl. Phys. Lett.*, 90, 242909(1) (2007).

[17] E. Z. Li, R. Sasaki, T. Hoshina, H. Takeda, T. Tsurumi. *Jpn. J. Appl. Phys.*, 48, 09KD11(1) (2009).

[18] E. Z. Li, H. Kakemoto, T. Hoshina, T. Tsurumi. *Jpn. J. Appl. Phys.*, 47, 7702 (2008).

[19] I. Kanno, K. Akama, K. Wasa, H. Kotera. *Appl. Phys. Exp.*, 2, 091402(1) (2009).

[20] H. P. Le. *J. Imaging Sci. Technol.*, 42, 49 (1998).

[21] IEEE Standards on Piezoelectricity ANSI/IEEE Standard. (1987). No. 176-1987.

[22] S. J. Zhang, E. F. Alberta, R. E. Eitel, C. A. Randall, T. R. Shrout. *IEEE Trans. Ultrason. Ferro. Freq. Control,* 52, 2131 (2005).

[23] M. Ahtee. *Acta Cryst.*, A34, 309 (1978).

[24] G. Shirane, R. Newnham, R. Pepinsky. *Phys. Rev.*, 96, 581 (1954).

[25] R. E. Jeager, L. Egerton. *J. Am. Ceram. Soc.*, 45, 209 (1962).

[26] S. Takahashi. *Ferroelectrics,* 41, 143 (1982).

[27] K. Carl, K. H. Hardtl. *Ferroelectrics*, 17, 473 (1978).

[28] J. B. Lim, S. J. Zhang, J- H. Jeon, T. R. Shrout. *J. Am. Ceram. Soc.*, 93, 1218 (2010).

[29] D. Viehland, Y. H. Chen. *J. Appl. Phys.*, 88, 6696 (2000).

In: Recent Advances in Ceramic Materials Research ISBN: 978-1-62417-729-3
Editors: Joan J. Roa Rovira and Merce Segarra Rubi

Chapter 3

HOW TO ASSESS THE CERAMIC SUITABILITY OF RAW CLAYS?

M. Hajjaji* and A. Khalfaoui

Laboratoire de Physico-Chimie des Matériaux et Environnement,
Unité Associée au CNRST (URAC 20), Département de Chimie,
Faculté des Sciences Semlalia,
Marrakech, Morocco

ABSTRACT

Clays are still considered as the main feeding materials for conventional ceramic products (bricks, tiles, faience, etc). As a matter of fact, the suitability of these materials depends on their inherent characteristics (chemical composition, mineralogical constitution, grain size and plasticity) as well as on extrinsic factors such as firing temperature, soaking time, heating rate and atmosphere, and shaping methods.

Given the numerous influencing factors, the ceramic suitability of clays is commonly approached by trial. In general, changes of the microstructure of ceramic samples as a function of operating factors, e.g. firing cycle parameters, composition of ceramic bodies, are investigated and the technical properties are evaluated and compared to standard data. Besides these investigations, some empirical diagrams involving grain size distribution and chemical or mineralogical composition are used. However, neither the firing conditions nor the technological magnitudes of the ceramic products can be predicted by the use of such diagrams.

The ceramic suitability of clay could be evaluated by adopting a new approach based on the response surface methodology (RSM).

In this chapter, the RSM results are compared to those obtained by the use of empirical diagrams, and discussed in relation to the clay firing transformations and some important properties (strength, water absorption and firing shrinkage).

* Corresponding Author's e-mail address: hajjaji@ucam.ac.ma (M. Hajjaji).

1. Introduction

Clays are natural blends of phyllosilicates (clay minerals) and non-clay minerals (tectosilicates, carbonates, oxihydroxides, etc). Because of their plastic behavior when mixed with water, clay minerals play a key role in the shaping process. Furthermore, when heated at higher temperatures, they may give rise to ceramic phases (mullite, glassy phase, etc), which in general contribute to the consolidation of ceramic products. Non-clay minerals also may have beneficial effects: quartz for instance facilitates drying, acts as a skeleton for green bodies and may react upon heating. Feldspars, particularly alkaline feldspars, are natural fluxing agents, which result in the formation of a viscous melt playing an active role in the sintering process [1].

As mixtures of plastic materials, fillers and fluxes, clays are considered as ideal raw materials for traditional ceramics. However, their suitability depends on the mineralogical and chemical compositions, the shaping method as well as on the firing conditions. Thereby, many studies have been carried out on clays in order to assess the effects of such factors on their firing transformations [2-12], and determine their ceramic suitability. Moreover, some diagrams were established and used as references to predict the ceramic suitability of clays [13-16]. Nevertheless, no information regarding the firing conditions (temperature, soaking time, heating rate, etc) and the technological magnitudes (water absorption, mechanical strength, shrinkage, etc) of the ceramic products can be known.

In this chapter, the ceramic suitability of clay is discussed in sight of the results of the firing transformations together with the ceramic properties measurements. Also, it is evaluated by the use of some reported empirical diagrams and the response surface methodology. In this purpose, it may be noted that the RSM is a statistical tool, involving mathematical models which allow to express a response as a function of natural variables. Thus, it allows to determine a response value for any parameters values belonging to the investigated experimental domains.

Because of their marked influence on ceramic properties and practical importance (time reduction and energy saving), the firing cycle parameters have received much attention in the use of RSM.

2. Clay Characteristics, Experimental Procedures and Techniques

The clay studied was from a Mio-Pliocene red deposit sited at the haut-atlas mountain (Morocco). As can be deduced from the X-ray diffraction patterns of the raw clay and its fine fraction (< 2μm) (Figure 1), it essentially consisted of illite, chlorite, serpentine, quartz and hematite.

The minerals relative amounts and the clay chemical composition are given in Table 1.

The clay grain size fractions (< 2 μm and > 20 μm) were isolation by sedimentation, applying the law of Stokes [17]. Considering the determined grain size distribution (Table 2) and the Shepard's diagram, which is a representative classification of sediments [18], the raw material is classified as a silty clay. As compared to Italian clays [15], it is convenient for structural products manufacturing.

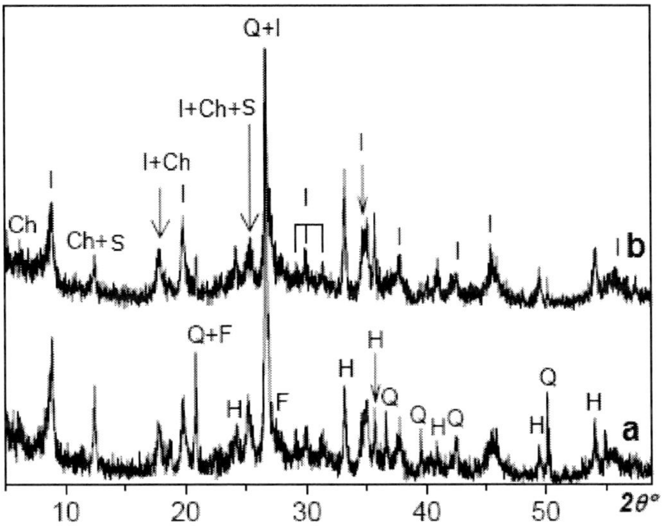

Figure 1. X-ray diffraction patterns of the studied raw clay *(a)* and its fine fraction (< 2µm) *(b)*. *I:* illite; *Ch:* chlorite; *S:* serpentine; *Q:* quartz; *H:* hematite; *F:* K-feldspar.

The Atterberg limits of the clay, namely the liquidity and plasticity limits (LL, PL) were conventionally determined, whereas the index of plasticity (PI) was deduced using the relation PI = LL – PL. The determined values are given in Table 2. As PI > 20, the clay is enough plastic to be shaped by extrusion [14]. Its drying sensitivity index (0.77) was as good as for structural clay products.

Table 1. Mineralogical and chemical compositions of the studied clay

Mineralogical composition					
Illite	Chlorite	Serpentine	Quartz	Hematite	[a]Balance
42	15	5	22	8	8
Chemical composition (wt. %)					
SiO_2	50.8	CaO	0.3	MnO	0.05
Al_2O_3	19.2	K_2O	4.3	[b]LOI	8.3
Fe_2O_3	9.4	Na_2O	1.6	-	-
MgO	4.3	TiO_2	0.8	-	-

[a] K-feldspar, calcite and calcium sulfate.
[b] Loss on Ignition.

Table 2. Grain size distribution and Atterberg limits of the studied clay

Grain size distribution (µm)			Atterberg limits (%)		
< 2	2-20	> 20	LL	PL	PI
62	32	6	47	21	26

LL: liquidity limit, PL: plasticity limit; PI: plasticity index.

Bars (4x10x40 mm^3) of the damped clay (26 wt.% humidity) were shaped by extrusion and kept at room temperature for about 24 hours. Then, they were oven-dried for 24 h at 105 °C, and fired at 800-1050 °C in an electric furnace, operating in open atmosphere. The fired samples were allowed to cool by switching off furnace. The investigated domains of soaking time and heating rate were 10-300 min and 1-20 °C/min.

The water absorption of fired samples was determined by weighting shards before and after immersion in boiled water for 2 hours. The standard deviation was < 0.3 wt. %. The bending strength of fired bars was measured by the three point's method, using an EX 150 DeltaLab apparatus. Its standard deviation varied in the range 1–5 MPa. For firing shrinkage determination, dimensions of green and fired bars were measured with a micrometer device. The standard deviation was 0.15%.

Fired samples were analyzed by X-ray diffraction (XRD) using a PHILIPS X'Pert MPD diffractometer equipped with a copper anticathode (Kα = 1.5418 Å). The amounts of crystalline phases were evaluated by the use of X'Pert HighScore program. Microstructure examinations were performed by a JEOL JMS 5500 scanning electron microscope (SEM) equipped with an EDAX Falcon spectrophotometer. For this aim, freshly fragmented pieces of fired samples were carbon coated.

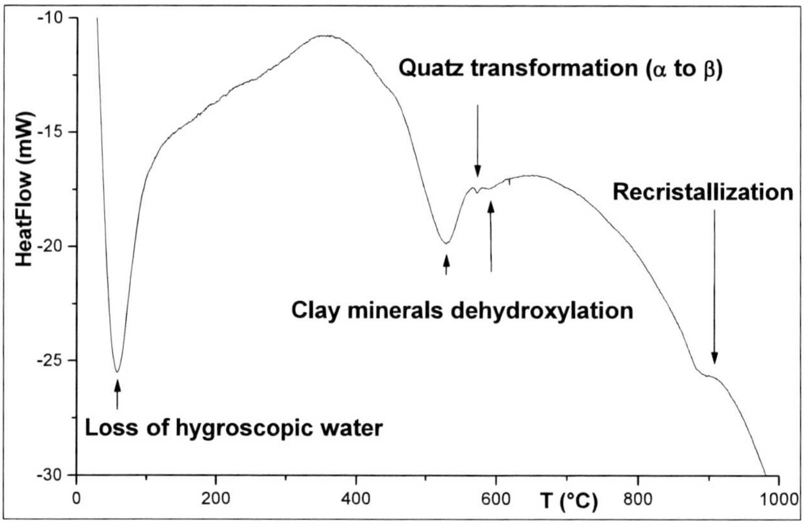

Figure 2. Thermal curve (DSC) relevant to the raw clay.

3. Firing Transformations and Ceramic Properties

The dehydroxylation processes of the clay minerals (serpentine, chlorite and illite) occurred at T < 900 °C, and a neoformation phenomenon manifested by 900 °C (Figure 2).

Based on the X-ray diffraction analyses (Figure 3), the later process was due to the formation of Mg-spinel. This phase persisted at higher temperatures, but its particles were too small (about 45 nm) to be distinguished by scanning electron microscope. The rise of the firing temperature resulted in the occurrence of olivine together with an intensive formation of the glassy phase, characterized by the X-ray line broadening in the range 25-35 ° (2Θ°). It also led to the partial disappearance of quartz, and to a slight increase of the content of hematite (Figure 4). Details relevant to the later results were reported in [19]. The rise of the amount of hematite was due to the ferric oxide formed by the decomposition of clay minerals with non-negligible amounts of iron.

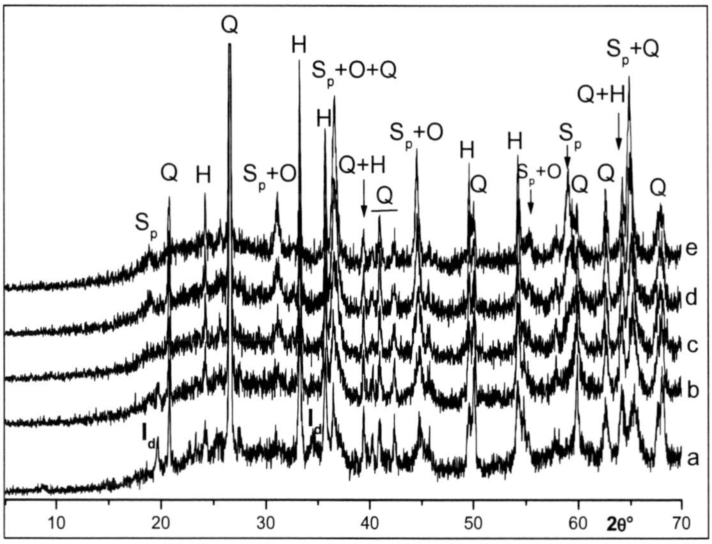

Figure 3. X-ray diffractograms of clay samples fired 8 h at 900°C *(a)*, 950°C *(b)*, 1000°C *(c)*, 1050°C *(d)* and 1075°C *(e)*. I_d: dehydroxylated illite; Sp: spinel; *Q*: quartz; H: hematite; *O*: olivine.

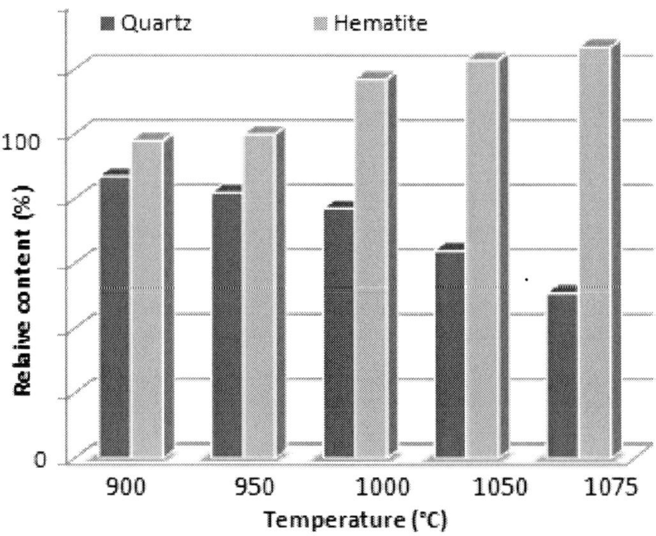

Figure 4. Relative contents of quartz and hematite in samples fired at different temperatures. The amounts of the minerals in unfired samples were taken as references.

Microscopic examinations of samples fired at 900 °C exhibited welded filaments identified to dehydroxylated illite (Figure 5a). As the firing temperature increased, featureless zones (glassy phase) developed at the expense of the filamentous microstructure (Figure 5b).

In view of these results, Mg-spinel essentially formed at lower temperatures from the breakdowns of magnesian minerals, namely chlorite and serpentine. The products of decomposed dehydroxylated illite, occurring at higher temperatures, contributed to the spinel formation. A part of the later products gave rise to the melt (glassy phase), which also derived from fused quartz. More details on the neoformation processes were reported elsewhere [19, 20].

The variations of the ceramic properties against firing temperature are shown in Figure 6. It is derived that beyond about 1020 °C, samples were the subject of marked expansion and absorption of water. Because of the intensive fusion happened at T > 1020 °C together with the escape of entrapped gas, samples were the subject of bloating pores and fissures. Such processes were undoubtedly responsible for the deterioration of the mechanical strength and intensively contributed to the increase of water absorption. In sight of these observations, it could be concluded that the optimum firing temperature approximated 1000 °C.

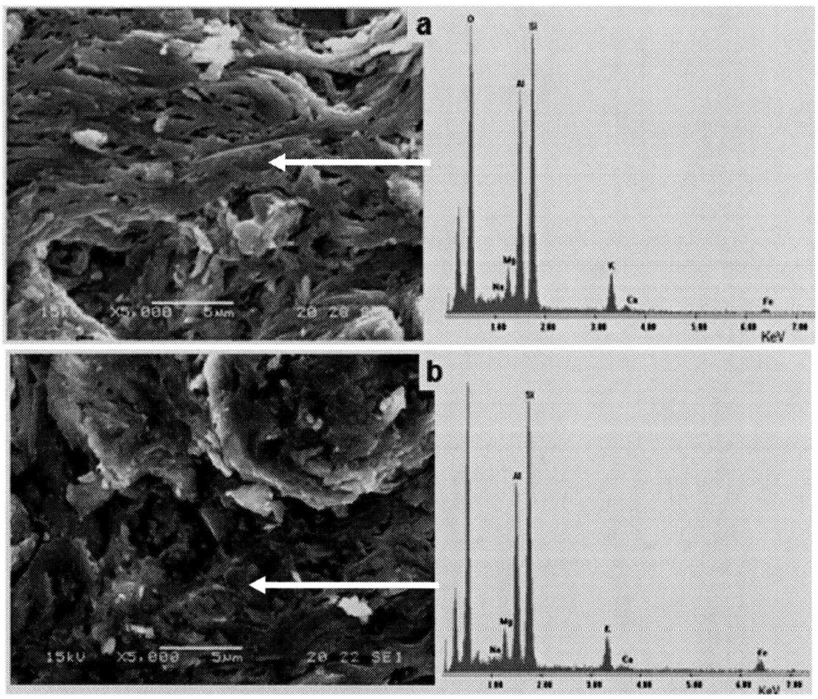

Figure 5. SEM micrographs of samples fired at 900 *(a)* and 950 °C *(b)*, and EDS spectra related to dehydroxylated illite *(a)* and amorphous glassy phase *(b)*.

Considering the reported values for the common ceramic properties (Table 3) and the magnitudes of the considered properties (Figure 6), the raw clay may be suitable for red stoneware tiles. The melt formation and pores elimination due to the sintering process were essential for the manufacture of the later products.

Table 3. Technological magnitudes of some conventional ceramic products

Ceramic property	Tiles	Bricks	Red stoneware tiles
WA (%)	15-20	15-32	0-6
S (%)	0-1	0-2	5-10
BS (MPa)	12-22	6-20	25-45

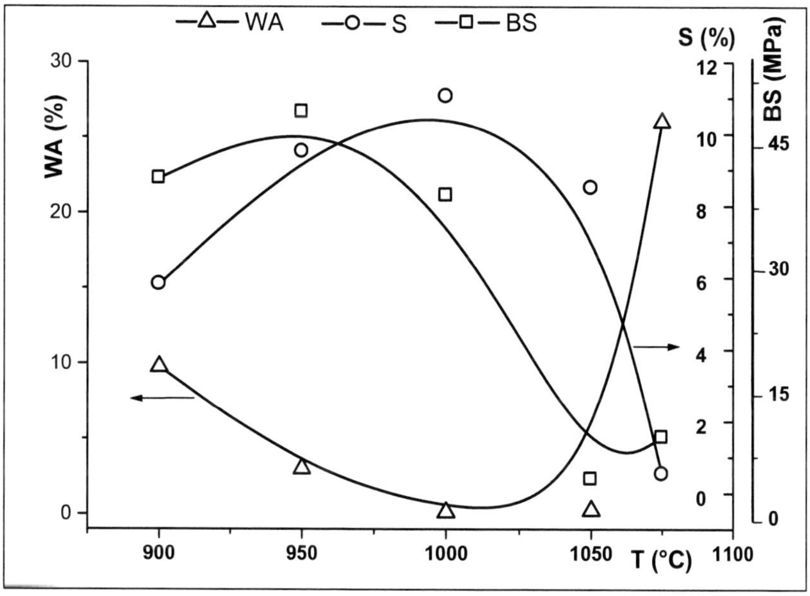

WA: Water absorption; S: Firing shrinkage; BS: Bending strength. Soaking time: 8 h.

Figure 6. Variations of the ceramic properties against firing temperature.

4. Empirical Diagrams and Clay Ceramic Suitability

The clay suitability was assessed based on some empirical diagrams defining adequate grain size distribution, mineralogical constitution and chemical composition for the production of conventional ceramics. Considering the position of the representative point of the clay grain size distribution in the Winkler's diagram (Figure 7a), the raw clay is neither convenient for bricks nor tiles. To manufacture such products, the raw clay should be amended by appropriate additions of suitable materials consisting of coarse particles, i.e. sand or grog. Without amendment, the clay is suitable for pottery.

In view of the clay mineralogy and the adequate argillaceous materials for Italian structural products (Figure 7b), the studied material would not be appropriate for roofing tiles or structural clay bodies. The later ceramic products may be produced by suitable additions of quartz.

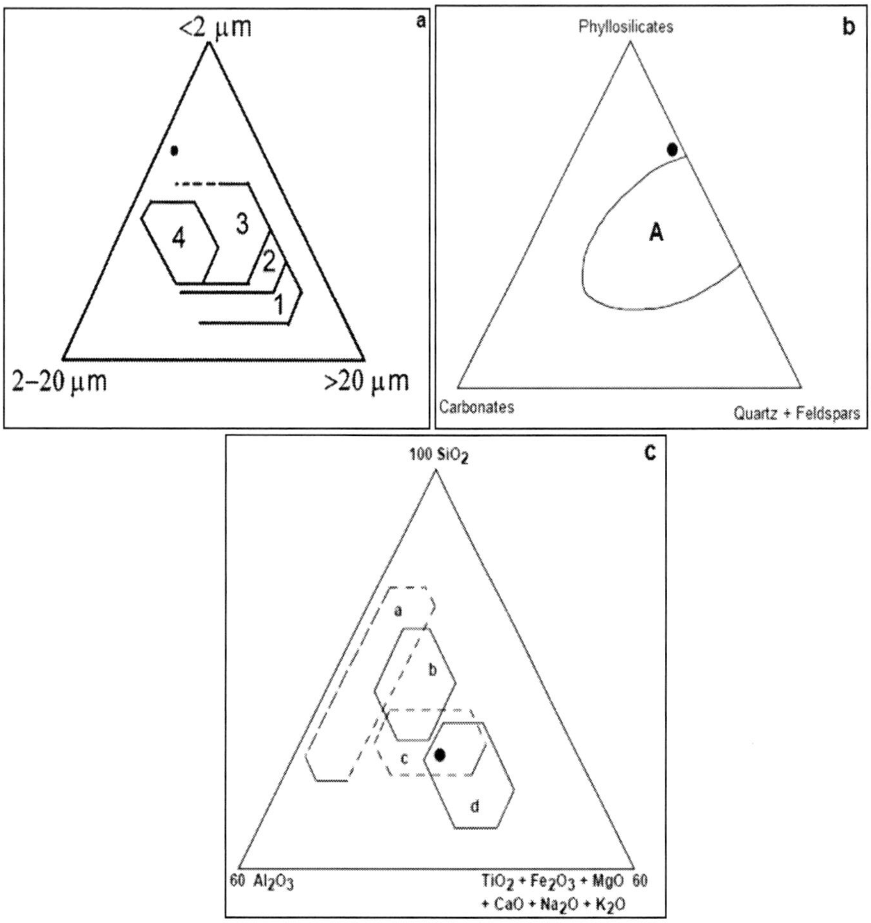

Figure 7. Empirical diagrams showing suitable domains for manufacturing conventional ceramic bodies. (•) Positions related to the studied clay. *(a):* Diagram of Winkler ((*1*) Solid bricks; (*2*) Vertically perforated bricks; (*3*) Roof tiles and lightweight blocks; (*4*) Hollow bricks). *(b):* Structural clay products (A). *(c):* Stoneware tiles (white *(a)* and red *(b)* bodies) and porous tiles (*c and d*).

The representative point of the chemical composition in the ternary diagram of Figure 7c allowed to deduce that the raw clay may be used as a feeding material for red porous tiles. This result is consistent with that deduced from the ceramic properties measurements.

5. Clay Ceramic Suitability Using the Response Surface Methodology

As previously mentioned, RSM is a statistical technique. It allows to determine the mathematical relation between a response and a set of variables. The graphical representation of the equation represents the response surface, which could be used to evaluate the effects of the variables and their mutual interactions on the response.

5.1. Model Presentation

As previously mentioned, the firing cycle factors (firing temperature (T), soaking time (t) and heating rate (h)) were taken as parameters (natural variables). The studied responses were bending strength (Y_1), water absorption (Y_2) and firing shrinkage (Y_3). The variation of a response Y_i against the coded parameters (X_1, X_2, and X_3), related to the natural variables as reported hereafter, was determined according to the polynomial model [21-24]:

$$Y_i = a_0 + a_1X_1 + a_2X_2 + a_3X_3 + a_{11}X_1^2 + a_{22}X_2^2 + a_{33}X_3^2 + \\ + a_{12}X_1X_2 + a_{13}X_1X_3 + a_{23}X_2X_3$$

a_o is a constant; a_1, a_2, and a_3 are the weights of the effects of firing temperature, soaking time, and heating rate, respectively. The quadratic term a_{ii} can be regarded as a curve shape parameter and a_{ij} expresses the interaction effect. $X_1 = (T-T_o)/\Delta T$, $X_2 = (t-t_o)/\Delta t$, and $X_3 = (h-h_o)/\Delta h$. The investigated domains of the natural variables (800-1050 °C; 10-300 min and 1-20 °C/min) were defined by taking into account the thermal behavior of the clay and by referring to some reported data [8]. T_o, t_o, and h_o are the values at the centers of the considered experimental domains (T_o = 925°C, t_o = 155 min and h_o = 10.5 °C/min). ΔT, Δt, and Δh are the variation steps of the natural variables values (ΔT = 125°C, Δt = 145 min, and Δh = 9.51°C/min).

Table 4. Doehlert experimental design in natural variables and the determined responses (Y_1, Y_2 and Y_3)

Run	T (°C)	t (min)	h (°C·min^{-1})	Y_1 (MPa)	Y_2 (%)	Y_3 (%)
1	1050	155	10.5	39.00	0.04	11.22
2	1050	155	10.5	45.56	1.95	10.61
3	1050	155	10.5	59.25	0.03	11.00
4	800	155	10.5	8.11	24.00	-0.40
5	800	155	10.5	7.94	23.95	-0.25
6	800	155	10.5	6.21	24.50	-0.50
7	988	281	10.5	43.40	0.23	10.87
8	988	281	10.5	39.00	0.00	10.68
9	988	281	10.5	43.40	0.13	11.05
10	863	29	10.5	22.80	17.56	3.36
11	863	29	10.5	33.00	14.00	3.43
12	863	29	10.5	24.90	17.15	3.37
13	988	29	10.5	34.30	2.15	10.28
14	988	29	10.5	39.00	0.21	10.10
15	988	29	10.5	47.15	0.93	10.24
16	863	281	10.5	40.03	7.93	4.88
17	863	281	10.5	42.40	9.93	5.05
18	863	281	10.5	39.80	9.30	5.46
19	988	197	18.3	41.20	0.27	10.48
20	988	197	18.3	52.00	0.23	9.75

Table 4. (Continued)

Run	T (°C)	t (min)	h (°C·min^{-1})	Y_1 (MPa)	Y_2 (%)	Y_3 (%)
21	988	197	18.3	43.40	0.24	10.75
22	863	113	2.7	42.60	9.80	5.57
23	863	113	2.7	45.15	8.47	5.50
24	863	113	2.7	40.00	9.73	5.03
25	988	113	2.7	36.70	0.10	11.27
26	988	113	2.7	44.90	0.02	10.96
27	988	113	2.7	40.80	0.06	10.97
28	925	239	2.7	29.60	6.41	8.12
29	925	239	2.7	37.90	5.68	8.25
30	925	239	2.7	32.60	6.20	8.40
31	863	197	18.3	36.70	14.20	4.99
32	863	197	18.3	26.70	11.00	4.67
33	863	197	18.3	26.70	12.60	4.92
34	925	71	18.3	47.20	7.60	7.50
35	925	71	18.3	37.00	6.18	7.44
36	925	71	18.3	42.10	7.62	7.73
37	925	155	10.5	35.00	6.50	7.85
38	925	155	10.5	47.20	5.42	7.85
39	925	155	10.5	42.00	6.45	7.63
40	925	155	10.5	35.40	7.43	7.65
41	925	155	10.5	31.85	5.63	8.35
42	925	155	10.5	31.46	6.88	7.96
43	925	155	10.5	27.90	5.67	8.14
44	925	155	10.5	31.46	5.34	8.19
45	925	155	10.5	23.88	5.50	8.18

5.2. Experimental Design and Model Validation

Because of its advantages (uniform distribution of the experimental points, possibility to extend the matrix, limited number of experiments, etc), the matrix of Doehlert [21] was adopted for the experimental design. In this case, the number of the planned runs (N) should be 13, since $N = K^2 + K + 1$ (K is the number of variables, i.e. 3). To validate the experimental error, the run at the experimental domain center was repeated two times, and each run was triplicated. The Doehlert experimental design in natural variables and the obtained responses (Y_i) are given in Table 4.

The coefficients a_i, a_{ii}, and a_{ij} of the model were calculated by least-square regression using the New Efficient Methodology for Research using Optimal Design (NEMROD) software [25]. For this purpose, the results of the planned experiments were taken into consideration (Table 4). The calculated coefficients composing the equations of Y_1, Y_2 and Y_3 are reported in Table 5.

To test the model validity, experiments were undertaken in arbitrary chosen firing conditions, and the experimental values of the responses were compared to the calculated ones (Table 6).

In view of these results the adopted model is valid. The model validation is supported by the higher correlation coefficients and the smaller standard deviations (Table 7).

Based on the absolute values of a_i , the weights of the effects of the parameters on the studied properties are classified as follows: firing temperature > soaking time > heating rate. A detailed study dealing with this subject was published elsewhere [23].

5.3. Clay Ceramic Suitability

For the clay ceramic suitability and the NEMROD program execution, the required values for the studied responses corresponding to tiles (floor, wall and roof), bricks and red stoneware tiles were taken into consideration [5, 26] (Table 3), and consequently desirability functions were established (Figure 8). In this purpose, it should be noted that each plot shows the smallest and the highest suitable values of the response as well as the most desired one (target).

Table 5. Calculated coefficients of the models relevant to the studied responses (*Y_1*: bending strength, *Y_2*: water absorption and *Y_3*: firing shrinkage)

Coefficient	Y_1	Y_2	Y_3
a_0	32.202	6.243	8.022
a_1	13.773	-11.775	5.796
a_2	1.624	-1.444	0.487
a_3	1.234	0.587	-0.355
a_{11}	-8.111	6.289	-2.726
a_{22}	8.343	-1.292	0.012
a_{33}	7.447	-1.228	0.289
a_{12}	-8.291	3.397	-0.497
a_{13}	16.603	-3.278	0.009
a_{23}	11.229	-3.430	0.696

Table 6. Experimental and calculated values of the responses determined in different firing conditions

T (°C)	t (min)	h (°Cmin^{-1})	Y_1 (MPa)		Y_2 (%)		Y_3 (%)	
			[a]Exp	[b]Cal	Exp	Cal	Exp	Cal
940	120	9.0	31	34	4.5	4.8	8.0	8.5
1020	30	9.3	49	50	0.0	0.0	11.0	10.9
900	60	12.0	30	29	10.6	10.4	5.8	6.0
860	90	6.0	28	32	15.0	13.0	3.1	4.0

[a]Experimental; [b]Calculated.

Table 7. Statistical data related to the correlation between the experimental and calculated values of the studied responses

Y_1 (MPa)		Y_2 (%)		Y_3 (%)	
R^2	ε	R^2	ε	R^2	ε
0.86	4.6	0.98	1.2	0.99	0.4

R^2 : Correlation coefficient; ε: Standard deviation.

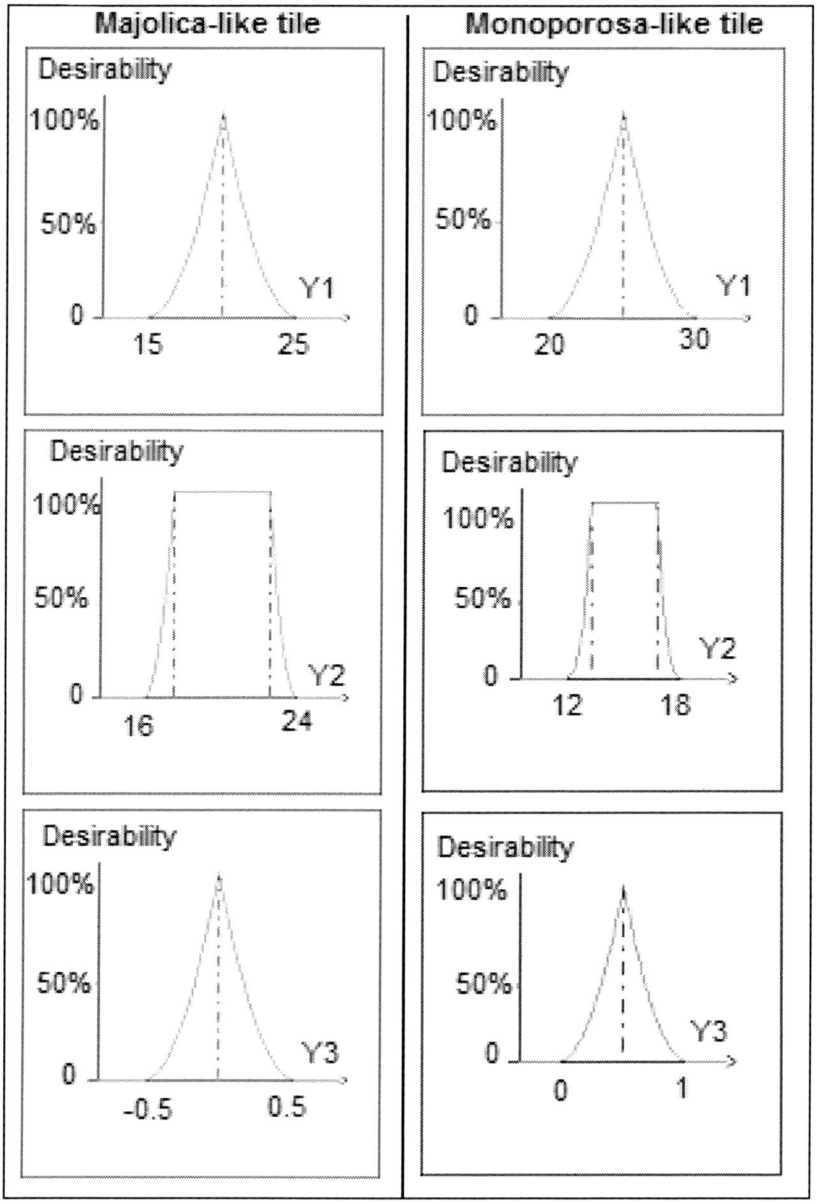

Figure 8. (Continued).

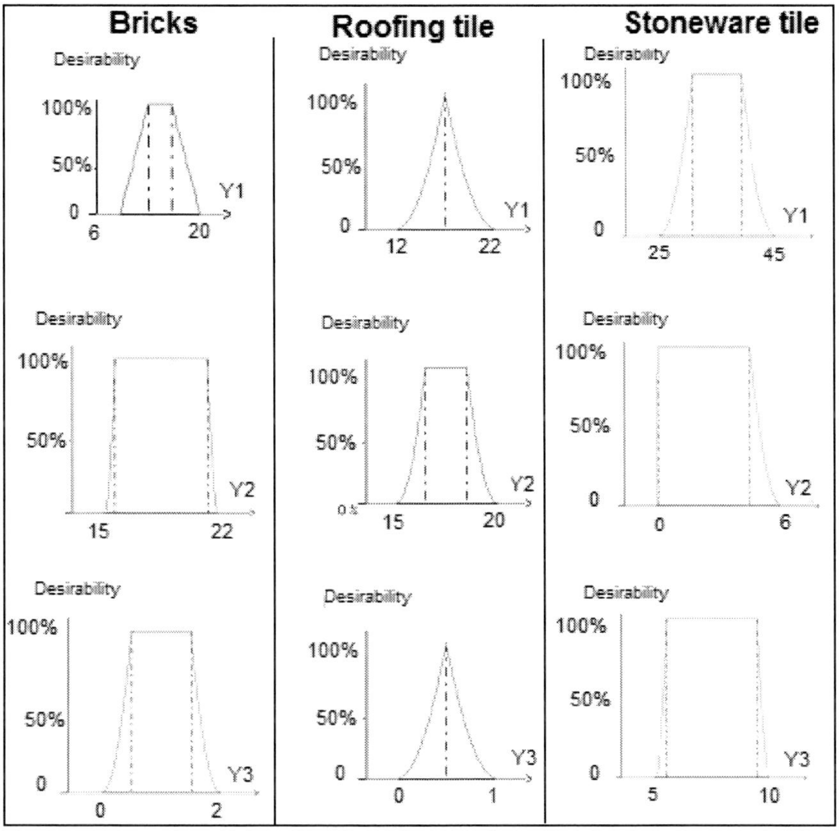

Figure 8. Desirability functions relevant to the values of the bending strength (Y_1), water absorption (Y_2) and firing shrinkage (Y_3) of different traditional ceramic products.

The desirability domains representing the suitable sets of parameters of the firing cycles to manufacture Majolica-like tiles are displayed in Figure 9. It is derived that this kind of products can be manufactured using lower temperatures (nearby 800 °C) and slow heating rates. As a matter of fact, there are very limited suitable firing parameters for their production. Under such firing conditions, the sintering process was not yet significant and the microstructure essentially consisted of dehydroxylated clay minerals.

As can be seen, Monoporosa-like tiles and roofing tiles (Figure 10) could not be made at all by the use of the adopted firing cycle parameters.

Bricks can be successfully fabricated using lower firing temperatures (Figure 11). As compared to Majolica-like tiles, bricks could be produced by rapid firing cycles. Referring to Figure 11(c), heating rates as high as 15 °C/min could be used to manufacture bricks. In contrast, heating rates for Majolica-like tiles should not exceed 10.5 °C/min (Figure 9 (c)).

Based on the plots of Figure 12, the studied clay was convenient for red stoneware tiles production. In this case, the firing temperature should be relatively high (around 940 °C). In fact, the use of higher temperatures engendered sufficient melt and consequently enough glassy phase to consolidate the ceramic matrix. The melt formation at such firing temperature was favored by the presence of high contents of clay fraction and sufficient amount of auto-fluxing clay minerals.

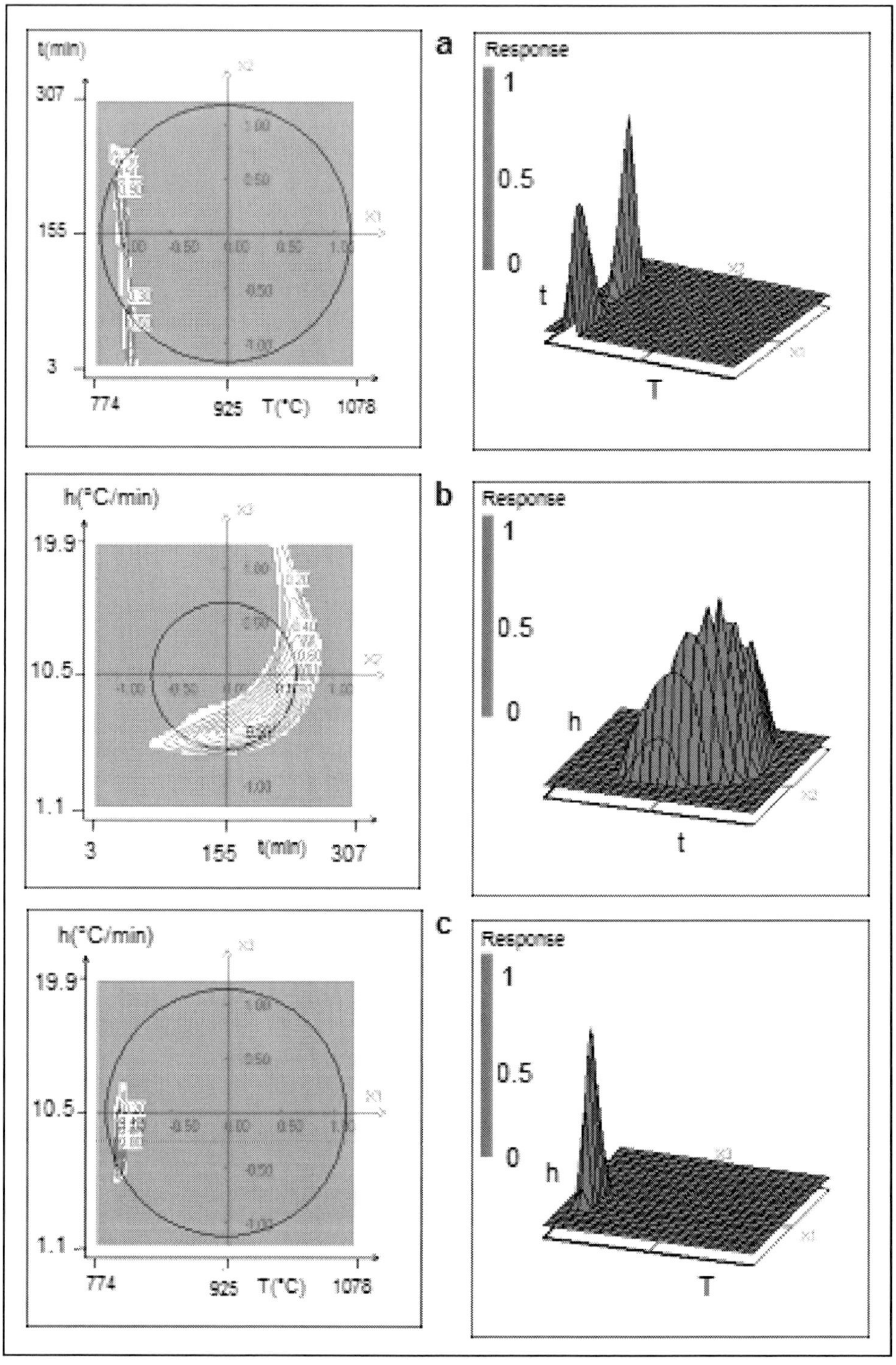

Figure 9. Desirability domains (two- and three-dimensional representations) for Majolica-like tiles. *(a):* h = 8.2 °C/min; *(b):* T= 800 °C ; *(c):* t = 210 min.

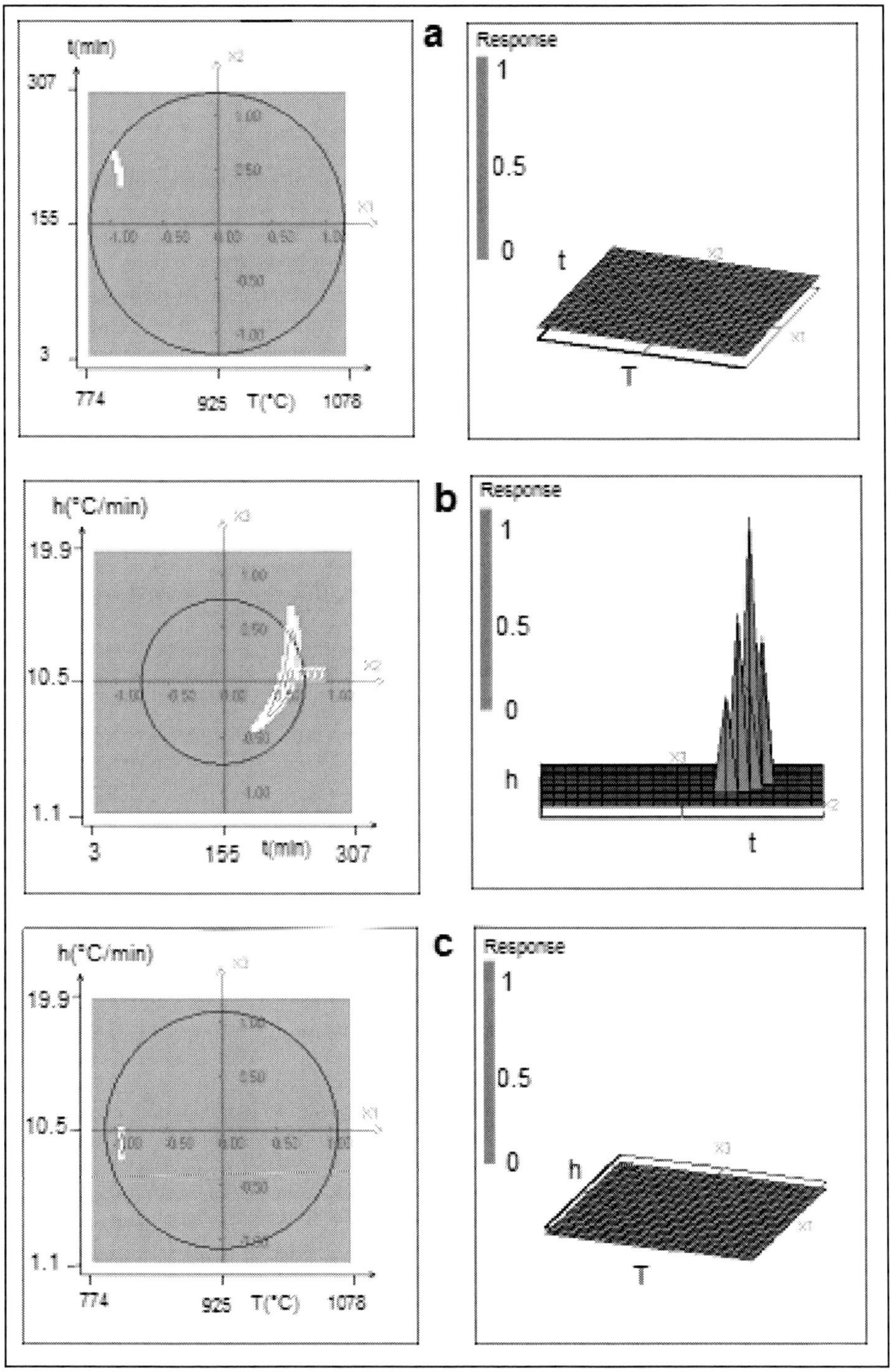

Figure 10. Desirability domains (two- and three-dimensional representations) for roofing tiles and Monoporosa-like tiles. *(a):* h = 9.5 °C/min; *(b):* T= 808 °C ; *(c):* t = 222 min.

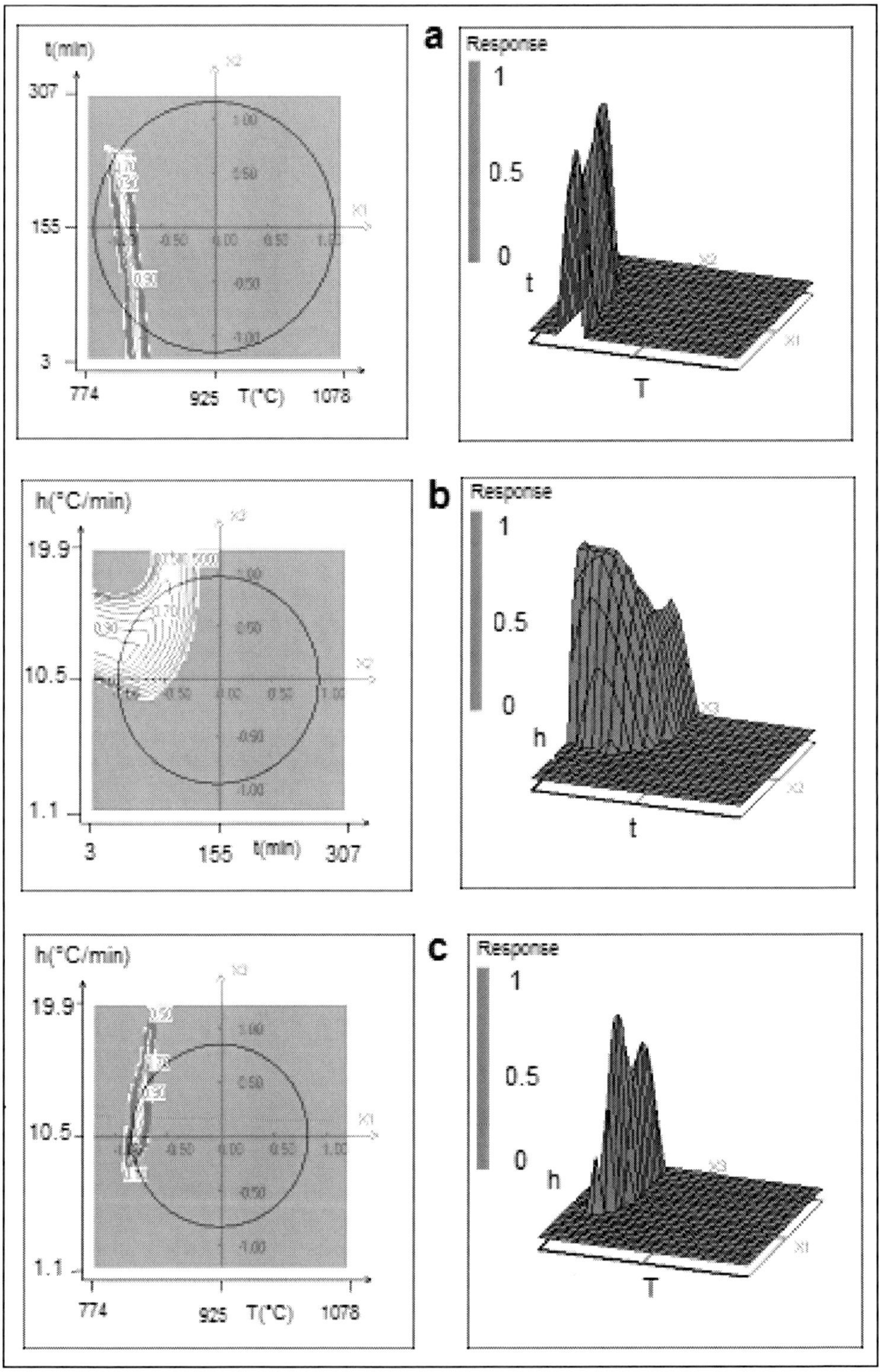

Figure 11. Desirability domains (two- and three-dimensional representations) for bricks. *(a):* h = 13.4 °C/min; *(b):* T= 834 °C ; *(c):* t = 46 min.

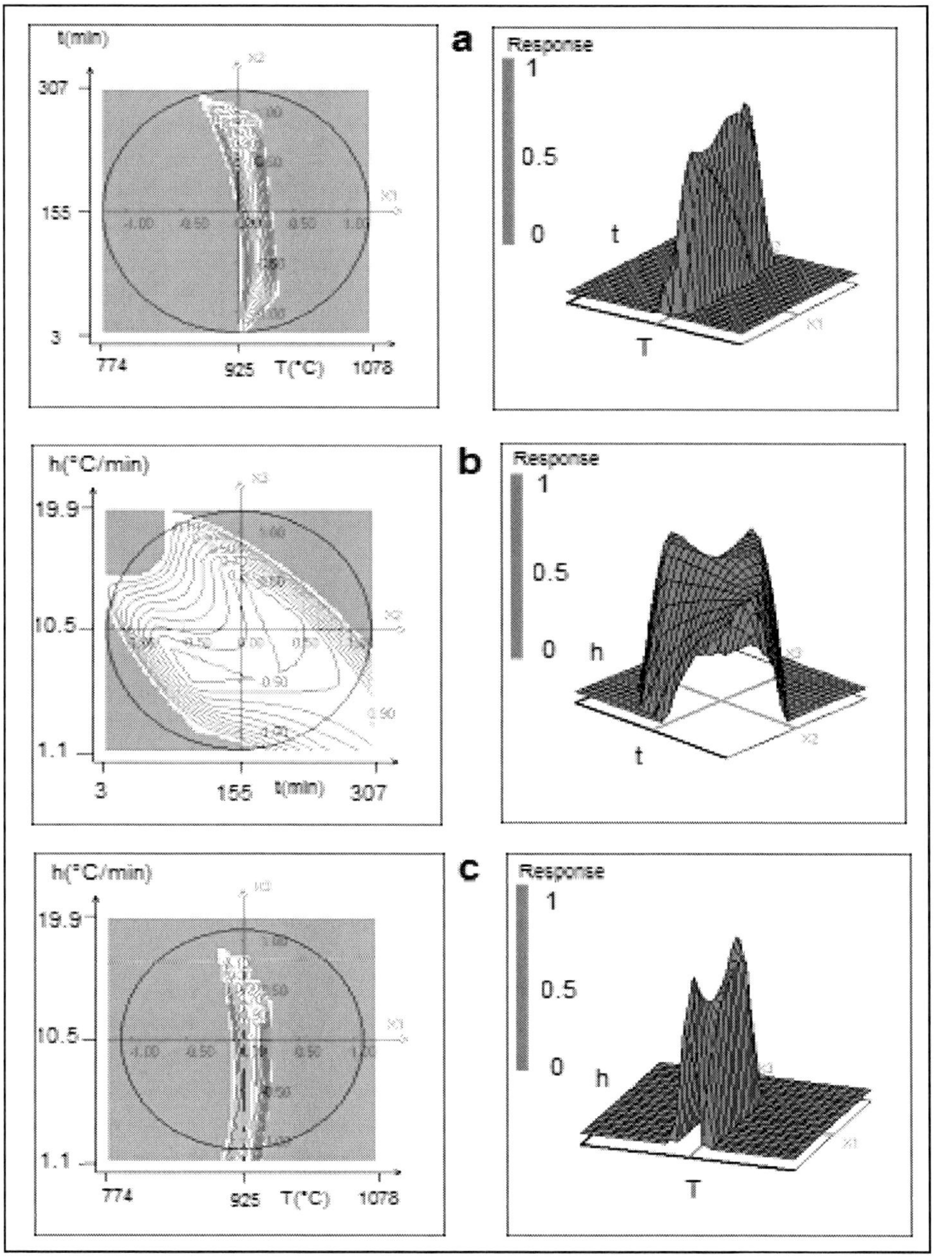

Figure 12. Desirability domains (two- and three-dimensional representations) for red stoneware tiles. *(a):* h = 11 °C/min; *(b):* T= 942 °C ; *(c):* t = 220 min.

CONCLUSION

The ceramic suitability of the chloritic-illitic raw clay was assessed in relation to its inherent and extrinsic characteristics and using response surface methodology. As a silty clay, the studied material seemed to be convenient for structural products manufacturing. However, in view of its mineralogical composition, it could not be suitable for bricks or tiles production. This result was supported by the use of the Winkler's diagram, based on the grain size distribution. Due to its good plasticity behavior, the clay was rather suitable for pottery.

Taking into consideration the gresification curves, the adequate firing temperatures was around 1000 °C. Higher temperatures resulted in intensive vitrification and the occurrence of defects. Taking into consideration the magnitudes of the ceramic properties, the raw clay could be suitable for red stoneware tiles. This concurred with the results involving the clay chemical composition diagram and the response surface methodology. According the later method, the appropriate temperature for red stoneware fabrication approximated 940 °C.

Unlike the above methods, the RSM results showed that the clay could be appropriate for Majolica-like tiles and bricks, and the adequate firing temperatures were as low as 800 °C. In this case, the sintering process, involving melt flow, was not yet significant.

ACKNOWLEDGMENTS

The authors appreciated the comments of Dr. Joan Josep Roa Rovira (Editor). Thanks to Mr. A. Bani for his technical help.

REFERENCES

[1] C. Zanelli, M. Raimondo, G. Guarini, M. Dondi. *J. Non-Cryst. Solids*, 357, 3251 (2011).

[2] J. Lemaitre, B. Delmon. *J. Mater. Sci.*, 12, 2056 (1977).

[3] T. Peters, R. Iberg. *Ceram. Bull.*, 57 (5), 503 (1978).

[4] Y. Maniatis, A. Simopoulos, A. Kostikas. *J. Am. Ceram. Soc.*, 66, 773 (1983).

[5] M. Dondi. *Appl. Clay Sci.*, 15, 337 (1999).

[6] S. G. Barlow, D. A. C. Manning, P. I. Hill. *Int. Ceram.*, 2, 5 (2000).

[7] M. Hajjaji, S. Kacim, M. Boulmane. *Appl. Clay Sci.*, 21, 203 (2002).

[8] M. I. Carretero, M. Dondi, B. Fabbri, M. Raimondo. *Appl. Clay Sci.*, 20, 301 (2002).

[9] M. Echajia, M. Hajjaji, S. Kacim. *Sil. Ind.,* 68, 61 (2003).

[10] A. Aras. *Appl. Clay. Sci.*, 24, 257 (2004).

[11] G. Stathis, A. Ekonomakou, C. J. Stournaras, C. Ftikos. *J. Eur. Ceram. Soc.*, 24, 2357 (2004).

[12] M. Echajia, S. Kacim, M. Hajjaji. *Industrial Ceramics* , 27, 185 (2007).

[13] H. G. F. Winkler. Bedeutung der Korngrossenverteilung und des Mineralbestandes von Tonen fur die Herstellung grobkeramischer Erzeugnisse. *Ber. Dtsch. Keram. Ges.,* 31, 337– 343 (1954).

[14] J. A. Bain, D. E. Highley. Regional appraisal of clay resources. A challenge to the clay mineralogist. *Proc. Int. Clay Conf.* AIPEA, Oxford. (1966).

[15] B. Fabbri, M. Dondi. La produzione del laterizio in Italia. *Faenza Editrice,* 169 p. (1995).

[16] C. Fiori, B. Fabbri, F. Donati, I. Venturi. *Appl. Clay. Sci.*, 4, 461(1989).

[17] T. Holtzapffel. Les minéraux argileux: Analyse diffractométrique et détermination. Société géologique du Nord, Publication N°12, 136 p. (1985).

[18] F. P. Shepard. *J. Sediment. Petrol.*, 24, 151 (1954).

[19] A. Khalfaoui, S. Kacim, M. Hajjaji. *J. Eur. Ceram. Soc.*, 26, 161 (2006).

[20] A. Khalfaoui, M. Hajjaji. *Appl. Clay Sci.*, 45, 83 (2009).

[21] D. H. Doehlert. *Appl. Statist.*, 19, 231 (1970).

[22] L. C. Ferreira, W. N. L. dos Santos, M. A. Bezerra, V. A. J. M. Lemos & Bosque-Sendra. *Anal. Bioanal. Chem.*, 375, 443 (2003).

[23] A. Khalfaoui, M. Hajjaji, S. Kacim, A. Baçaoui. *J. Am. Ceram. Soc.*, 89, 1563 (2006).

[24] D. Njoya, M. Hajjaji, A. Baçaoui, D. Njopwouo. *Mater. Character.*, 61, 289 (2010).

[25] D. Mathieu, R. Phan Tan Luu. *Software NEMROD*. Université d'Aix-Marseille III, France. (1980).

[26] B. Strazzera, M. Dondi, M. Marsigli. *Appl. Clay Sci.*, 12, 247 (1997).

In: Recent Advances in Ceramic Materials Research
Editors: Joan J. Roa Rovira and Merce Segarra Rubi
ISBN: 978-1-62417-729-3

Chapter 4

AGING OF DENTAL ZIRCONIA

C. A. M. Volpato[1*], R. F. Kikko[1], D. K. Vasconcellos[1], I. M. Zani[1] and M. C. Fredel[2]

[1] Department of Dentistry, Federal University of Santa Catarina, Brazil
[2] Department of Mechanical Engineering, Federal University of Santa Catarina, Brazil

ABSTRACT

Research focused on the microstructure and processing of ceramic materials has considerably increased in dentistry, due to biological, mechanical and aesthetic demands. Currently, zirconia is considered one of the most promising materials for biomedical applications because it combines excellent mechanical properties with biocompatibility. In dentistry, this ceramic has been widely used to make prosthesis substructures and implant abutments.

To be applied for structural purposes, it must be stabilized in the tetragonal phase at room temperature.

However, when stabilized zirconia contacts water or steam, it may undergo a considerable degradation of its mechanical properties. The presence of moisture interferes with its metastability, creating an unfavorable condition for the longevity of the material, characterized by macro- and microcracks, or even fractures.

This phenomenon is known as "low temperature degradation" or "aging". The present study evaluated the degree of fit between ceramic abutment/implant and its resistance to fracture before and after thermal aging.

The results presented no statistical differences in the degree of fit abutment/implant; however, the best numerical results were found with the control group.

The results were statistically significant when evaluating resistance, suggesting that aging decreased the fracture resistance of zirconia abutments.

* Corresponding Author's e-mail address: claudiamvolpato@hotmail.com (C. A. M. Volpato).

1. Introduction

The development of technologies for the production of biocompatible materials is motivated by the demand for materials that perform both old and new functions more adequately [1,2]. Currently, ceramics are an important alternative for the reconstruction of lost tooth structure, as these materials reproduce the optical properties of enamel and dentin, such as fluorescence, opalescence and translucency [3-6]. However, although ceramics have excellent optical properties, they are brittle materials when subjected to stress and may fracture due to the propagation of microscopic defects that may be present on the surface [7]. When considering these aspects, the use of structural ceramics based on zirconia has been suggested [8-10].

The application of zirconia as a material of choice for the construction of prosthetic substructures and abutments for implants occurs because zirconia is a biocompatible and bio-inert material and provides superior mechanical properties. Ceramic structures are currently obtained by machining CAD/CAM (Computer Aided Design / Computer Aided Manufacturing) blocks of pre-sintered zirconia-based tetragonal ceramic that is partially stabilized by yttria (3Y-TPZ). The ceramic structures are machined with a dimensional increase of 25 to 30% to compensate for sintering shrinkage [11]. The ceramic microstructure is mostly formed of micrometric particles of ZrO_2 and approximately 2-3% of Y_2O_3. After sintering, a structure is formed, composed of nanometric tetragonal grains, that is metastable at room temperature (3Y-TZP) [9].

However, a large number of failures with zirconia orthopedic prostheses has drawn the attention of the scientific community. Whilst studying these failures, researchers have demonstrated that 3Y-TPZ is destabilized in the presence of water or steam [12-14]. Fractures of orthopedic prostheses were found to be related to a large amount of the monoclinic phase present in the zirconia. The 3Y-TPZ is a partially stabilized tetragonal zirconia at room temperature. The presence of moisture favors the transformation from the tetragonal to monoclinic phase ($t \rightarrow m$), increasing the content of the monoclinic phase and, consequently, the surface roughness and subcritical crack growth (SCG). This phenomenon, described as "low temperature degradation *(LTD)* or aging", was found to be responsible for the failure of orthopedic prostheses [15-17].

Thus, with the current application of zirconia in dentistry, it is important to understand the aging phenomenon, with its main influences on the microstructure of the material, the clinical consequences, and possible actions that are being studied to control it. With this knowledge, the dentist is able to carry out dentistry aimed, not only on aesthetic concepts, but also on longevity.

2. Cristallograph Phases of Zirconia

Zirconium is commercially extracted from zirconate and baddeleyita. The zirconate ($ZrSiO_4$) is found in greater amounts in the terrestrial crust and contains about 67.2% ZrO_2 and 32.8% SiO_2. The main reserves of this material are in Australia, Ukraine, South Africa and the United States of America. In 1892, Hussak discovered badeleyita in Brazil. It is almost pure ZrO_2, with approximately 2% HfO_2; however, it is a much more rare material than zirconate. Thus, most zirconia is produced from zirconate, with its purity controlled depending on the technique used in its decomposition [9,18].

Pure zirconia exhibits three polymorphic phases, also known as allotropic or crystalline phases, which are classified according to different temperatures: monoclinic *(m)*, tetragonal *(t)* and cubic *(c)*. At low temperatures (up to 1170°C), the material has a monoclinic phase that is thermodynamically stable. A tetragonal structure may be observed at intermediate temperatures (1170-2370°C), and the material exhibits a stable cubic structure at high temperatures (> 2370°C) (Figure 1). [8,18-19].

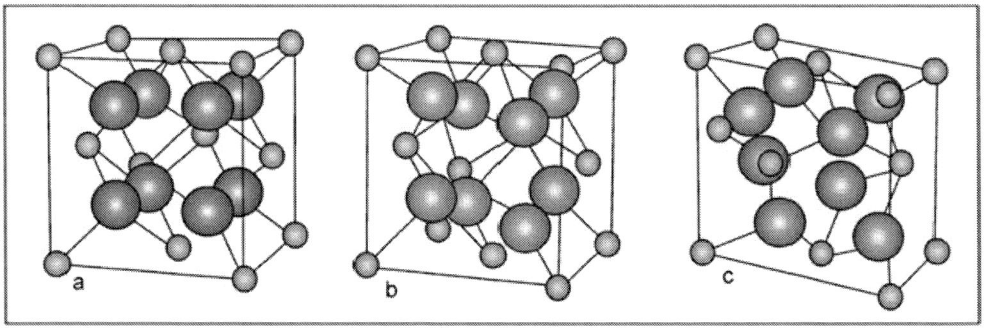

Figure 1. Crystal structure of monoclic (a), tetragonal (b) and cubic zirconia(c). Ref. [19].

The industry has great interest for advanced ceramics because of their favorable mechanical behavior with regards to resistance, fracture toughness, coefficient of thermal expansion and chemical stability [8,20]. However, the polymorphic nature of zirconia can restrict its use.

Pure zirconia is predominately found in the monoclinic phase. During the heating process, it undergoes phase transformation, also causing changes in volume, which may cause external stresses and fractures of the material, thereby precluding the use of pure zirconia. To enable zirconia to be used as a structural material, it is important that the tetragonal or cubic phases are stabilized at room temperature. Oxides (CaO, MgO) and rare earths (CeO_2, Y_2O_3) are used as stabilizers of tetragonal or cubic phases, avoiding phase transformations which may occur during heating and/or cooling. Early studies involving the addition of oxides revealed that the temperature of transition from the tetragonal to monoclinic phase is very low [15], which allows for a stabilized tetragonal phase at room temperature [21]. Ideal stabilizers are oxides whose cations are polyvalent to the zirconia, ie which have a low valent cubic crystal structure. Examples of oxides which are incorporated into the crystalline structure of zirconia include alkaline earth cations: Mg^{2+}, Ca^{2+}, and cations with rare earths: TR^{3+} and Y^{3+}. Also, the ratio of the difference between the radius of the cation in relation to the radius of zirconium is ideally less than 40% [21,22].

For zirconia in biomedical applications, yttrium oxide (Y_2O_3), extracted from a xenotime mineral, is the most commonly used stabilizer [23]. In most cases, ceramics based on zirconia are stabilized with 2.5% to 3.5% Y_2O_3 by heating the material up to 1400-1500°C, followed by rapid cooling, generating tetragonal zirconia polycrystals. Partially stabilized zirconia can also be produced with Y_2O_3 content between 3 and 6%. When yttria is added to zirconia, some of the zirconium ions are substituted by yttria ions, generating a large number of oxygen voids, which guarantees the neutrality of the ionic group. The amount of Zr^{4+} ions substituted for Y^{3+} ions determines the final polymorphic phase of the material [24].

The maintenance and degree of stabilization of the tetragonal zirconia are directly dependent on the type and amount of stabilizer oxide that is added, which is classified as fully or partially stabilized zirconia. Generally, partially stabilized zirconia consists of a mixture of cubic, monoclinic and tetragonal structures. After stabilization, the tetragonal or cubic phases become metastable at room temperature, due to the presence of these oxides [18,25].

3. CHARACTERISTICS

Ceramic materials are naturally inert, they do not cause adverse local or general tissue reactions when in contact with the human body. The biocompatibility of zirconia has been confirmed by several *in vitro* and *in vivo* studies. Zirconia is a chemically inert material, having a low toxicity when compared with titanium oxide and a similarly low toxicity when compared to alumina. However, its radioactive content must be purified to achieve a bio-inert form [26]. Another important factor to be considered is that zirconia can obtain an excellent surface finish, facilitating good tissue interaction after delivering ceramic abutments and prostheses [27].

Substructure prostheses made of zirconia offer a good masking of unfavorable substrates (discolored teeth, pins and metal cores), due to a suitable level of opacity. Additionally, zirconia provides light control after lamination, due to its homogeneity and high density (residual porosity <0.05%), even with a minimal thickness (0.5 mm) [4]. The opaque behavior of this ceramic is caused by the grain size being greater than the wavelength of light. Furthermore, zirconia has a high refractive index (1.8 – 2.17), a low coefficient of absorption (dependent on the microstructure and the type of dopant) and high opacity in the visible and infrared spectra [4,28].

However, zirconia cannot be used alone as a restorative material. It should be covered with a ceramic that has a higher degree of translucency to provide for a natural looking restoration [9].

Ceramic materials are brittle in nature. They do not have the capacity for deformation, and thus may crack under stress. In the case of zirconia, the addition of oxides allows for stabilization at room temperature, yielding a ceramic material with greater fracture toughness. However, all ceramic materials, at some point, may crack due to microstructural defects (such as voids, porosities and imperfections in the ceramic crystals). In stabilized zirconia, the tension generated by a crack can induce a reverse transformation (metastable tetragonal to monoclinic phase [$t \rightarrow m$]) at the crack propagation site. This transformation is of the martensitic type, and is accompanied by an increase in volume of 3-5% [29]. In this situation, the volume expansion that occurs as a function of the $t \rightarrow m$ transformation induces a transformed zone at the crack tip and a compressive stress that, in addition to reducing the driving force of crack propagation, generates an opposition force to its opening and a resistance to crack propagation. This mechanism is called "toughening by transformation" [30]. (Figure 2).

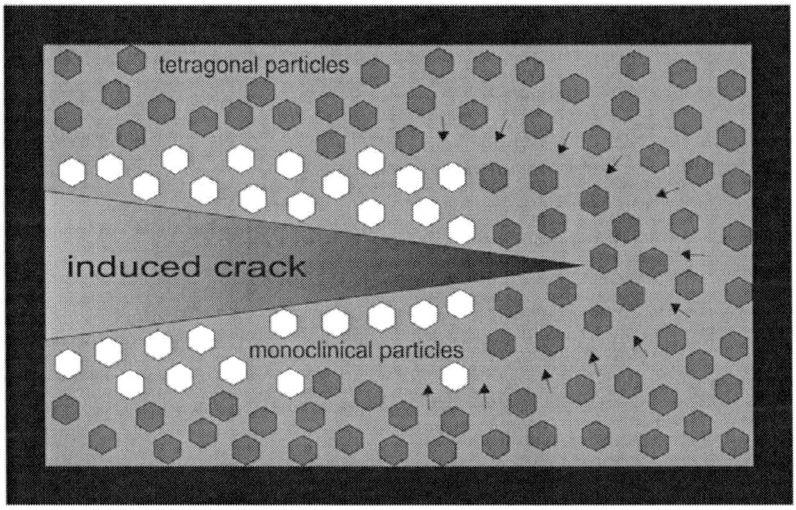

Figure 2. Mechanism to increase toughness: phase transformation of zirconia from tetragonal to monoclinic. Adapted from Vagkopoulou et al. 2009 [10].

The increased strength of the material occurs because energy that would be available for propagating the crack is partially dissipated in the area formed by the transformed monoclinic zirconia. [30].

However, if the swelling of the particle exceeds the yield strength of the material, it may promote crack propagation, resulting in new microcracks and possibly catastrophic fracture. Another important fact is that, as the toughening mechanism of zirconia is directly related to the $t \rightarrow m$ transformation, good flexural strength is guaranteed when a large amount of the monoclinic phase is detected on the surface of the ceramic [31].

However, a considerable reduction in the resistance is related to the total amount of the monoclinic phase, and the flexural strength decreases as the bulk of the structure becomes transformed [32].

Some requirements are necessary for the toughening process so that the $t \rightarrow m$ transformation is effective [30]. There must be a metastable phase, as the transformed zone can generate a compressive field of stresses that oppose the tension needed for crack propagation. Furthermore, the process must be virtually instantaneous, non-time dependent, and there must be a change in form and/or volume so that it can be induced by stress. It has also been observed that the speed of the process is high and there is no change in composition [22,32].

When in the tetragonal phase, the $t \rightarrow m$ transformation results in increased fracture toughness and is influenced by several factors. Although the factors do not act in isolation, each factor directly influences the transformation to the other crystalline phases. Among the factors which influence the $t \rightarrow m$ transformation are:

- *Defects:* the presence of defects and/or oxygen voids hinders the transformation to other crystalline phases by preventing atomic movement [8,24].
- *Grains or particles:* the size, shape and location (inter- or intragranular) of the ZrO_2 particles influence the $t \rightarrow m$ transformation. Large grains are more prone to transformation than are small grains. There is a critical grain size, below which the

phase transformation occurs (≈ 0.2 μm); change occurs spontaneously for larger grain sizes [33,34]. This critical size is dependent on the type of matrix that is contained in the zirconia.

- *Type and amount of stabilizers:* different dopants will act to stabilize zirconia individually. Currently, the stabilizer widely used in Dentistry is yttria. Ceria has shown a lower solubilization in humid conditions; however, when it is used as a dopant, it results in a zirconia with an unfavorable color [20]. Adequate amounts of stabilizer should be added to zirconia to optimize this metastable tetragonal phase [22,24]. The amount used depends on the type of microstructure, the route of synthesis of ceramic powders and, uniformity and distribution of the stabilizer.
- *Power system:* the transformation of the particle is governed by the energy of the particle-matrix system [8]. With the addition of stabilizers, the free energy changes in such a way that the tetragonal phase can be stabilized at room temperature. The energy (stress) applied to the material contributes to the phase transformation of the particle.
- *Temperature:* the phase transformation is temperature dependent. A few grains (greater than 0.2 μm) start at room temperature, and other grains, mainly due to their smaller size, require a temperature increase to form [35].

4. Factors that Influence the Behavior of Aging

Although zirconia may have good mechanical properties, the phenomenon of aging decreases its lifetime, and has represented a serious problem in the application of zirconia as a biomaterial [20,36].

As reported by Kobayashi in 1981 [37], tetragonal zirconia is susceptible to low temperature degradation, a phenomenon also known as "aging". [38]. This susceptibility of zirconia to aging depends on a number of factors, including: grain size, quantity, distribution and type of stabilizer, as well as the presence of residual stresses [39].

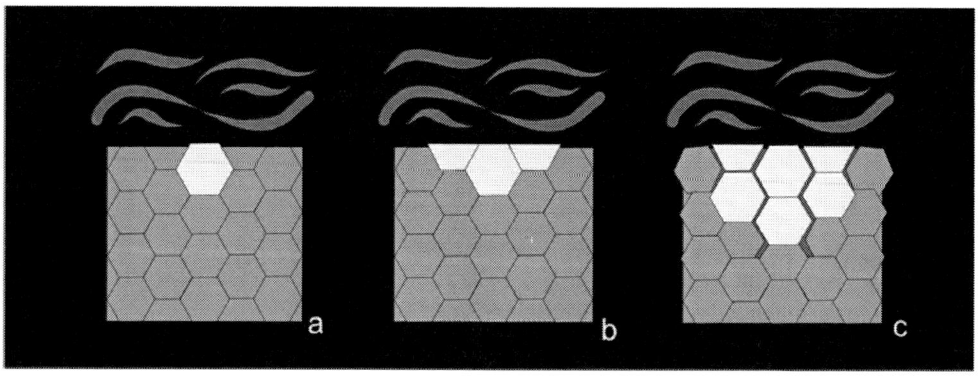

Figure 3. a) Schematic of the aging process in cross section. b) specific grain nucleation on the surface, leading to the formation of microcracks and stress in the neighboring grains. c) growth of the transformed area, leading to extensive formation of microcracks and surface roughening. Thus, water penetrates into the material due to the formation of microcracks around the processed grains. Adapted from Chevalier, 2006 [16] and Lughi, Sergo, 2010 [20].

It is known that the $t \rightarrow m$ transformation can be initiated by external stresses. In aging, this transformation is initiated when zirconia contacts water or steam. In a wet environment, spontaneous and progressive transformation begins on the zirconia surface by means of a mechanism of stress corrosion, and thereafter, advances into the material by nucleation and growth mechanisms in some grains, followed by propagation to the neighboring grains, surface degradation, roughness, release of particles, and micro- or macrocracks [17,20]. As the grain size increases with the created monoclinic phase, resulting in subcritical crack growth (SCG), a path for water to penetrate into the material is created, increasing susceptibility to aging. Due to failures that occurred with ceramic prosthetic femoral heads, Chevalier detailed the phenomenon of aging, which can be seen in Figure 3. [16].

Several mechanisms have been proposed to explain the LTD in a wet atmosphere; however, there is no clear experimental evidence for these models, due to the absence of rigorous scientific studies [9,37,39]. Some study designs have been proposed:

1) the presence of water vapor reacts with yttria to form clusters of $Y(OH)_3$, which removes the tetragonal structure of yttria, transforming from tetragonal to monoclinic grains [40];
2) water reacting with Zr-O-Zr connections, forming Zr-OH bonds, which induces the transformation [41];
3) the dissociation of water on the surface of zirconia and bond formation of Zr-OH and Y-OH, resulting in the deformation of the network surface, promoting the $t \rightarrow m$ transformation [42]; and
4) the creation and annihilation of oxygen voids, resulting in the phase transformation, ie, with the reduction of oxygen voids and their occupation by OH- anions, the tetragonal phase is not stable and transformation would occur [43]; the filling of oxygen voids by O^{2-}, and not OH^-, could be responsible for this degradation. The O^{2-} ions would be generated by the dissociation of water on the surface of Y-TZP ceramics [16,44].

The detailed model by Guo *et al.* [45] has been used worldwide, and uses sequential steps to describe the phenomena observed during the aging process:

1. Initial water chemical adsorption on the particle surface;
2. Reaction of water with oxygen to form surface hydroxyl groups;
3. Penetration of the hydroxyl groups into the material by diffusion across the grain;
4. Formation of proton defects by filling the oxygen voids with hydroxyl ions; and
5. $t \rightarrow m$ transformation caused by reducing the concentration of oxygen voids.

After surface modifications, water can penetrate into the material. The changes in the configuration of the nanograins are caused by defective protons migrating into the mass. The defective protons (step 3 and 4 of the Guo model) diffuse into the interior of the defects, especially in the oxygen voids derived from the stabilizers. As a result, the reduction in the concentration of oxygen voids causes a gradual transformation of the solid structure from a tetragonal to a monoclinic structure, reflecting the stabilization of the zirconia [45].

Regardless of the model, the consensus of the current authors is that the following observations on aging should be considered:

- The $t \rightarrow m$ transformation is influenced by the presence of water or steam, starting at the surface and progressing to the interior of the material;
- Aging effects are due to an increase of the monoclinic phase and reduced mechanical strength, hardness and density;
- The deterioration of mechanical properties takes place due to the $t \rightarrow m$ transformation by means of a propagation failure, known as subcritical crack growth (SCG);
- The phenomenon is strongly influenced by grain size, content and temperature of the stabilizer;
- The aging of zirconia can be observed at room temperature; however, the aging occurs more rapidly at 200-300 °C; and
- A reduction in grain size and/or an increase in the concentration of stabilizing oxides reduce the rate of transformation.

In vitro tests have been performed to evaluate the rate of zirconia aging. Chevalier et al. [14] performed artificial aging in an oven at 134°C for 1 hour, which is equivalent to 3-4 years in vivo. Those authors found that a temperature of 130°C for 7 hours caused the monoclinic content to increase by 30%, and the nucleation rate reached a high level of saturation [16,17]. However, many questions exist with regards to artificial aging. Evaluations at high temperatures are not representative of the physiological environment, and extrapolating results from a temperature higher than the human body may not be valid. It is also important to remember that the rate of strength degradation is not the same for all Y-TZP ceramics [46]. This variation in degradation is related to differences of the microstructural parameters, such as concentration, type and distribution of the stabilizer, grain size, failure distribution, residual stress and density. Another important finding is that the substructure of zirconia (used in the manufacture of ceramic prostheses) does not directly contact moisture in the oral environment, because it is covered with other ceramics; although recent studies have shown that the cement used in fixing these prostheses can absorb water through the dentinal tubules. Thereby, zirconia could interact with that water and undergo aging [47].

Using new techniques, and working at a nano scale, studies have been performed to find solutions to limit the phenomenon of aging. Additions of small traces of alumina (Al_2O_3), silica and/or ceria (CeO_2) have been reported [48-50]. Good results have been found with the addition of alumina particles, since this prevents disruption of the network of tetragonal zirconia by decreasing the particle size and providing a higher elastic modulus (250 – 350 GPa), which maintains the tetragonal zirconia phase [50]. However, adding a large amount of alumina can greatly enhance the degree of opacity, which could be a problem in dentistry [51].

The addition of ceria has attracted strong interest from researchers as a stabilizer of zirconia, having shown positive results against aging. It is needed in greater quantities than yttria, and must present a typical composition (Ce-TZP) with 8% ceria to provide toughness, since the system is non-transformable at concentrations above 12%. Another interesting feature of ceria-zirconia is that it presents a pseudo-plastic behavior and may even bend before fracture [52]. For some authors, this is because Ce-TZP can present hysteresis cycles in the stress-strain curve and even provide a shape memory effect, in the form of shape recovery after heating [53]. However, the color of ceria is an issue, because it varies in color between yellow and brown, even presenting as dark gray [50].

5. Aplications in Dentistry

Three types of zirconia have been applied in Dentistry [9]:

- *ZTA (zirconia toughened alumina):* an alumina matrix that is reinforced by zirconia grains. A similar compound was developed using this philosophy (InCeram Zirconia, VITA Zahnfabrik, Germany) by adding 33% stabilized zirconia and 12% Ceria (TPZ-12Ce) to InCeram Alumina (70-80% aluminum oxide). Initially, the material was created using a ceramic infiltration technique (slipcasting) and later, the same material was available in pre-sintered blocks to be finished by machining.
- *Mg-PSZ (Magnesia partially stabilized zirconia):* the microstructure is a cubic zirconia that is partially stabilized from 8 to 10% magnesium oxide. Due to the difficulty in obtaining precursors of Mg-PSZ that are free of silica (SiO_2), the present magnesium silicates may form a low concentration of magnesia, which favors the $t \rightarrow m$ transformation, resulting in a decrease in the mechanical properties and stability of the material.
- *3Y-TZP (Yttria tetragonal zirconia polycrystalline):* initially used in the medical field, 3Y-TZP, also known as tetragonal zirconia polycrystalline, contains from 2 to 4% yttrium oxide (Y_2O_3). The microstructure of this material consists of nanosized tetragonal grains, often combined with a small fraction of the cubic phase. In dentistry, 3Y-TZP nanoparticles (from 0.2 to 0.5μm in diameter) are compacted in blocks for machining in CAD/CAM systems and further sintering at high temperature (1350 up to 1550°C). The blocks are machined with a volume that is 25 to 30% higher than desired (depending on the batch of material), to counter the sintering shrinkage. In the variation called TZP-A, small amounts of alumina are added to 3Y-TZP.

6. Study of the Degree of Adaptation and Flexural Strength of Dental Abutments

Structural ceramics based on zirconia have enabled the production of abutments for implants and substructures for crowns and fixed prostheses in aesthetic and masticatory areas by combining high mechanical strength with biocompatibility [29].

During the fabrication of ceramic abutments for implants, a proper fit to the implant is essential; failure of achieving an adequate fit between these structures may favor the accumulation of plaque. Additionally, mechanical overloading of the system may result in loosening or fracture of the screw, peri-implantitis, bone loss and even loss of osseointegration [54-56].

When abutments are installed on implants, they are readily subjected to stress for several reasons: the geometry is complex, as the customized pieces have some thicker and thinner areas, generating a different distribution of stress and ceramic abutments are attached to the implants using metal screws, and these two materials exhibit different behaviors when subjected to mechanical stresses.

These pieces must withstand mastication; when they are positioned in the anterior region, they are at a disadvantage, because the dissipation of the generated forces depends on the anterior guidance of each patient [57].

In a study conducted at the Federal University of Santa Catarina, the fit of ceramic abutments to implants and the fracture toughness of the ceramic before and after thermal aging were evaluated *in vitro*.

Twenty abutments were made of 3Y-TPZ (Zirkonzahn, Italy) using the machining technique, and tested in a simulated region of the central incisors. The abutments were divided into two random groups: G1, which received no treatment (control) and G2, which was subjected to thermal aging in a thermocycler. Five hundred cycles were performed with alternating water baths at temperatures ranging from 5°C to 55°C (± 5°C) with a 15 dwell time for each temperature. The samples were then immersed in distilled water and stored in identified bottles, where they remained for 14 days at 37°C in an oven for evaluation using a scanning electron microscopy (SEM) (JEOL JSM-6390LV, JEOL, USA).

The abutments of both groups were joined to the implants (Osseotite, 3i, USA) with a new metal screw and positioned on a metal matrix to prevent movement during SEM evaluation. The tightening of the screw was performed (20N) using a manual torque wrench (Neodent, Brazil).

All of the specimens were then placed in the SEM with the palatal face (with the opening for the screw) facing the reader. Measurements of the implant/abutment interface were made in three regions (L: left side, M: middle and R: right side) at a magnification of 700X. The images obtained were analyzed with vertical linear measurements using a digital program (MeasureIt, Olympus, USA) to quantify the data (Figures 4 and 5).

After SEM evaluation, the abutments were included in PVC cylinders, ensuring that the coronal portion and the implant platform were visible. The positioning of the teeth in relation to the PVC cylinder was controlled using a parallelometer. The specimens were coupled to a universal testing machine (model 4444, Instron, USA) and subjected to a progressive static load at a speed of 1mm/min.

A special device attached to the lower portion of the universal testing machine allowed the specimens to stay positioned at an angle of 45° to the horizontal plane (135° to the long axis of the abutment), simulating a load with a direction similar to that found clinically in anterior regions (Figure 6).

The test was performed at a speed of 1mm/min until fracture of the abutment, as determined by visual observation of the fracture, a distinctive sound with or without visual observation, or a sudden reduction in the value of force recorded by the universal testing machine. The values (N/m^2) were recorded with the machine software.

The results of the fit to the implant platform were recorded as an average of the three local readings (*L:* left side, *M:* middle and *R:* right side) and are presented as mean, standard deviation, median, minimum and maximum values in Table 1. There was a wide variation in the values of the degree of cervical adaptation between groups and within each group tested, noting that G1 had lower mean values when compared to G2.

Figure 4. Specimen placed in the SEM.

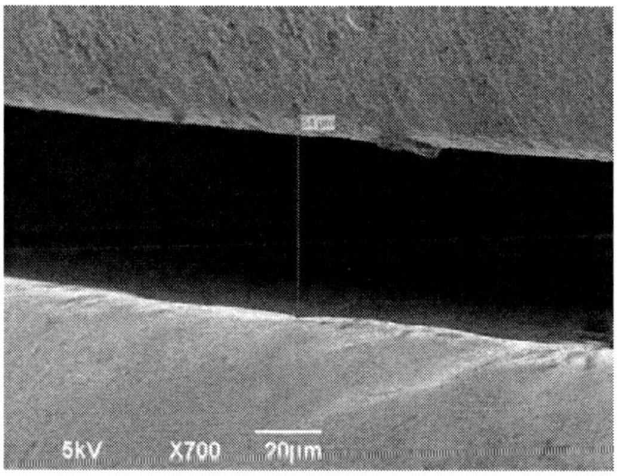

Figure 5. Measurement of the implant/abutment interface at 700X magnification.

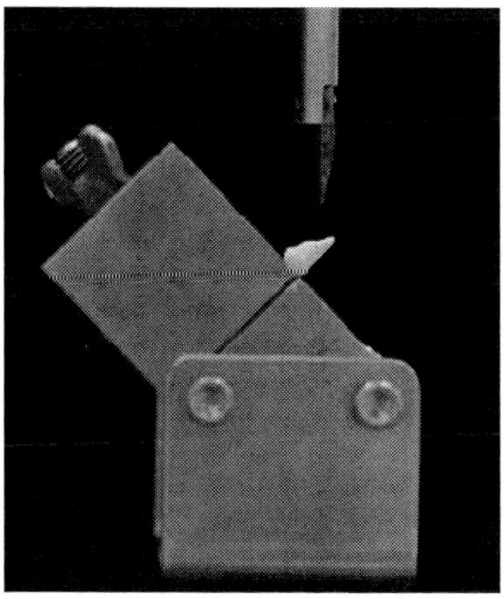

Figure 6. Specimens positioned at a 45° angle to the horizontal plane in a special device attached to the lower portion of the universal testing machine.

Table 1. Values of mean, median, standard deviation, minimum and maximum degree of cervical adaptation (μm) among the groups

Group	X(+SD)	Median	Minimum	Maximum
G1	30.73 (+36.28)	12.46	2.10	86.28
G2	57.19 (+46.73)	63.82	1.26	141.67

Table 2. Values of mean, median, standard deviation, minimum and maximum fracture strength resistance (N/m^2) among the groups

Group	X(+SD)	Median	Minimum	Maximum
G1	470.60 (+77.56)	465.5	377	612
G2	399.10 (+74.56)	403.5	271	532

The average values of marginal gaps were 57.19 μm (± 46.73μm) and 30.73 μm (±36.28 μm), respectively, for G1 and G2, with the results similar to those found by Scarano *et al.* [54] Although the control group presented a lower degree of misfit when compared with the experimental group, there was no statistical difference between the groups when applying the unpaired *t* test at a confidence interval of 95% ($\rho = 0.175$). Therefore, thermal aging did not interfere with the abutment/implant misfit. However, a high standard deviation was found, which may be the result of the sum of the possible variables: a small number of samples used and manual and visual adjustment of the cervical region after sintering the abutments.

The values of fracture strength resistance are presented as means, standard deviation, median, minimum and maximum in Table 2. The results showed higher mean values for G1 when compared to G2.

The values of fracture strength for G1 (470.60 ± 77.56 N) were lower than those reported in the literature [58-61]. Statistical differences were found between the groups ($\rho > 0.05$), with G1 (470.60 ± 77.56 N) superior to G2 (399.10 ± 74.56 N). The fracture strength values of G2 were lower than those found by Att *et al.* [59], with values of fracture strength of up to 443.6 N in thermocycled samples. The high values of standard deviation for the results reinforce the hypothesis that human variables are strongly linked to the differing results. However, it is important to note that the values found in this current study exceeded the maximum occlusal forces for humans in both groups tested (370 N).

These present results can be explained by the fact that aging occurred due to the presence of water contacting zirconia. Aging did not statistically affect the degree of adaptation for the abutment/implant; however, it compromised the mechanical behavior, reducing its fracture strength. It is possible that a change may have started on the surface of zirconia. Water penetrates into the material from the surface, providing for the $t \rightarrow m$ transformation and leading to the formation of microcracks (SGC), which might have compromised the fracture strength of the material.

Although the clinical situation is not adequately reproduced, static tests have been used in the majority of investigations on fracture strength of ceramic abutments [58-63]. The predominance of this type of test is related to a shorter execution time and ease of access to the continuous compression load transmission equipment. However, equipment that performs cyclic loading are the most suitable for assessing the strength of materials, as they are able to simulate an environment that allows results that are similar to the clinical situation [60,63].

Another important aspect to be discussed is that, although there are limitations with *in vitro* investigations, they are essential for identifying the quality of the materials prior to clinical studies, while also allowing for studies that are impossible to perform *in vivo*, such as those designed to study the fracture strength [61]. It is noteworthy that the results shown in laboratory studies do not necessarily reflect the clinical behavior of the material evaluated, since *in vitro* studies are limited in the reproduction of the factors that act *in vivo*. Therefore, it is not possible to affirm that ceramic abutments will present the same clinical behavior as observed in this current study when subjected to masticatory loading, because other variables found in the oral cavity were not simulated.

The literature presented studies that had a great variation in the number of thermocycles (100 to 1,000,000 cycles); as there is no consensus for this variable. Because thermocycling is a laboratory simulation of thermal oscillations that occur in the oral cavity, it is extremely difficult to determine precisely the number of these oscillations *in vivo* and reproduce them *in vitro*. In the absence of accurate data, it is suggested that, if such cycles occur between 20 and 50 times a day, 10,000 cycles would represent 1 year in the mouth [64]. However, it was found that 500 cycles were enough to promote changes in the mechanical behavior of thermocycled samples, resulting in a statistically significant difference between the groups. It is likely that increasing the number of cycles would further potentiate this difference.

CONCLUSION

Since its introduction in dentistry, zirconia has attracted attention because of its biocompatibility and excellent mechanical properties, such as good strength and fracture toughness. However, as confirmed by previous studies with artificial limbs, zirconia that has been partially stabilized by yttria can age when it comes into contact with water or steam, resulting in deterioration of its mechanical properties. In this current study, the results showed no significant differences in degree of cervical adaptation between the groups; however, the best numerical results were found in the control group. The results for the fracture strength resistance were statistically significant, suggesting that aging decreased the fracture strength of the abutments. Therefore, it has been a major goal of the scientific community to quantify this phenomenon and find solutions to minimize it, because zirconia has been demonstrated as a promising structural material for clinical use.

ACKNOWLEDGMENTS

Center for Research in Ceramics and Glass / Federal University of Santa Catarina (CERMAT/UFSC).

REFERENCES

[1] R. G. Craig. *J. Oral Rehabilitation*. 26, 841 (1999).

[2] R. G. Craig, J. M. Powers, J. C. Wataha. Dental materials: properties and manipulation. 9th ed. St Louis: Mosby, 2007.
[3] M. J. Heffernan, S. A. Aquilino, A. M. Díaz-Arnold, D. R. Haselton, C. M. Stanford, M. A. Vargas. *J. Prosthetic* Dentistry. 88, 4 (2002).
[4] M. J. Heffernan, S. A. Aquilino, A. M. Díaz-Arnold, D. R. Haselton, C. M. Stanford, M. A. Vargas. *J. Prosthetic* Dentistry. 88, 10 (2002).
[5] M. S. Cho, B. Yu, Y. K. Lee. *Dental Materials*. 25, 695 (2009).
[6] C. A. M. Volpato, S. Jr. Monteiro, M. A. C. Andrada, M. C. Fredel, C. O. Petter. *Dental Materials*. 25, 87 (2009).
[7] M. Guazzato, M. Albakry, S. P. Ringer, M. V. Swain. *Dental Materials*. 20, 449 (2004).
[8] J. R. Kelly, I. Denry. *Dental Materials*. 24, 289 (2008).
[9] I. Denry, J. R. Kelly. *Dental Materials*. 24, 299 (2008).
[10] T. Vagkopoulou, S. O. Koutayas, P. Koidis, J. R. Strub. *European Journal of Esthetic Dentistry*. 4, 130 (2009).
[11] T. Miyazaki, Y. Hotta, J. Kunii, S. Kuriyamas, Y. Tamaki. *Dental Materials*. 28, 44 (2009).
[12] B. Cales, Y. Stefani, E. Lilley. *Journal of Biomedical Materials Research,*28, 619 (1994).
[13] C. Piconi, W. Burger, H. H. G. Richter, A. Cittadini, G. Maccauro, V. Covacci, N. Bruzzese, G. A. Ricci, E. Marmo. *Biomaterials*. 19, 1489 (1998).
[14] J. Chevalier, S. Deville, E. Munch, R. Jullian, F. Lair. *Biomaterials*. 25, 5539 (2004).
[15] C. Piconi, G. Maccauro. *Biomaterials*. 20, 1 (1999).
[16] J. Chevalier. *Biomaterials*. 27, 535 (2006).
[17] J. Chevalier, L. Gremillard, A. V. Virkar, D. R. Clarke. *Journal of the American Ceramic Society*. 92, 1901 (2009).
[18] S. O. Koutayas, T. Vagkopoulou, S. Pelekanos, P. Koidis, J. R. Strub. *The European Journal of Esthetic Dentistry*. 4, 348 (2009).
[19] E. H. Kisi, C. J. Howard. *Key Engineering Materials*. 153, 1 (1998).
[20] V. Lughi, V. Sergo. *Dental Materials*. 26, 807 (2010).
[21] W. D. Richerson. *Modern Ceramic Engineering*. Properties, processing and use in design. Ed. Marcel Dekker, 2006.
[22] R. G. Luthardt, M. Holzhuter, O. Sandkuhl, V. Herold, J. D. Schanapp, E. Kuhlisch, M. Walter. *Journal of Dental Research*. 81, 487 (2002).
[23] A. S. Kumar, A. R. Durai, T. Sornakumar. *Materials Letters*. 58,1808 (2004).
[24] S. Fabris, A. T. Paxton, M. W. Finnis. *Acta Materialia*. 50, 5171 (2002).
[25] J. W. McLean. *The Journal of Prosthetic Dentistry*.85, 61 (2001).
[26] M. Gahlert, T. Gudehus, S. Eichhhorn, E. Steinhauser, H. Kniha, W. Erhardt. *Clinical Oral Implants Research*. 18, 662 (2007).
[27] L. Rimondini. *International Journal of Oral and Maxillofacial Implants*. 17, 793 (2002).
[28] F. F. Lange. *Journal of the American Ceramic Society*. 69, 240 (1986).
[29] P. F. Manicone, P. R. Iommetti, L. Rafaelli. *Journal of Dentistry*. 35, 819 (2007).
[30] P. M. Kelly, L. R. Francis Rose. *Progress in Materials Science*. 47, 463 (2002).
[31] M. Guazzato, L. Quach, M. Albakry, M. V. Swain. *Journal of Dentistry*. 1, 9 (2005).
[32] K. Shimizu, M. Oka, P. Kumar, Y. Kotoura, T. Yamamuro, K. Makinouchi, T. Nakamura. *Journal of Biomedical Materials Research*. 27, 729 (1993).

[33] F. F. Lange. *Journal of the American Ceramic Society*. 69, 240 (1986).
[34] W. Z. Zhu. *Ceramics International*. 22, 389 (1996).
[35] S. Schmauder, U. Schubert. *Journal of the American Ceramic Society*. 69, 534 (1986).
[36] S. Lawson. *Journal of the European Ceramic Society*. 15, 485 (1995).
[37] K. Kobayashi, H. Kuwajima, T. Masaki. *Solid State Ionics*. 3-4, 489 (1981).
[38] I. L. Denry, J. J. Peacock, J. A. Holloway. *Journal of Biomedical Materials Research B: Applied Biomaterials*. 93B, 236 (2010).
[39] J. W. Kim, N. S. Covel, P. C. Guess, E. D. Rekow, Y. Zhang. *Journal of Dental Research*. 89, 91(2010).
[40] F. F. Lange, A. G. Evans. *Journal of the American Ceramic Society*. 62, 62 (1979).
[41] T. Sato, M. Shimada. *Journal of the American Ceramic Society*. 68,356 (1985).
[42] F. F. Lange, G. L. Davis, B. I. Davis. *Journal of the American Ceramic Society*.69,237 (1986).
[43] B. R. Lawn, M. Yoshimura, T. Noma, K. Kawabata, S. Somiya. *Journal of Materials Science Letters*. 6, 465 (1987).
[44] J. Chevalier, B. Cales, J. M. Droguet. *Journal of the American Ceramic Society*. 82, 2150 (1999).
[45] X. Guo. *Journal of Physics and Chemistry of Solids*. 60, 539 (1999).
[46] J. Chevalier, C. Olangnon, G. Fantozzi. *Journal of the American Ceramic Society*. 82, 3129 (1999).
[47] R. Shahin, M. Kern. *Dental Materials*. 26, 922 (2010).
[48] A. J. Feighery, J. T. S. Irvinie. *Solid State Ionics*. 121, 209 (1999).
[49] L. Gremillard, T. Epicier, J. Chevalier, G. Fantozzi. *Acta Materialia*. 48,4647 (2000).
[50] V. Sergo, C. Schmid, S. Meriani, A. G. Evans. *Journal of the American Ceramic Society*. 77, 2971 (1994).
[51] D. Y. Lee, D. J. Kim, Y. S. Song. *Material Science and Engineering*. 289, 1 (2000).
[52] P. E. Reyes-Morel, J.- S. Cherng, W. Chen. *Journal of the American Ceramic Society*. 71, 648 (1988).
[53] C. Schmid, O. Sbaizero, V. Sergo, S. Meriani. *Journal of the American Ceramic Society*. 75, 2003 (1992).
[54] A. Scarano, B. Assenza, M. Piatelli, G. Iezzi, G. C. Leghissa, A. Quaranta, P. Tortora, A. Piatelli. *Journal of Oral Implants*. 6, 269 (2005).
[55] P. Vigolo, F. Fonzi, Z. Majzoub, G. Cordioli. *The International Journal of Oral and Maxillofacial Implants*. 20, 108 (2005).
[56] B. Yuzugullu, M. Avci. *Clinical Implant Dentistry and Related Research*.10, 113 (2008).
[57] S. Deville, J. Chevalier, L. Gremillard. *Biomaterials*. 27, 2186 (2006).
[58] M. Yildrin, D. Edelhoff, O. Hanisch, H. Spiekermann. *The International Journal of Periodontics & Restorative Dentistry*.20, 81 (2001).
[59] W. Att, S. Kurun, T. Gerds, J. R. Strub. *The Journal of Prosthetic Dentistry*. 95, 116 (2006).
[60] A. Sundh, G. Sjorgren. *Dental Materials*. 22, 778 (2006).
[61] R. B. Kerstein, J. Radke. *International Journal of Oral and Maxillofacial Implants*. 23, 1029 (2008).
[62] F. Butz, G. Heydecke, M. Okutan, J. R. Strub. *Journal of Oral Rehabilitation*, 32, 838 (2005).

[63] N. D. Adatia, S. C. Bayne, L: F. Cooper, J. Y. Thompson. *Journal of Prosthodontics*.18,17 (2009).
[64] M. S. Gale, B. W. Darvell, G. S. P. Cheung. *Journal of Dentistry*. 22, 370 (1994).

In: Recent Advances in Ceramic Materials Research
Editors: Joan J. Roa Rovira and Merce Segarra Rubi
ISBN: 978-1-62417-729-3

Chapter 5

PECULIARITIES OF THE STRUCTURE AND PHONON SPECTRUM IN NANO- AND MICROSTRUCTURE CERAMICS

E. N. Khazanov **[*]** ***and A. V. Taranov***

Institute of Radioengineering and Electronics Russian Academy of Sciences, Moscow, Russia

ABSTRACT

The experimental results on transport characteristics of diffusion propagation of slightly non-equilibrium thermal phonons (NP) at Helium temperatures are reviewed for a number of dense dielectric nano- and microstructured ceramics and composites. The relations between the diffusion coefficient of NP and grain boundaries (GB), grain size and structure, and process-dependent parameters of synthesis are analyzed. The model of GB considered as a finite-depth flat layer and of the NP transport in the polycrystalline structure is suggested under assumption $qR >> 1$ (where q is the phonon wave vector and R is the mean grain size). Under these conditions, the temperature dependence of the diffusion coefficient $D(T)$ is shown to be determined by spectral properties of GB. The model allowed us to estimate the mean values of thickness and acoustic impedance of GB. It is shown that the gap in the phonon spectrum may arise at $qR \sim 1$ This kind of gap has been revealed in the composites based on polycrystalline Yttrium-stabilized Zirconium (YSZ) matrix with nanocrystalline fraction of corundum (r = 20-30nm). The gap characteristics are controlled by both commensurability of the phonon wavelength with fragment size and GB structure. Inclusions of secondary phases and defects with characteristic sizes smaller than the grains of the main ceramic material shift the upper edge of the gap to higher frequencies. The inclusion of nano-sized fragments of α-Fe (<5%) into the polycrystalline corundum matrix substantially (over two orders of magnitude) suppresses the NP diffusion. In this case the phonon kinetics is determined by the size of metal fragments as well as by the surrounding spinel layer. It is shown that transport properties of NP in the $Ce_{1-x}Gd_xO_{2-y}$ ceramic solid solutions may depend on amount of impurities in the grain as well as on formation energy of a defect (a vacancy) in the anion sub-lattice. The value of relaxation energy can be experimentally determined

[*] Corresponding Author's e-mail address: khazanov@cplire.ru (E. N. Khazanov).

when it is comparable to the energy of NP. The better quality of GB and of grain structure (and hence the better acoustical, optical and thermophysical properties) of the optically transparent ceramics on the basis of cubic oxides and fluorides can be achieved when the material structure is formed by the twinning process as the main mechanism of plastic deformation.

The peculiarities of phonon transport in the ferroelectric ceramics having both sharp and diffused phase transition, corresponding to the crystal and glassy state are investigated. The impact of domain structure to the kinetics of non-equilibrium phonons in $BaTiO_3$ single crystal is examined. The value of reflection coefficient of sub-terahertz NP on domain boundaries is estimated.

1. INTRODUCTION

Different ceramics and ceramic-based composites are widely used both as novel construction materials and active elements in high-technology devices. Translation of the attractive properties of the basic material to the relevant ceramic prototype having polycrystalline structure as well as synthesis of the ceramic samples with large dimension as compared with common single crystals open new perspectives for developing energy-consuming and wide-aperture devices impossible otherwise [1]. Modern ceramic materials are produced from appropriate nano-powders. Therefore, all thermal and other physical properties of the materials are defined by their internal structure. Varying dimension, orientation, packing, and grain-to-grain boundaries through modifying parameters of the used technology process allows fabrication of the ceramic materials with desired unique properties. As a result, reliable control of the microstructure and chem ical content becomes extremely important [2, 3]. At present the efforts are focused on optimization of the grain structure, grain size, and grain boundaries (GB). The problem is solved partially by common X-ray methods, providing information on the averaged grain size from the width of the ray diffraction line, and the transmission electronic microscopy revealed an important property of the ceramics – the continuous nanometer intermediate layer, existing between neighbor grains [4]. However, neither X-ray methods, nor the electronic spectroscopy is not suitable for probing large volumes of the test materials and, thereby, they are not applicable for estimating an averaged condition at the grain-to-grain boundaries [5].

The kinetic characteristics of phonon system always contain information on structure peculiarities of the material. As a rule, the thermophysical properties of dielectric materials at temperatures in the range $T \geq 50$K are determined by the inelastic processes of phonon–phonon interaction. This circumstance complicates the study of the transport characteristics of phonons, because the measured quantities in this case are averaged over the phonon spectrum and scattering mechanisms. As the phonon temperature (energy) decreases, there arise conditions for the competition between elastic and inelastic phonon scattering (with times τ_0 and τ^*, respectively) and the phonon transport in materials with structural defects can occur according to a large number of different regimes [6]. The regime of classical diffusion, when a heat pulse of weakly non-equilibrium phonons (NP) propagating in the sample under investigation is elastically scattered by structural defects in the absence of inelastic processes of phonon–phonon interaction, is most informative for investigating structural defects and phonon transport processes.

The appropriate conditions for this regime are provided by the relationship for the times $\tau_0 << \tau_b << \tau^*$ (where τ_b is the ballistic propagation time of NP in the sample under investigation), which ensures a constant number of phonons for an observation time $t < \tau^*$ and makes it possible to exclude the influence of boundary scattering. As a rule, these conditions at liquid-helium temperatures (below 4K) are satisfied for almost all dense dielectric oxide materials with defects. It should be noted that the wavelengths of thermal phonons λ_{ph} = 10–50 nm are comparable to characteristic sizes of structural units of ceramic materials (grain boundaries, pores, nanofragments of another phase). Owing to this specific feature, thermal phonons can serve as an effective tool for studying micro-/nano-inhomogeneous materials. Among these materials are ceramics, solid solutions, any materials containing point or extensive defects, excitations with comparable energy.

2. EXPERIMENTAL TECHNIQUE

In the study of the ceramic structure, we employ the method that originates from the method of thermal pulses [7] and is based on the measurement of the temperature dependences of the transport characteristics of NP in the diffusion regime at helium temperatures. The heating of the metal film of the NP injector to temperature T_h, such that $\Delta T = T_h - T_0 << T_0$ makes it possible to study the temperature dependences of NPs using variations in thermostat temperature T_0. Under such conditions, the inelastic phonon–phonon processes can be disregarded and the scattering efficiency of NPs is determined only by the structural features of the material. The NP pulses that are detected on the opposite (with respect to the injector) end surface of the sample by the superconducting bolometer (Sn, In) are well described with the aid of the solution to the diffusion equation.

$$\frac{\partial^2 T}{\partial x^2} = \frac{1}{D}\frac{\partial T}{\partial t} \tag{1}$$

where $D=K/c_v$ is the diffusion coefficient, K, c_v are thermal conductivity and heat capacity respectively. The experimentally measured value is the time of the maximum diffusion signal (plane source),

$$t_M = \frac{L^2}{2D(T)} \tag{2}$$

where L is the dimension of sample, $D(T) = lv/3$, l is the NP free path length, and v is the polarization-averaged velocity of sound.

The temperature varies by pumping the helium vapors; the bolometer operating point varies by applying small magnetic field. The experimental technique has been proved when studying the phonon kinetic in yttrium-rare earth garnet single crystals (YAG:Re) [8]. Such a material serves as model because it allows controlled isomorphic substitution of Y ions by Re ions in the crystal lattice. The maximum of experimentally observed signal in YAG:Re solid solutions is formed by elastically scattered phonons of frequency $\hbar\omega \cong 2.8k_BT$ (k_B is the Boltzmann constant) [8], so we can use the single-frequency model. The thermal conductivity experiments indicate that the temperature dependence $K(T)=c_V D(T) \sim T$ in the YAG ceramic at 8-16K is controlled by the phonons of these frequencies [9]. The diffusion mode of phonon propagation in ceramics at helium temperatures has been proved and observed previously in [10]. The method allows to estimate the acoustic impedance and GB thickness that are averaged with respect to the volume of the sample [11].

3. NP Propagation Model in Ceramic ($QR >> 1, l/R >> 1$)

Figure 1 represents the cleavage of different ceramic samples made by scanning electronic microscopy. All samples have dense polycrystalline structure. At helium temperatures, the NP mean free path lengths in dielectric single crystals are several fractions of centimeter. When the condition $l/R >> 1$ is met the NP propagation can be represented as the ballistic propagation of phonons in the grain with probability f_ω of the transmission through GB to the neighboring grains. So we can assume that NP scattering is determined only by the structure and properties of GBs.

When the condition $qR >> 1$ is met (q is the phonon wave vector), the phonon spectrum of the material is similar to the spectrum of the vibrational excitations of the grain material and GB can be represented as a plane layer with finite thickness d whose acoustic impedance differs from the acoustic impedance of the grain material [11].

For the NP diffusion in the ceramic sample on the time scale $t >> t_0$ (t_0 is the time of the photon passage through the grain), the diffusion coefficient is $D \sim R^2/t_0$. In the high-density ceramics, we have:

$$t_0 \sim \frac{R}{v \cdot f_\omega} \tag{3}$$

and,

$$D \sim R\, v\, f_\omega \tag{4}$$

Expression (3 and 4) shows that $D \sim R$ in the framework of the proposed model and f_ω is the main quantity that determines the NP spectral characteristics (i.e., $D(T)$). Figure 2 demonstrates the dependence of diffusion coefficient D on the mean size of the grain for high-density oxide ceramics with different compositions at $T = 3.8$ K under the condition $qR >> 1$.

For each sample, a solid line connects the experimental points. Note that the NP diffusion coefficients increase with increasing mean size of the grain in accordance with expression (4). Point A corresponds to the YAG single crystal [8]. Each series of samples has been fabricated under the same conditions. The varied parameters are sintering temperature T_s and time, t. The dependency $D(R)$ is close to linear in the range of 3-4 orders of magnitude, thus indicating that GB properties are stable in wide range of fabricating technology parameters. It seems fair to suppose that spectral characteristics of diffusion coefficient (f_ω value) show resonant behavior when the wave vector projection component is comparable to the GB thickness.

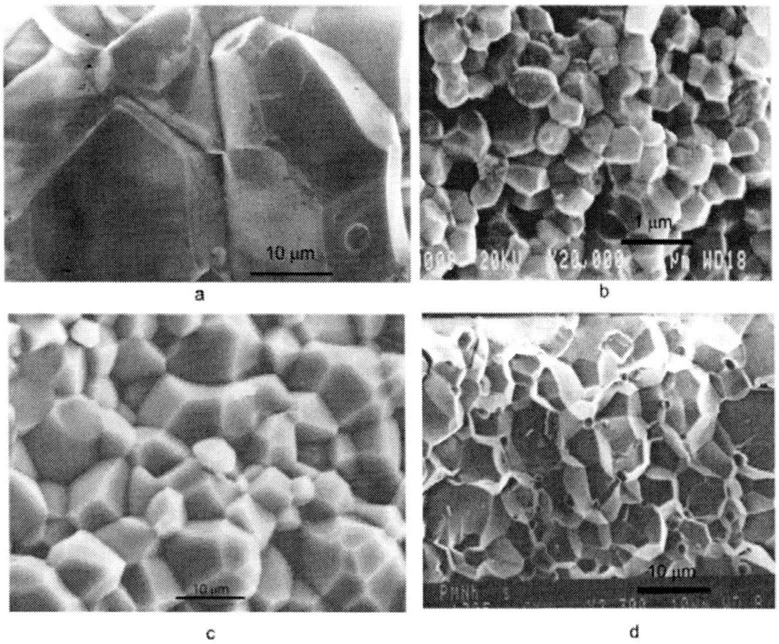

Figure 1. The cleavage of ceramic samples: *a)* Al_2O_3; *b)* ZrO_2:Y_2O_3 (YSZ); *c)*$Y_3Al_5O_{12}$ (YAG); and *d)* ferroelectric PMN ceramic.

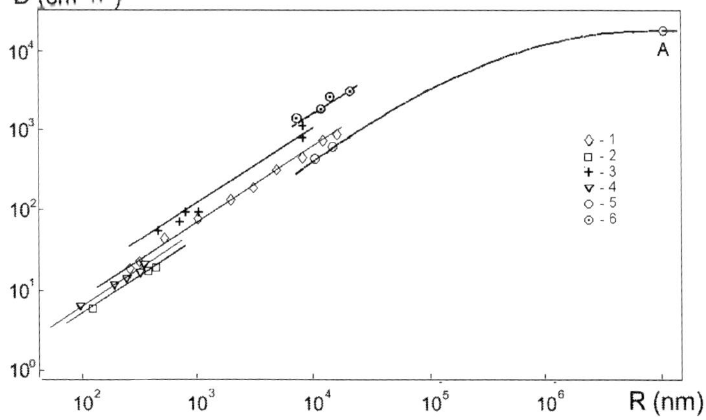

Figure 2. $D(R)$ dependences in the range $qR >> 1$ for the oxide ceramics at $T = 3.8$ K: (*1*) Al_2O_3 [12], (*2*) YSZ [11], (*3*) Y_2O_3 : Nd [13],(*4*) TiO_2 [14], (*5*) YAG [15], and (*6*) YAG : 1% Nd [15].

It may lead to change of sign of derivative $\frac{\partial t_m}{\partial T}$, and hence $\frac{\partial D}{\partial T}$, observed in AlN ceramic samples exposed to supplementary sintering at T_s = 1750C (Figure 3) [16]. The structure and size of ceramic grains have no visible differences, but time of signal maximum at T = 3.8K reduced more than 20 times (see inset Figure 3). The value of thermal conductivity at room temperature increased only by 20-25%.

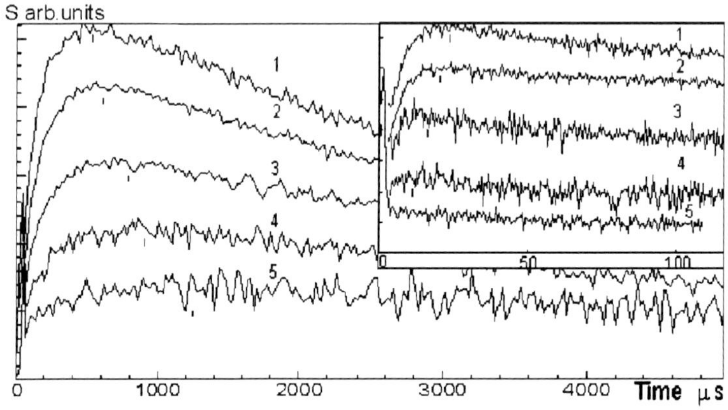

Figure 3. The bolometer signal in AlN ceramics L = 0.138 cm: T = 3.83K (1); T = 3.44K (2); T = 2.98K (3); T = 2.57K (4); T = 2.28K(5) Inset – repeated sintering at T_s = 1750C.

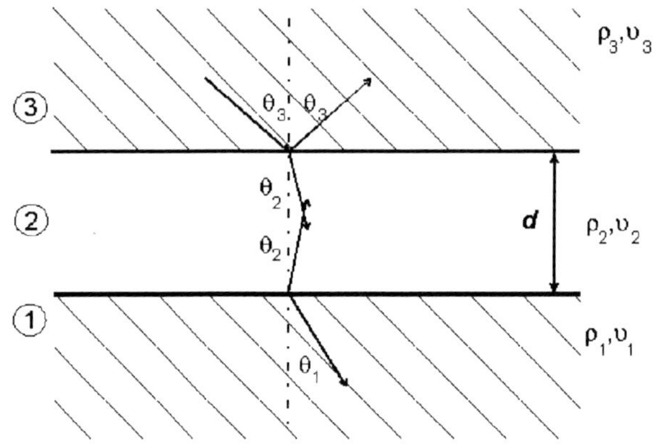

Figure 4. The acoustic model of phonon transmission from grain to grain via plane boundary layer.

The experimentally revealed increasing and decreasing forms of the temperature dependence for the arrival time t_m of the maximum for the signal of a pulse of nonequilibrium phonon radiation, are illustrated by the example of the structure shown in Figure 4, which is a simplified model of a phonon (high-frequency acoustic wave) propagating from grain *3* to grain *1* via the boundary layer *2* [12]. The materials of the structure of Figure 4 are assumed to be isotropic and homogeneous with preassigned values of density ρ_i and group velocity of sound v_i (i = 1, 2, 3), with the grains *1* and *3* assumed to be semibounded plane layers of thickness d. We follow the method of acoustic matching of media [17, 18] to estimate the probability f_ω of phonon transmission from grain *3* to grain *1* via the boundary layer *2* as the ratio of the power of radiation transmitted to grain *1* to the power of radiation incident on the boundary layer, with due regard for all angles of incidence θ_3 and transmission θ_1 (Figure 4).

This determination gives;

$$f_\omega = \frac{\int_0^{\pi/2} \alpha(\theta_3)\cos\theta_3 \sin\theta_3 d\theta_3}{\int_0^{\pi/2} \cos\theta_3 \sin\theta_3 d\theta_3} \quad (5)$$

Here, the coefficient $\alpha(\theta_3)$ is written as $\alpha(\theta_3)=\left(Z_3/Z_1\right)|W|^2$, where $Zi = \rho_i v_i/\cos\theta_2$ is the acoustic impedance of the medium, $i = 1, 2, 3$, and W is the coefficient of transparency of the boundary layer with respect to the wave amplitude, that is equal to:

$$W = \frac{4Z_1Z_2}{(Z_1+Z_2)(Z_2+Z_3)} \cdot \frac{1}{\exp(-i\gamma_2 d)+V_{12}V_{23}\exp(i\gamma_2 d)} \quad (6)$$

with the quantities V_{12} and V_{23} having the meaning of the coefficients of wave reflection at the boundaries of the medium 2, 1 and 3, 2, i.e.,

$$V_{12} = \frac{Z_1 - Z_2}{Z_1 + Z_2}; V_{23} = \frac{Z_2 - Z_3}{Z_2 + Z_3} \quad (7)$$

$\gamma_2 = q_2 \cos\theta_2$, is wave number of wave propagation inside the boundary layer in the direction transverse to its boundaries, ($q_2 = \omega/v_2$).

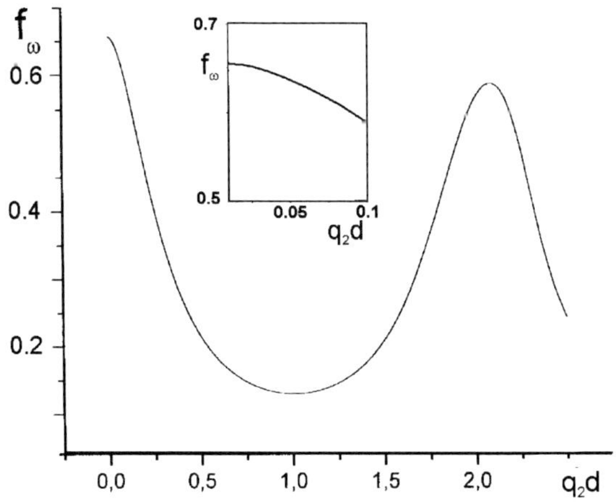

Figure 5. Predicted curves of probability f_ω of the transmission of a phonon of frequency ω from grain *3* to grain *1* via boundary layer *2* of thickness *d* as a function of the layer thickness q_2d, where $q_2 = \omega v_2$. The other parameters: $v_1/v_2 = 1.83$, $v_2/v_3 = 0.6$, $\rho_2 v_2/\rho_1 v_1 = 0.32$, $\rho_3 v_3/\rho_2 v_2 = 3.5$. Given in the inset is the value of f_ω for small values of q_2d.

The results of numerical integration and the dependence of the coefficient of transmission of phonon flux through the boundary between grains on the acoustic thickness of the boundary q_2d are given in Figure 5. For a thin (perfect) boundary between grains, ($q_2d \ll 1$)

and $f_\omega \cong 0.7$ (see the inset in Figure 5), the value of f_ω decreases as the frequency (or the phonon temperature) increases, and, therefore, the value of t_m increases. The absolute value of f_ω for this case agrees with our estimates based on the experimental curve *1* in Figure 2.

In the case of a fairly thick boundary layer, $q_2 d > 1$, for the preassigned parameters of the layer $\rho_2 / \rho_{1,3} \cong 0.6$ and $v_2/v_{1,3} \sim 0.6$, the transparency of the boundary to the phonon flux decreases, $f_\omega \cong 0.1$, and may exhibit an increasing temperature dependence, and, consequently, t_m will decrease as the temperature increases. The estimates based on the condition $q_2 d \sim 1$ give $d > 5$ nm for the thickness of the intergrain layer, which appears to be a reasonable value.

4. Analysis of GB Structure in YSZ and YAG: Nd Based Ceramics

A comparative analysis of acoustic impedance and GB thickness has been performed on high-density $ZrO:Y_2O_3$ ceramic samples prepared according to two technological schemes:

1) Dry static compacting of nanopowders obtained by plasmo-chemical synthesis under simultaneous action of high-intensity ultrasound. Subsequent sintering is carried out in vacuum at a temperature $T = 1650$C. The power of ultrasonic action in compacting varied from 0 to 3 kW (samples 1-4 in table 1). The contacts between crystallites or subgrains (according to the terminology used in [19]) are very close and poorly resolved by the electron microscope, the subgrains are bound into finite formations with c $R >> r$, which can be referred to as agglomerates. In electron micrographs, such agglomerates appear as well-facetted grains (see Figure 1b).
2) Magnetic-pulse compacting followed by annealing in a resistance furnace in air [20]. Owing to the high density of compacts, the ceramic was fritted at lower temperatures (1100–1450C) with short holding times (about 20 min). X-ray diffraction analysis of the samples being studied made it possible to determine the crystal modification of $ZrO_2:Y_2O_3$ samples and the mean size of r - regions of coherent scattering. According to the results of X-ray measurements, the crystallite size in samples 5–11 [20] prepared from weakly agglomerated powders corresponds to the grain size on electron micrographs; the boundaries between the grains are clearly manifested.

The method of estimation of GB's acoustic impedance and thickness based on temperature dependences *D(T)* was developed in [11] for the case of thin and thick GB. The experimental results were compared with the following expression:

$$l = l_0 \frac{f_\omega}{1 - f_\omega} \tag{8}$$

from [21], where l is the effective phonon free path in the lamellar periodic structure having period R. The NP scattering in the grain material is neglected. The mean free path of the phonon inside grain between scattering on GB $l_0 = 0.6R$ [22]. We assumed that for all samples from table 1 $\rho_1 = \rho_3 =$ density of basic material and a $v_1 = v_3 = 4.33 \cdot 10^5 \text{cm} \cdot \text{s}^{-1}$. Figure 6 represents the family of dependences of $f_\omega/1\text{-}f_\omega$ value on q_2d. It can be noted that the theoretical curves in Figure 6 indicate a strong dependence of $f_\omega/1\text{-}f_\omega$ value on q_2d.

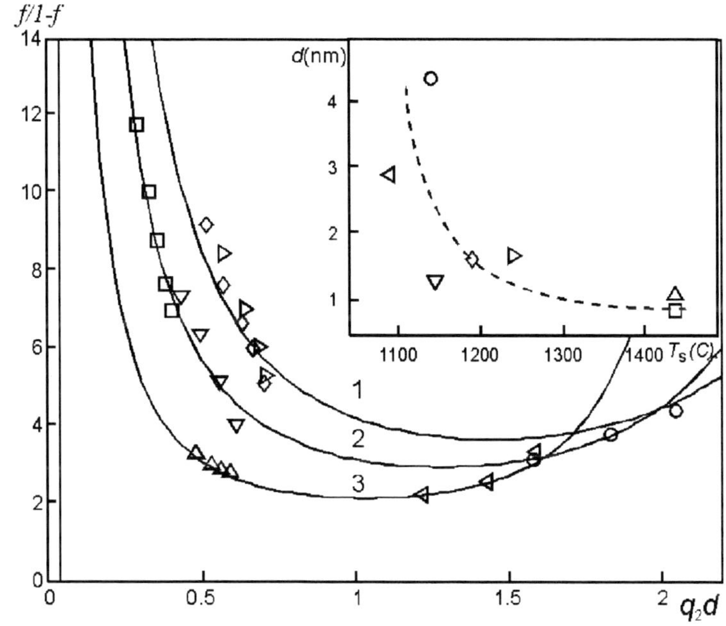

Figure 6. Curves describing quantity $f_\omega/1\text{-}f_\omega$ as a function of q_2d for ρ_2v_2/ρ_1v_1=0.6 (1); 0.7 (2), 0.75 (3). Symbols correspond to the results of experiments for samples obtained by magnetic-pulse compacting: O - sample 798.3, Δ - 528.1, □ - 722.1, ◊ - 798.4, ∇ - 712.2, < - 236, > - 799.5. The inset shows the dependence of d on the compact annealing temperature T_s.

This is because of the resonance mechanism [11] that emerges due to the difference between ρ_2v_2 and ρ_1v_1, and congruence between the grain boundary thickness q_2d and the projection of the wave vector of injected phonons. Thus, the intergranular layer, which operates as a resonant structure, ensures that passage of the heat pulse depends on temperature; the derivative of this dependence can be either positive or negative (see Figure 6). Figure 6 represents also the experimental data (points) for samples 5-11 from table 1.

The calculated dependences $f_\omega/1\text{-}f_\omega$ were compared with experimental values of $l(T)/l_0$ according to the following scheme: for each sample we find the section of the theoretical curves at T = 3.8K and corresponding values of $l|_{3.8K}/l_0$. ; the points of sections determine the values of q_2d; assuming $q_2 \approx \dfrac{3kT}{\hbar \cdot v_2}$ we calculate the values of d. The coordinates of other points of temperature dependency for that sample correspond to experimental values of $l(T)/l_0$ and $q_2(T)d$. (for the value of d calculated earlier).

The curves corresponding to the best fitting between theory and experiment makes it possible to determine the thickness d of the intergranular layer and its acoustic parameters (i.e., the ratio ρ_2v_2/ρ_1v_1.). The results of precisely such data processing are presented in Figure 6.

Table 1. Parameters of $ZrO::Y_2O_3$ ceramic samples studied

No	Sample No.	Phase composition (mol. %)	Powder compacting regime		Sinterin g tempera ture Ts,°C	Density g·cm-3	Micro hardness GPa	Coherence region size r (nm)	Grain size R nm	mean free path l (nm)	*l/R*
			Pressure MPa	Ultra sound kW							
1	1-57	ZrO2:2.8%Y2O3	350	0	1650	5.45	11.6	77	430	930	2.16
2	953	ZrO2:2.8%Y2O3	350	1	1650	5.58	12.85	54	350	2140	6.11
3	9-58	ZrO2:2.8%Y2O3	350	2	1650	5.58	12.82	64	345	950	2.75
4	3-63	ZrO2:2.8%Y2O3	350	3	1650	5.24	8.9	79	420	2040	4.86
5	236	ZrO2:2.8%Y2O3	135	-	1100	6.1	17	63	63	126	2.0
6	798.3	ZrO2:4.1%Y2O3	135	-	1150	6.1	16.5	78	78	206	2.6
7	798.4	ZrO2:4.1%Y2O3	135	-	1200	6.1	16.5	84	84	256	3.0
8	722.1	ZrO2:4.1%Y2O3	135	-	1450	5.595	16	120	120	504	4.2
9	799.5	ZrO2:4.1%Y2O3	135	-	1250	6.1	17	95	95	286	3.2
10	712.2	ZrO2:9.8%Y2O3	135	-	1155	5.95	16	115	115	328	2.43
11	528.1	ZrO2:9.8%Y2O3	135	-	1450	5.89	16	100	100	319	1.77

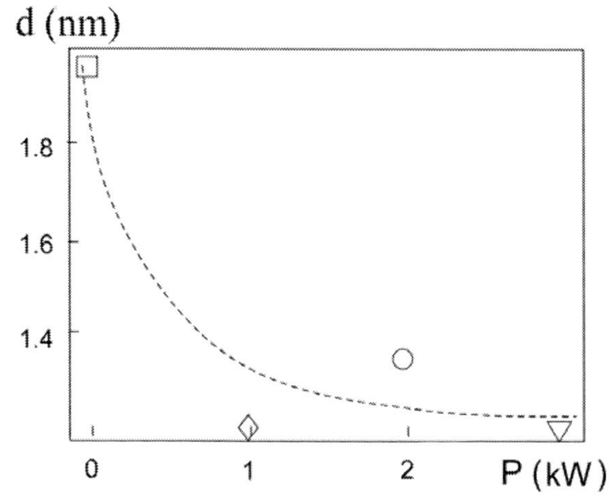

Figure 7. The dependence of GB thickness *d* on ultrasound power for samples 1-4 from table 1.

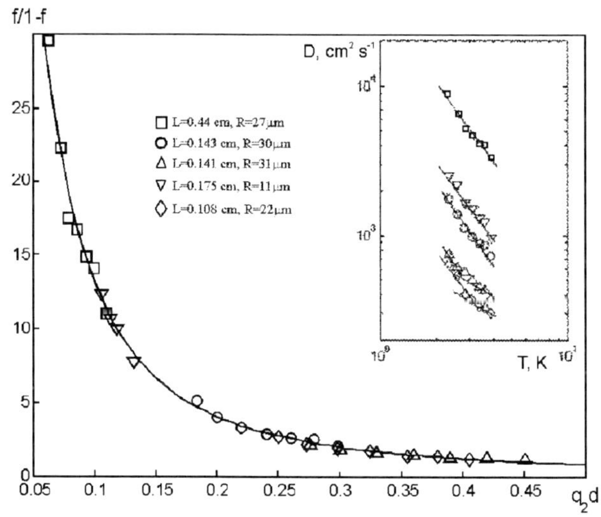

Figure 8. plots of $f\omega/(1-f\omega)$ vs. GB acoustic thickness q_2d in the YAG : 1% Nd samples obtained from the experimental results shown on inset◊ - L=0.108 cm, R=22μm; Δ - L=0.141 cm, R=31μm; ○- L=0.143 cm, R=30μm; v - L=0.175 cm, R=11μm; □- L=0.44 cm, R=27μm.

The inset shows the dependences of the value of *d* determined here on sintering temperature. A decrease in the value of *d* to about one-third of the initial value is also observed for samples 5–11 upon an increase in their fritting temperature from 1100 to 1450°C. Figure 7 shows the results of the same procedure for samples 1–4 obtained under the ultrasound action on the stage of compacting. The thickness of GB becomes smaller when ultrasound power increases. The ratio v_2/v_l is estimated to be 0.6-0.8. It means that the GB layer can be amorphized like in [23]. The additional investigations of the structure of samples 1-4 by scanning probe microscopy, X-ray diffraction, X-ray photoelectron spectroscopy [24] show that ultrasound leads to disperse structure alterations and activates the diffusion of Y ions in the interface regions. So the experimental data spread for 1 and 2 kW samples can be explained by that effect. Figure 8 demonstrates the theoretical dependence of the ratio on q_2d

in the samples of the optically transparent YAG:1%Nd ceramics from [25] as an example of the $qR >> 1$ case. The inset represents the experimental results for YAG:1%Nd samples. The dependence represents the left-hand slope of the resonance curve for the impedance ratio of the grain material (YAG, ρ = 4.55 g $\cdot$cm^{-3}, and v = 5.6$\cdot$10^{5} cm$\cdot$s^{-1}) and GB material (SiO_2 glass phase, ρ_2 = 2.02 g $\cdot$cm^{-3} , and v_2 = 4.05$\cdot$10^{5} cm$\cdot$s^{-1}) z_2/z_1 = 0.32. It was supposed that GB consists of amorphous SiO_2. In the beginning a small addition of SiO_2 was assumed during powder activation process to neglect the porosity [26]. The experimental data are in agreement with the calculated dependence, which indicates the quality of the proposed model. The calculated values of GB thickness varied between 0.2 and 0.5 nm depending on sintering conditions. In the Y_2O_3:Nd^{3+}, Lu_2O_3 ceramics that are synthesized in the absence of SiO_2 [13, 27] the estimated thicknesses of GB are close to the grain lattice constant. When the condition $qR >> 1$ is satisfied, the general regularity is observed: the NP diffusion coefficient increases with increasing grain and decreases with increasing GB thickness.

5. Peculiarities of the Structure of LiF Ceramics and Single Crystals

Optical ceramics are one of the most serious recent innovations in the field of laser materials [28-30]. In addition to oxide ceramics, significant progress has been made in the field of fluoride ceramics [31, 32]. In [33, 34], the authors obtained lasing in optical lithium fluoride ceramic with dye centers. Lithium fluoride is a classical object for studying the dislocation structure [35, 36] and the effect of mechanical deformation on the real structure [37, 38]. The samples of LiF optical ceramic [39] was produced by hot formation by the method described in [33]. Ceramic samples were chopped off from single crystals in the form of cubes with edge size of 1 cm with natural crystallographic faces along the planes (100) and then subjected to hot pressing at T = 973 K under different static loads. The samples obtained were characterized by the degree of deformation from 30 to 65% of the original size of the single crystal. The ceramic samples under test had the shape of plates with a thickness of 1.6 – 5 mm. Possible mechanisms of phonon scattering in the range of temperatures $T << \Theta$ (Θ is the Debye temperature, T_D) in LiF single crystals were considered in [35].

In [36], the authors showed that, in the range of liquid helium temperatures, the scattering of NPs in LiF single crystals with the density of dislocations N = 10^{6}– 10^{7} cm^{-2} is determined by the competition of mechanisms related to the deformations that arise in the crystal around dislocations and to the flutter effect due to the oscillation of dislocation segments in the field of an elastic plane wave followed by the emission of a phonon. When the phonon wavelength λ is much less than the average length l_d of dislocation segments, then $\tau^{-1} \sim T^{-1}$, $D = 1/3\ v^2\tau$, ($D(T) \sim T$), and a flutter effect is observed. When $\lambda >> l_d$ (lower temperatures), phonons take a dislocation as a linear defect and are scattered by it due to additional stresses in its neighborhood by the law $\tau^{-1} \sim T$ ($D(T) \sim 1/T$).

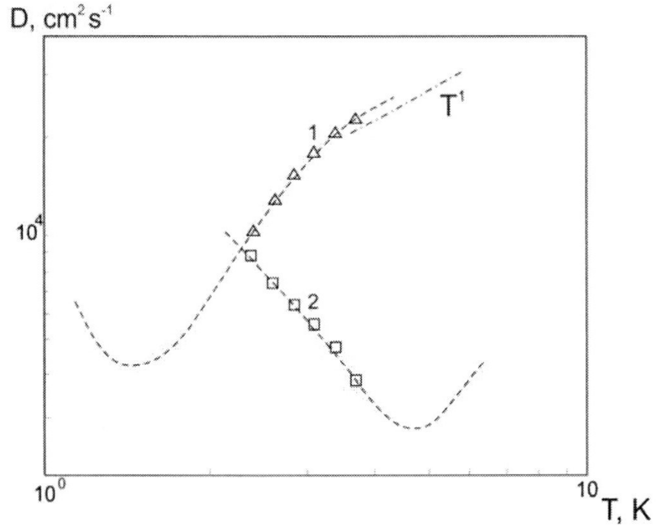

Figure 9. Experimental function D(T) in LiF single crystals: (1) unirradiated sample with dislocation density of $N = 10^6$ cm^{-2} and (2) sample exposed to γ-irradiation with a dose of 10^6 rad. The dashed curves illustrate a qualitative behavior of the functions that is characteristic of the flutter mechanism of scattering of NPs.

Thus, in the low temperature region, there may exist a resonance at $\lambda \cong l_d$. Figure 9 shows the diffusion coefficients *D(T)* of NPs in a single crystal with the dislocation density N = 10^6 cm^{-2} (1) and in a similar sample exposed to γ-radiation with a dose of 10^6 rad (2). The emergence of additional radiation defects in the irradiated sample leads to the pinning of dislocations, i.e., to a decrease in the length of dislocation segments and, as a consequence, to a shift of the resonance maximum to shorter wavelengths of NPs. In Figure 9, in the range of temperatures considered, the functions *D(T)* for $\lambda << l_d$ (1) and $\lambda >> l_d$ (2) correspond, to the right and left slopes of the resonance curve respectively.

It is possible to estimate the dislocation segments length supposing that the maximum of phonon signal is formed by NPs corresponding to the maximum of spectral density of Planck's distribution in the phonon injector film $\omega \approx \frac{3kT}{\hbar}$. At T = 3K $v \cong 4 \cdot 10^5$ cm·s^{-1} $\lambda_{ph} \cong$ 20nm $l_d \geq$ 40nm in unirradiated sample. In irradiated sample $l_d \lesssim$ 10nm. Figure 10 shows the results of electron microscopy of the chips of strained LiF ceramic samples. In Figure 10a (the deformation degree of the original material is 30%), the grains of ceramic have the shape of large blocks with a size of R = 200–300 μm, whose boundaries mainly coincide with the crystallographic planes of the single crystal. It allows suggesting that the process of deformation reflects the dislocation mechanism of relaxation of stresses in the ceramic material. In the second case (Figure 10b), $R \cong 100$ μm (the deformation degree is 65%), and the grains are rotated relative to each other; this is associated with the relaxation of larger static stresses and, possibly, with a change of the mechanism of plastic deformation of the material when the dislocation mechanism cannot guarantee any longer the relaxation of increasing stresses.

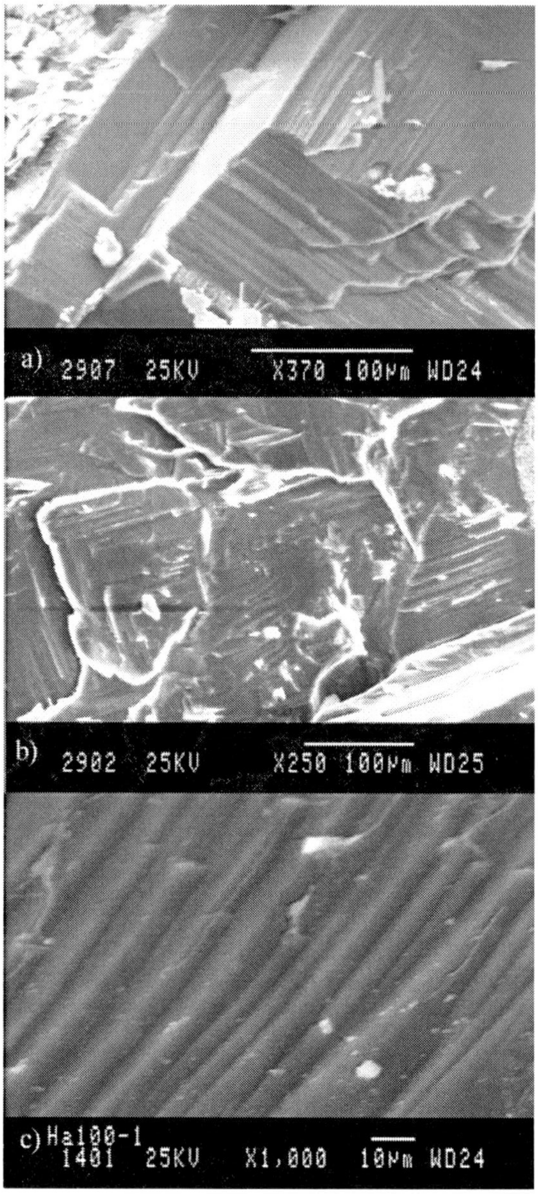

Figure 10. Micrographs of LiF ceramic samples for different deformation degrees of: (a) 30%, (b) 65%, and (c) >50%.

In the first case, the destruction of samples occurs both along the boundaries of grains (blocks) and through the grains. In the second case, the destruction occurs mainly along the boundaries of grains, which exhibit a streak pattern. For higher resolution, the grain surface in the samples deformed under hot pressing conditions by more than 50% exhibits ordering of structural elements with a size of $h \leq 10$ μm, which is similar to twinning regions (Figure 10c). The AFM spectroscopy results of the surface of a LiF ceramic chip with intermediate degree of deformation (46%) are presented in Figure 11. They show that the elements of the grain structure observed in Figure 10c have a layered arrangement. The analysis of larger scale images shows that each element of the structure consists of oriented layers with a mean width of 50 – 100 nm. One can see that the width of the layers decreases as the degree of deformation increases and is beyond the resolution of the device for the maximal degree of deformation (65%).

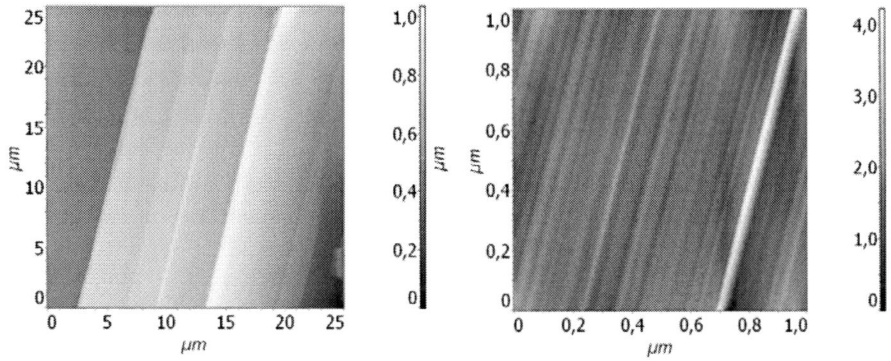

Figure 11. Atomic Force Microscopy, AFM (topographic mode)images of LiF ceramic samples with a deformation degree of 46% at different scales.

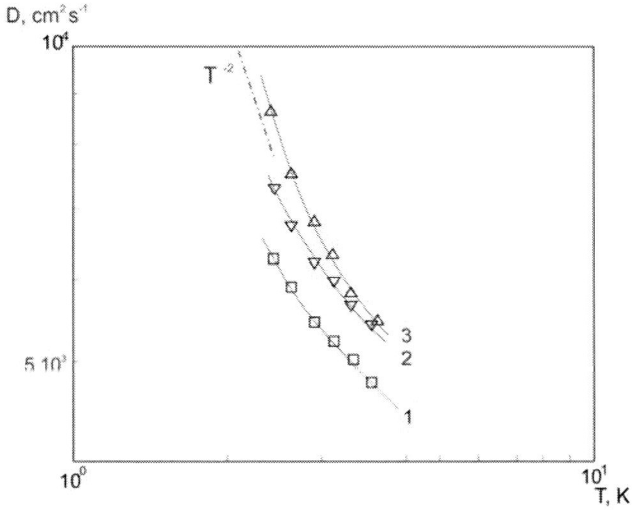

Figure 12. Experimental function $D(T)$ in LiF ceramic samples for different deformation degrees of the original single crystals: (1) 30%, (2) 45%, and (3) 65%.

The intermediate-thick layer shows the same behavior. Figure 12 represents the temperature dependence of the diffusion coefficient *D(T)* of subterahertz phonons in ceramic samples obtained under hot pressing conditions with deformation degrees of 30% (1), 40% (2), and 65% (3). The higher the degree of deformation, the greater the diffusion coefficient for a given temperature, and the steeper the temperature dependence *D(T)*. It was mentioned above that, as the GB thickness decreases ($q_2d << 1$), the function *D(T)* tends to T^{-2}, which corresponds to curve 3 in Figure 12.

The fact that the cleaved surface in a sample in Figure 10b passes along the grain boundaries of the ceramic follows the reduced defectiveness and, hence the enhanced strength of the sample. Due to the twinning in the structure of grains, defects that induce stresses of opposite sign annihilate on the boundaries of the twinning regions [39]. In the present case, this fact may lead to the cleaning of the grain structure of the ceramic from dislocations and to improve the thermophysical and optical properties. In all the cases, the ratio of the mean free path of NPs to the grain size, $l/R \geq 3$-7; for the structure of broad layers, $l/R > 50$. The latter result shows that NPs are predominantly scattered by the boundaries of grains and that there is hardly any scattering inside the grains by the boundaries of the structures shown in Figure 11. This fact may provide evidence for the natural crystallographic character of the boundaries that arise between the layers of the material in the grain structure during plastic deformation. The mean value of the thickness of the intergrain boundary in a ceramic sample obtained from a single crystal by deformation of 65% (curve 3 in Figure 12) at different temperatures (energies) of NPs is $d = 1.8 \pm 0.3$ nm, which is typical for oxide ceramic samples and for the natural CaF_2 ceramic [40].

6. The Structure Characteristics of YAG-Based Ceramics

The initial phonon spectroscopy experiments on optically transparent YAG and YAG:Nd ceramics showed that the GB thickness in less than lattice constant of ceramic material $d < a$ ($a = 1.2002$) [15]. So small value (when compared to the phonon wavelength) makes it possible to use such ceramics as laser material. From the other hand, it follows from Figure 2 that the inhomogeneous distribution of activator ions is possible due to displacement of heavy Nd atoms to GB region during fabrication. So the correct calculations of GB acoustic impedance are needed. As it was shown above (Figure 8), the experimental results satisfy the GB model [12].

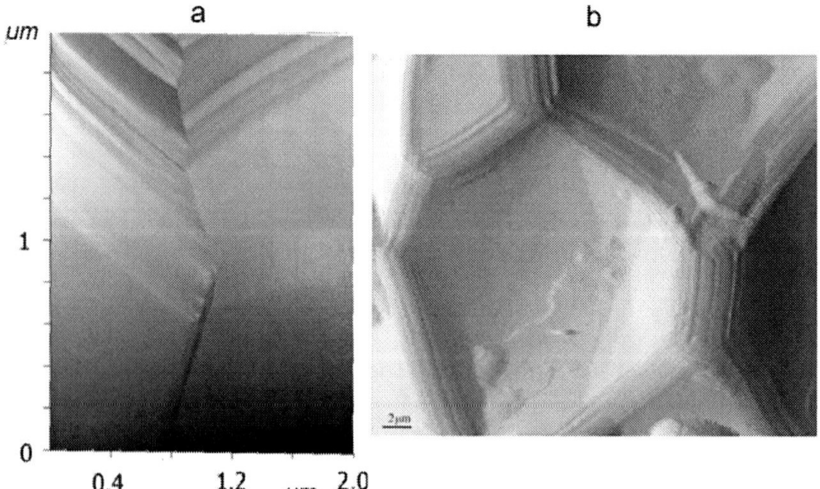

Figure 13. AFM images of the ceramic samples chipped along (a) GB and (b) grain.

Table 2. Characteristics of YAG ceramic samples

Sample	L, cm	R, μm	$D\|_{3.86K}$, $cm^2 s^{-1}$	$l\|_{3.8K}$, cm	$l\|_{3.8K} R^{-1}$
1	0.310	1.0	$4.41 \cdot 10^3$	$2.36 \cdot 10^{-2}$	236.0
2	0.900	7.0	$2.37 \cdot 10^3$	$1.27 \cdot 10^{-2}$	18.0
3	0.310	6.5	$1.46 \cdot 10^3$	$0.78 \cdot 10^{-2}$	12.0
4	0.310	6.5	$1.25 \cdot 10^3$	$0.67 \cdot 10^{-2}$	10.3
5	0.145	13.5	$0.86 \cdot 10^3$	$0.46 \cdot 10^{-2}$	3.4
6	0.160	11.0	$5.60 \cdot 10^3$	$0.30 \cdot 10^{-2}$	3.0
7	0.240	1.0-2.0	$1.50 \cdot 10^3$	$0.80 \cdot 10^{-2}$	54.0
8	-	3.0	-	$1.00 \cdot 10^{-2}$	25.0
9	-	4.0	-	$0.75 \cdot 10^{-2}$	25.0
10	-	7.5	-	$1.50 \cdot 10^{-2}$	20.0
11	0.175	11.0	$0.96 \cdot 10^3$	$0.51 \cdot 10^{-2}$	4.6
12	0.440	27.0	$3.20 \cdot 10^3$	$1.73 \cdot 10^{-2}$	5.8
13	0.143	30.0	$0.73 \cdot 10^3$	$0.39 \cdot 10^{-2}$	1.3

As the fabrication technology is improved, the phonon mean free path in YAG ceramics become comparable to that in the single crystals of YAG:Re solid solutions [8]. Meanwhile l/R ratio reaches 200 and more [41]. It conflicts to the model of GB as a finite thickness layer. The maximum of phonon mean free path was reached in the samples synthesized by the technology of vacuum sintering without external pressure [42]. The comparative analysis of the kinetic characteristics and the structure of samples demonstrate several differences that correlate with the grain size. The study of the chips of samples using the raster electron microscopy and AFM shows the twinning in the entire volume of samples with $R = 1 - 2$ μm. The distance between the twinning planes is about 100 nm (Figure 13a). The twinning is observed only in the GB region or is absent at all at $R > 10$ μm (Figure 13b) [41].

Table 2 presents the parameters of samples and the experimentally determined kinetic characteristics of NPs in the YAG ceramic samples at $T = 3.86$ K. Samples 1–10 were produced at the Konoshima Chemical Company (1–3, YAG:1% Nd; 4–7, YAG (results from [15]); and 8–10 YAG (results from [9]). Samples 11–13 were produced at the Institute of Radio Engineering and Electronics, Russian Academy of Sciences (YAG:1% Nd) [25]. In the samples under study, R ranges from 1 to 30 μm. The temperature dependences $D(T)$ are presented in Figure 14. It is mentioned above that the twinning processes lead to the structural improvement. The moving twinning boundary provides the elimination of defects in the crystal and insignificantly affects the NP scattering. We can present several results that indicate an increase in the number of defects and the corresponding strains in the grain structure with an increase in the grain size.

In particular, the destruction of the samples with the grain size close to 1–2 μm predominantly occurs along the GB. In the ceramic samples with a larger grain size ($R > 10$ μm) we mainly observe the destruction of the grains. This result indicates the presence of strains in the structure of large grains and, hence, lower strength. Note that the problem of the structural defects and the corresponding strains and anisotropy of grains remains unsolved. The optical anisotropy (about 5%) in the neighboring grains was observed in the course of annealing in the Y_2O_3 ceramics [43]. Figure 15 shows the asymptotes of the trailing edge of the diffusion signal $S(t)$ for several samples from Table 2.

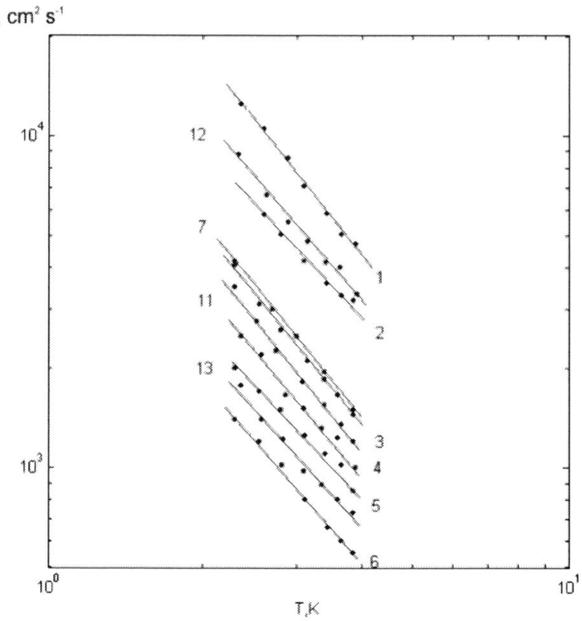

Figure 14. Temperature dependences of diffusion coefficient *D(T)* in YAG ceramics for samples from table 2.

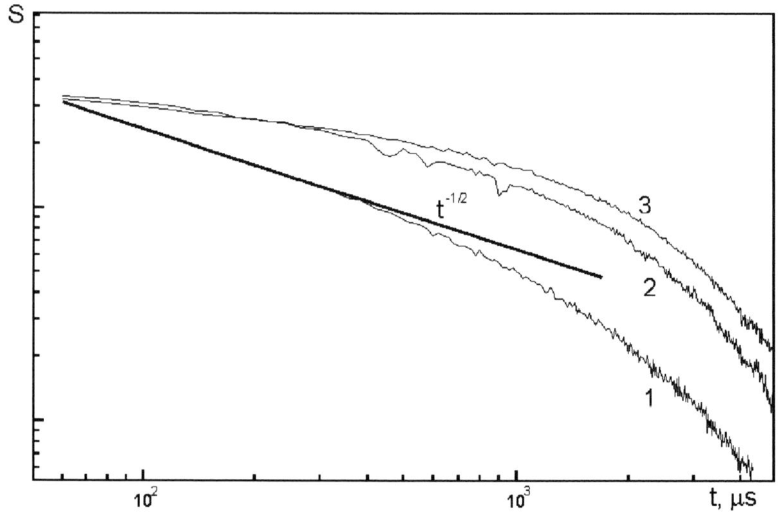

Figure 15. Asymptotes for the trailing edge of the NP diffusion signal measured by the bolometer in the YAG : 1% Nd samples (Table 2): (1) sample no. 1, (2) sample no. 3, (3) sample no. 12.

The dependence $S(t) \sim t^{-1/2}$, which is observed for samples 1 and 7, is typical of the classical diffusion and plane source of NPs. Longer trailing edges of the signals for samples 3 and 12 ($R > 10$ μm) can be due to the NP reflection from GB because of the anisotropy of the neighboring grains.

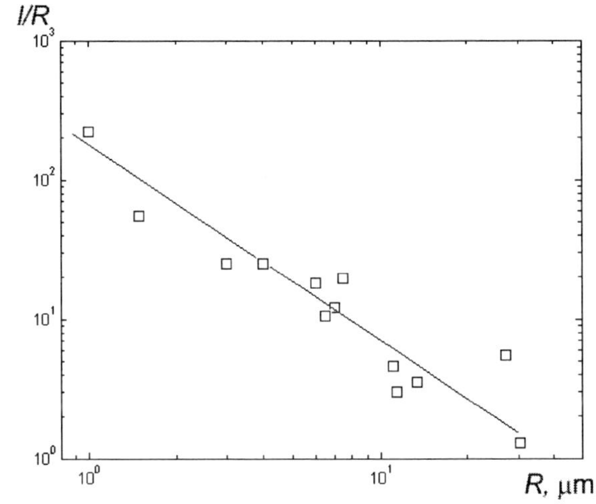

Figure 16. Plot of the ratio l/R vs. grain size in the YAG samples (Table 2) at $T = 3.86$ K.

The dependence of the ratio l/R on the grain size (Figure 16) additionally proves an increase in the number of defects and the corresponding strains with an increase in the grain. The theoretical analysis of the diffusion propagation of the NP with a frequency of 0.874 THz in the high density α-Al_2O_3 ceramics with zero thickness of GB, where the NP scattering is determined only by disordering of crystallites (grains) and their elastic anisotropy, can be found in [22]. The NP free path length is given by;

$$l - l_0 \frac{1}{1-\langle\cos\alpha\rangle} \tag{8}$$

where α is the mean angle of the NP single scattering. The expression (8) illustrates NP scattering due to the anisotropy of neighboring grains and provide an explanation for the difference of the experimental data D(T) ~ T^{-2} on Figure 14.

The above facts indirectly prove an increase in the strain of the structure of grains with an increase in the grain size, which causes the worsening of the thermophysical and optical (depolarization) parameters of material.

7. Propagation on Non-Equilibrium Phonons in Nano-Structured Ceramics (The Gap in the Phonon Spectrum $qR \sim 1$)

The assumption of a planar boundary indicates that this model can be applied only to coarse grain ceramics. Within this model, one cannot describe the dependence of the phonon diffusion coefficient on the grain size, which occurs when the average grain size of the ceramic becomes comparable with the wavelength of injected phonons. To take into account the finite boundary curvature, spherical shells randomly distributed in space and having elastic properties differing from those of grains were considered as intergranular regions in

[44, 45]. It was found that, at certain parameters, leading to resonant phonon scattering from these shells, a gap arises in the wavelength range λ~*R* in the phonon spectrum of such a system; this gap was observed in [46].

Let us consider an elastic medium with a density ρ_0 whose elastic properties are described by one modulus of elasticity K_0 (scalar model) with the dispersion relation $\omega(q)=v_0q$, for phonons of all polarizations. As main scattering centers, we will consider spherical shells with an external radius R_g, thickness *d*, and the elasticity parameters v_1, ρ_1 ($K_1=v_1^2\rho_1$), which model grain boundaries. The material within a shell and beyond is assumed to be that of ceramic grains. It is shown that resonant scattering from spherical shells may occur when the condition $K_1/K_0 << 1$ is satisfied. The resonant frequency is;

$$x_r = \sqrt{\frac{K_1 R_g}{K_0 d}} \tag{9}$$

The formal analysis of the frequency behavior of the diffusion coefficient with respect to the position of two remaining model parameters: resonant level x_r and the defect concentration c_g was made in [46] and present on Figure 17. For all curves, Rayleigh scattering (D ~ $1/q^4$) is observed at small x; at large x, geometric scattering occurs: $D=lv_0$; $l=2R_g/3c_g$. In the intermediate region (x ~ 1), arises a gap. The upper curve in Figure 17, is related to a system containing absolutely solid spheres (well-stabilized boundaries). This curve shows no resonant behavior.

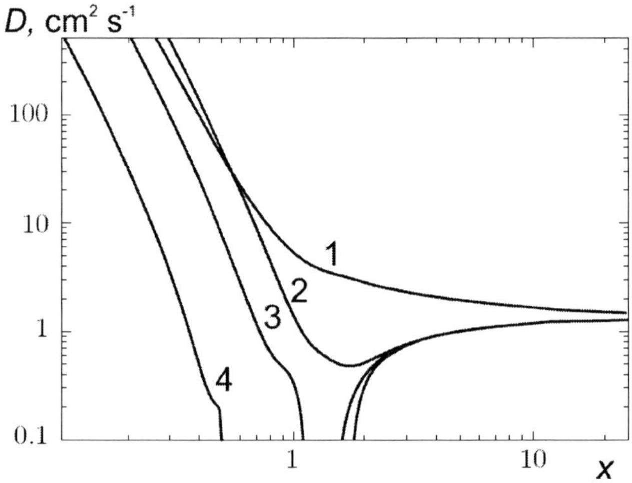

Figure 17. Diffusion coefficient *D*(*x*) ($x=qR_g$) calculated at $c_g = 0.5$ for different resonant frequencies:.(2): $x_r = 0.5$, (3) $x_r=1.0$, (4) $x_r = 1.5$; the curve 1 corresponds to the case of absolutely solid spheres.

Figure 17 indicates that, in the range of x close to unity, the behavior of the phonon diffusion coefficient may exhibit qualitative changes, including object opacity for phonons and a change in the sign of the derivative $\partial D/\partial T$, with a change in the elasticity parameters of the boundary layer of the ceramics or the concentration of scattering centers. Along with grain boundaries, pores are also effective scattering centers in ceramics. It was shown in [46] that a factor of fundamental importance is that inclusions with a characteristic size smaller than the grain size ($R_g/R_p > 1$) can play an important role even at low concentrations, shifting the gap (or minimum) to higher frequencies. Formation of a gap in the phonon spectrum of a material in the presence of several types of scattering centers is considered below in more detail (Figure 20) by a specific example. Figure 18 shows the dependences $D(R)$ at $T = 3.8$ K for two types of ceramics: Al_2O_3 and the YSZ solid solution. The curves 1 and 2 are continuations of corresponding dependences of Figure 2 to the region $R < 1000$nm. In both cases, sharp falloff in the dependence $D(R)$ is observed at $qR < 20$ (for Al_2O_3 $v = 7.4 \cdot 10^5$ cm·s^{-1}; for YSZ $v = 4.33 \cdot 10^5$ cm·s^{-1}, which is indicative of the onset of effective scattering of NP by ceramic grains at $R \sim 100$ nm).

The presence of a gap in the phonon spectrum suggests that the parameter $qR \sim 1$ can be implemented; i.e., at $q \sim 10^6$cm^{-1} the grain size should be 10–20 nm. The current absence of such single-phase nanostructured ceramics based on the noted oxides does not exclude the possibility of forming composites on their basis, which, along with the main fraction of stable grains (crystallites), contain a certain amount of the nanostructured phase of another material. Such composites (YSZ + Al_2O_3) were synthesized in [47].

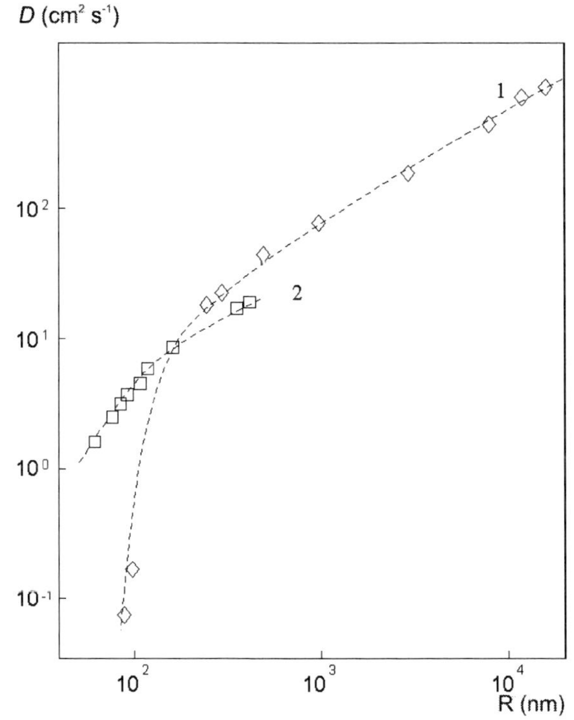

Fig.18. The dependence of effective diffusion coefficient D on mean grain size at $T = 3.8$K for: Al_2O_3 (1) and YSZ (2) ceramic samples.

The feature of this structure is the presence of a fine-grained metastable phase of corundum ($R = 20 - 40$ nm), which is partially transformed into a denser α-Al_2O_3 (ρ=3.97g/cm^3) with an increase in the synthesis temperature T_s. This transition leads to the formation of nanoscale shrinkage pores, whose number is related to the content of the metastable Al_2O_3 phase. The presence of additional nanoscale scattering centers (metastable corundum phase, shrinkage pores), even with relatively low concentrations, may lead, as was shown above, to a significant reconstruction of the phonon spectrum formation of a gap and shift of its top edge to higher frequencies. Along with the 9.8% YSZ + 14.3% Al_2O_3 composites, we investigated the 9.8% YSZ single-phase ceramic (samples 712, 713, 720, and 721), synthesized at the same temperature T_s. The parameters of the samples studied are listed in Table 3. The samples were plane-parallel polished plates with a thickness of 0.1 – 1 mm and an area of about 0.5 cm^2.

Table 3 Parameters of YSZ ceramic samples

Sample	T_s, C	ρ, g·cm^{-3}	R, nm	L, 10^{-2} cm	Content of Al_2O_3 phases	
					α-phase, wt.%	(β+γ)-phases, wt.%
721	1092	5.49	84	2.65	0	0
713	1140	5.74	110	3.2	0	0
775	1140	5.22	92	2.0	25	75
712	1155	5.75	135	2.1	0	0
769	1155	5.21	64	2.1	32	68
720	1200	5.84	132	3.3	0	0
776	1200	5.26	84	2.75	62	38

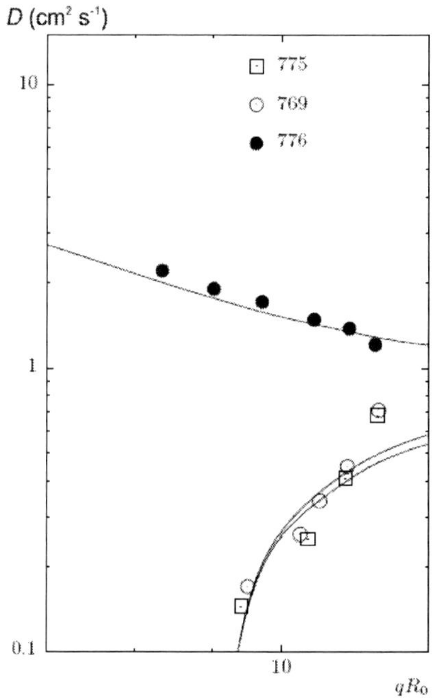

Figure 19. Diffusion coefficient of NP in the YSZ-Al_2O_3 composites. R_0=100nm (table 3).

It was shown in [46] that at $T_s \geq 1155°C$ the phase boundaries in YSZ are well-stabilized and phonon scattering is temperature-independent. The dependence $D(T)$ for composites differs both in magnitude and behavior from the corresponding dependence for the YSZ samples even when phase boundaries in YSZ are well-stabilized (T_s=1155°C, sample 769). For all single phase YSZ samples $\partial D/\partial T < 0$. The $D(T)$ dependence for 9.8 % YSZ+14.3% Al_2O_3 composites (samples 775, 769) is $\partial D/\partial T > 0$ (Figure 19). The reason is that these composites, along with YSZ matrix grains, contain other nanoscale defects: grains of the metastable Al_2O_3 phase and pores. It was shown in [47] that the growth of particles of metastable Al_2O_3 phases in a cubic oxide matrix is hindered, and, at T_s < 1200°C, the size of metastable phase particles does not exceed 20 – 40 nm. At the same time, in the temperature range T_s = 1100–1200°C, the α-Al_2O_3 phase grows most intensively. This process is accompanied by shrinkage with formation of nanoscale shrinkage pores, which grow with an increase in T_S. The formed α-phase grains had the size R = 80 nm at Ts =1200°C. Thus, the pattern of formation of the structure of 9.8% YSZ + 14.3% Al_2O_3 composites, synthesized at Ts = 1140–1200°C, can be represented as follows. A matrix composed of grains (crystallites) of the cubic YSZ phase with R > 100 nm contains nanoscale fragments of metastable Al_2O_3 phases, which, with an increase in T_s, are partially transformed into the α-Al_2O_3 phase, undergoing shrinkage. As a result, nanoscale shrinkage pores are formed, which are localized near Al_2O_3 grains and increase in size with an increase in T_S. An increase in the annealing temperature to T_S =1200°C leads to a decrease in the concentration of nanostructured defects, which actively scatter non-equilibrium phonons. Therefore, the dependence $D(T)$ with a sharp rise (derivative $\partial D/\partial T>0$, see Figure 19), which corresponds to the top of the gap in the phonon spectrum in samples 775 and 789, is transformed for sample 776 (synthesized at 1200°C) into a dependence with $\partial D/\partial T<0$, and scattering of non-equilibrium phonons in this sample begins to be controlled predominantly by grain boundaries, as in samples 712 and 720 ($T_s \geq$ 1155°C).

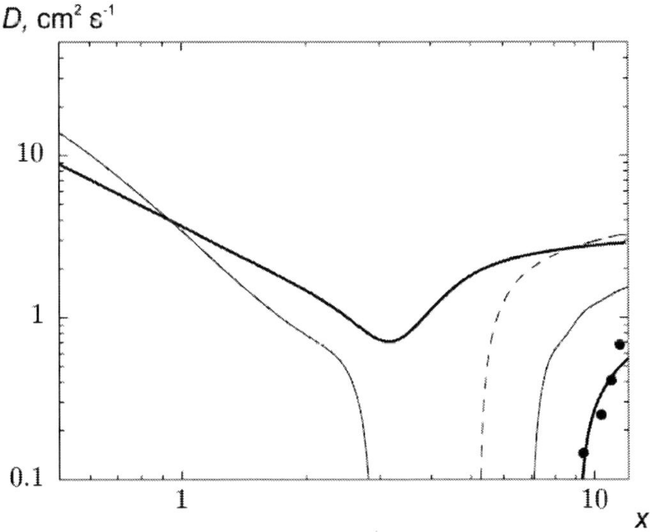

Figure 20. Diffusion coefficients related to pores (dashed line), shells (semibold line), and shells around Al_2O_3 grains (thin line) and the total (experimental points) diffusion coefficient for sample 775.

The presence of nanoscale inclusions in multiphase ceramics leads to the formation of a gap in the phonon spectrum, which is due to the resonant scattering of phonons from these defects. The formation of a gap in the phonon spectrum of samples 775 and 769 is clarified in Figure 20, which shows the diffusion coefficients due to individual defects: pores ($c_g = c_i = 0$, $c_p = 0.05$), shells ($c_g = 0.74$, $c_i = c_p = 0$), and shells around metastable Al_2O_3 phases ($c_g = c_p = 0$, $c_i = 0.21$), as well as the total diffusion coefficient. Thus, the technique of ceramic synthesis may be used to create phononic crystals with an effective periodicity parameter determined by the average grain size (which is a well-controlled parameter in fabrication of ceramics). The features of the phonon spectrum of ceramics in the range $qR \sim 1$ (the presence or absence of a gap in the spectrum) are determined to a large extent by the elastic properties of grain boundaries. The presence of inclusions of another phase or pores with a characteristic size smaller than the grain size of the main material (matrix), even with relatively low concentrations, shifts the gap in the phonon spectrum to higher frequencies.

8. Phonon Kinetic Peculiarities in Corundum-Based Cermets

Cermets, i.e. composites based on polycrystalline corundum matrix and nanostructured iron fraction, are another example of phonon kinetic peculiarities at $qR \sim 1$. A high melting temperature, a high hardness, high wear and chemical resistance, and a low density represent the properties of ceramics that excel those of metals. However, the absence of plasticity (brittleness) and low fracture toughness restrict the use of ceramics as a structural material. The introduction of a metallic fraction into a dielectric ceramic matrix makes it possible to create composite materials (cermets) that combine the advantages of ceramics and metals. As plastic-metal inclusions, researchers usually use Ag, Al, Ni, Fe [48, 49]. Important requirements imposed on cermets consist in the facts that metallic inclusions should closely bind a ceramic matrix and that their characteristic sizes should not exceed a critical size at which thermal stresses can induce cracks around metallic grains [50]. The presence of a metallic fraction in nanocomposites can substantially change the transport properties of subterahertz phonons due to both resonance scattering by structural fragments and an effective electron–phonon interaction in metallic grains at low temperatures, where $ql_{el} >> 1$ [51] (q is the phonon wavevector and l_{el} is the electron mean free path in the metal). Al_2O_3+xFe (x = 5, 10, 15 wt. %) cermet samples represented thin plates $(1.3\text{-}1.5)\cdot 10^{-2}$ cm thick with an area of about 0.5 cm^2. AFM (atomic force microscopy) image of the sample containing R_{Fe} = 30-80 nm fragments is presented in Figure 21. Table 4 gives the structural parameters of the samples, the technological conditions, the diffusion coefficients measured at $T = 3.86$ K, and the exponents in the temperature dependences $D(T) \sim T^h$ for the temperature range 2.3–3.86 K.

It can be seen that for average Al_2O_3 grain size in the ceramics under study, the resonance scattering of NPs begins to manifest itself apart from geometrical scattering. When analyzing the experimental results [52], we revealed the following three main features:

1) As the fraction of metallic phase increases, the diffusion coefficient at $T = 3.86$ K decreases.

2) The diffusion coefficients are very low (they are two orders of magnitude lower than those calculated using the $D(R) \sim R$ dependence in the geometrical-scattering range (see Figure 17).
3) The temperature dependence has the form $D(T) \sim T^h$ with an exponent $h = 1–3$, which decreases as the amount of the metallic phase increases.

Obviously, as the fraction of the metallic phase (whose thermal conductivity is higher than that of corundum at liquid-helium temperatures) increases, the thermal conductivity of the composite and, hence, the diffusion coefficient of phonons in a macroscopic dielectric sample should increase. However, in our studies, the diffusion coefficient of NPs decreases with increasing fraction of the metallic phase.

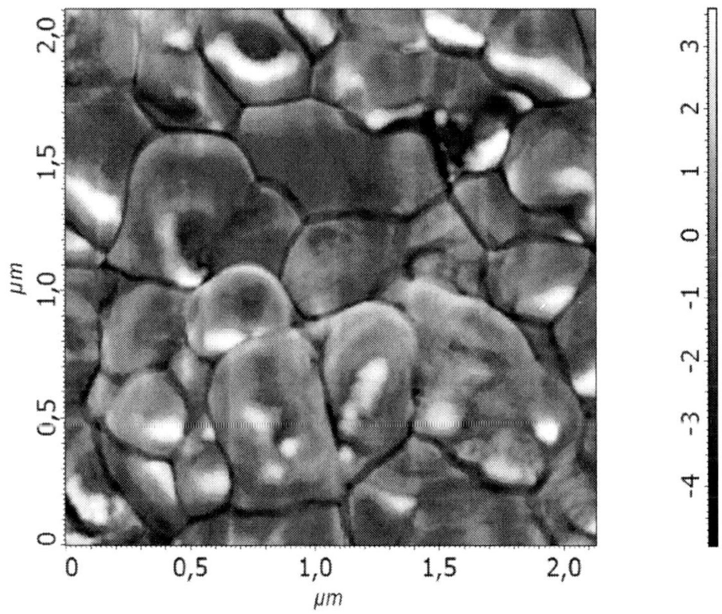

Figure 21. AFM phase-contrast image of a sample with 10 wt % Fe. Bright regions correspond to a softer metallic phase.

Table 4. Parameters of cermet samples

N	Phase composition, wt. %	US power, kW	Sintering temp/sint. time, °C/hour	CSA size Al_2O_3	D, $cm^2 \cdot s^{-1}$ at T=3.86K	h
1	Al_2O_3:5%Fe	0	1400/0.5	155	$4.24 \cdot 10^{-2}$	3
2	Al_2O_3:10%Fe	0	1400/0.5	94	$3.26 \cdot 10^{-2}$	2
3	Al_2O_3:15%Fe	0	1400/0.5	79	$2.30 \cdot 10^{-2}$	1
4	Al_2O_3:15%Fe	0	1450/2	>300	$2.27 \cdot 10^{-1}$	1

As follows from Figure 17, the diffusion coefficient measured at $T = 3.86$ K ($D \leq 0.05$ $cm^{2} s^{-1}$) can only exist near the gap edge (where the diffusion coefficient increases exponentially with temperature), which is in conflict with the detected temperature dependences $D(T) \sim T^{1-3}$.

The presence of fine inclusions (metallic fraction) in the system changes the gap (minimum) position but cannot qualitatively change the behavior of the $D(T)$ dependence; that is, for the results obtained to be explained, we have to assume an additional mechanism. For example, metallic grains can be considered as the centers of NP entrapment. The fact that metallic nanofragments can serve as NP traps follows from the relation between the time of electron–phonon interaction $\tau_{e\text{-}ph}$ and the time τ_R of the ballistic NP path in the bulk of an iron grain, $\tau_R = R_{Fe}/vc$ (where c is the coefficient of exit of NPs from a grain, $10^{-2} < c < 1$). If $\tau_{e-ph} << \tau_R$ metallic grains can serve as NP traps. For $ql_{el} >> 1$, we have [51], where n and ρ are the iron concentration and density, respectively; m is the electron mass, and v_F is the Fermi velocity. Estimates demonstrate that, under the experimental conditions, inequality is fulfilled at $\omega \geq 10^{12}$ (c=0.1). However, at $T = 4$ K. Thus, the NPs with a frequency $\omega > (2\text{-}3)\,\frac{kT}{\hbar}$ from the phonon spectral region corresponding to the maximum spectral density of the Planck distribution at $T = 4$ K can interact with electrons in the metal.

The diffusion coefficient of NPs under conditions of an effective electron–phonon interaction is one to two orders of magnitude lower than that in an analogous polycrystalline Al_2O_3 matrix. The propagation of NPs in the presence of trapping centers was considered in [53, 54], where experimental results on the propagation of NPs in solid solutions with $Y_{3-x}(Er,Ho)_xAl_5O_{12}$ paramagnetic centers were analyzed. According to those results, the propagation of a δ-like thermal pulse in an infinite one-dimensional medium is described by two diffusion coefficients, which characterize "fast" and "slow" processes.

According to [52], we take into account the specific heats of the matrix and metallic inclusions and write the dependence of D_{eff} on the iron weight concentration p in the form:

$$D_{eff} = \frac{D_0}{1 + \dfrac{pc_{Fe}}{(1-p)c_{Al_2O_3}}} \tag{10}$$

Since the electronic specific heat of iron at a concentration of about 10 wt % and liquid-helium temperatures is approximately two orders of magnitude higher than that of the phonon subsystem. This expression demonstrates that D_{eff} decreases as $D \sim 1/p$ as the iron concentration increases. At liquid-helium temperatures, we have $c_{el}/c_{ph} \sim T^{-2}$; therefore, we obtain $D_{eff} \sim D_0 T^{2}$. In dielectric ceramics, D_0 is controlled by elastic phonon scattering by grain boundaries. As is seen from Figure 17, the detected temperature dependence $D \sim T^{h}$ ($h = 1 \div 3$) can be obtained for phonons with qR of several units (to the right of the gap in the phonon spectrum), which corresponds to the experimental data.

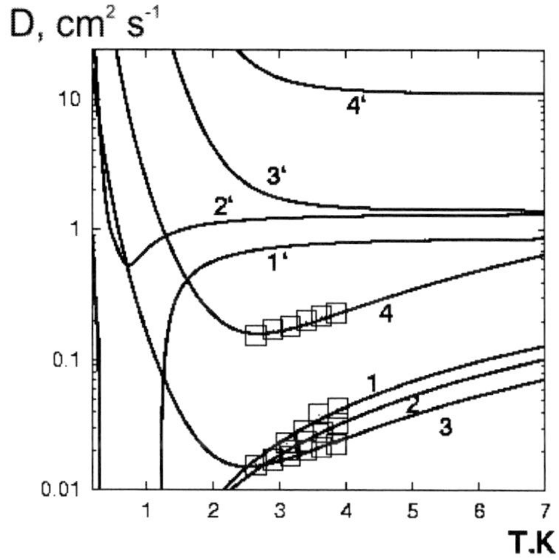

Figure 22. Temperature dependences $D(T)$ for the cermet samples: (*1'–4'*) theoretical dependences calculated for a polycrystalline matrix without regard for the fraction of iron [46], (*1–4*) dependences calculated by Eq. (10), and (squares) experimental data.

Figure 22 shows the experimental and calculated $D(T)$ dependences, and their numerals correspond to those in the table 4. Numerals with primes indicate the temperature dependences of the diffusion coefficient in the structure of a polycrystalline Al_2O_3 matrix that were calculated according to [46] without regard for the entrapment of NPs by fine iron inclusions. The calculated dependences that describe the experimental data were obtained using Eq. (10), which takes into account the ratio of the specific heats in the given temperature range. As the iron-fraction concentration increases, the absolute values of D at $T = 3.86$ K decrease and the temperature dependence $D(T)$ weakens. According to [46], this behavior indicates an increase in resonance frequency ω_r, which is caused by NP scattering by the structural fragments in the sample under study. Curves *4'* and *4* in Figure 22 correspond to a sample with larger grains in the polycrystalline matrix ($T/t = 1450/2$, where t is the sintering time) and to the fraction of nanodispersed iron that is identical to that for curves *3'* and *3* (which is 15 wt %). In this case, the diffusion coefficient increases by an order of magnitude because of the weakening of the resonance character of NP scattering by the larger grains of the Al_2O_3 fraction.

9. Subterahertz Phonons Transport in $Ce_{1-x}Gd_xO_{2-y}$ Sub-Micron Ceramics

Phonon spectroscopy provides information both on the structure of a disordered system and on relaxation processes when the relaxation energy is comparable to the phonon energy. In particular, our earlier investigations of the structural features of substitutional solid solutions of the yttrium rare-earth aluminum garnets $Y_{3-x}Re_xAl_5O_{12}$ (Re=Er, Ho) which were performed using phonon spectroscopy in the subterahertz frequency range [55,56], revealed that the kinetic characteristics of phonons can depend not only on the amount of substitutional impurities but also on the formation energy of vacancies in the anion sublattice, which in solid solutions is determined by the vacancy position with respect to impurity atoms. The calculations for the energy characteristics of vacancies in the anion sublattice of YSZ, i.e., the solid electrolyte with a crystal structure identical to that of the $Ce_{1-x}Gd_xO_{2-y}$ were performed in [57,58]. According to the results obtained, the formation energy of a vacancy at the oxygen site depends on the position relative to yttrium impurity ions. It turned out in [58] that the configuration in which two yttrium ions are contained in the second coordination sphere relative to the vacancy is the energetically most favorable configuration (the ground state with the energy E_0). The energy of all the other configurations separated from each other by the energy barrier with a height of the order of $E_b \approx 1$ eV is higher than E_0 by 200–300 K.

The presence of two-level systems (TLS) with the activation energy $\Delta = E_1 - E_0 \approx 200$ K in the structure leads to an additional contribution to the heat capacity, which has a maximum at the temperature $T \approx 0.4\Delta$. The heat capacities calculated in [58] are in good agreement with the experimental results obtained in [59]. For such large values of $\Delta \approx 200$ K, the TLS in the YSZ compound do not affect the propagation of subterahertz phonons [11]. (At $T = 3$ K, the energy of thermal phonons corresponding to the maximum of the spectral density of the Planck distribution is estimated as $\hbar\omega \cong 3k_BT \cong 0.7$meV.

The $Ce_{1-x}Gd_xO_{2-y}$ system have the same structure as YSZ, so under the same experimental conditions, there can arise a similar situation with other values of the activation energy Δ for which the interaction of phonons with two-level systems will play a decisive role. The initial nanopowders were synthesized through the evaporation of a target under irradiation by light from a CO_2 laser and through the chemical combustion (sample 997). The investigations performed revealed that all powders were single-phase solid solutions of the $Ce_{1-x}Gd_xO_{2-y}$ composition with a varying molar fraction of gadolinium in the range $x = 0.09$–0.30 and had a cubic lattice (a_0 =0.5424 nm at $x = 0.20$). The technological conditions used for the synthesis of ceramic materials based on the $Ce_{1-x}Gd_xO_{2-y}$ solid solutions and the results obtained from X-ray structural phase analyses of the composition, electron microscopy examination, and electrical conductivity measurements of the samples under investigation were described in [60, 61]. The main technological and structural parameters are presented in the table 5. The samples under investigation were prepared in the form of plane-parallel polished plates with a thickness L=0.02-0.03 cm and an area of 0.5 cm^2. Figure 23 from [62] shows the NP signals detected by the bolometer in the $Ce_{0.909}Gd_{0.091}O_{2-y}$ ceramic sample at different temperatures.

Table 5. Parameters of $Ce_{1-x}Gd_xO_{2-y}$ ceramic samples

Sample	T/t C/min	X (-)	R nm	$D_{T=3.76K}$ cm^2s^{-1}	l, cm $\cdot 10^{-6}$	l_0, cm $\cdot 10^{-4}$	v, $cm\,s^{-1}$ $\cdot 10^5$
911	1300/0	0.091	200	0.570	4.62	1.33	3.71
913	1300/0	0.126	200	0.425	3.48	1.23	3.66
904	1300/0	0.2	200	0.260	2.18	1.23	3.57
905	1300/0	0.2	200	0.365	3.07	1.65	3.57
801	1300/0	0.3	≥200	0.195	1.70	1.15	3.43
519	1300/0	0.2	110	0.152	1.28	0.68	3.57
997	1460/60	0.2	>>200	0.690	5.80	2.87	3.57

As the NP temperature increases, the time of the signal maximum t_m decreases (i.e., the phonon diffusion coefficient increases). The mean free path of thermal phonons at $T = 3.76$ K has the absolute value $l = 4.6 \cdot 10^{-6}$ cm, which decreases with decreasing of NP energy.

For an average grain size $R \approx 200$ nm, the condition $l \ll R$ obeys. This indicates an intense phonon scattering inside the grains. Under the experimental condition $qR \gg 1$, the phonon spectrum of the ceramic material is similar to the phonon spectrum of the grain material. Therefore, the phonon velocity averaged over the polarizations in solid solutions of different compositions was determined from the expression

$$v^{-3} = \frac{1}{3} \cdot \left(v_l^{-3} + 2v_t^{-3}\right) \tag{11}$$

The longitudinal v_l and shear v_t velocities of acoustic waves were calculated from the data reported in [63] on the measured elastic modules with allowance made for the changes in the density for samples of the compositions CeO_2:10-20%Gd_2O_3 with extrapolation of the dependence $v(x)$ to $x = 0.3$. The data on v and the kinetic characteristics of NP scattering at $T = 3.76$ K are presented in the table 5.

Figure 24 a,b shows the temperature dependences of the mean free path of thermal phonons in the temperature range $T = 2.4$–3.8 K. In all cases, the mean free path of thermal phonons obeys the inequality $l \ll R$. The observed dependences exhibit an identical behavior ($\partial l/\partial T > 0$), which may indicate the same mechanism of phonon scattering for the samples synthesized under different conditions from powders prepared by different methods [10].

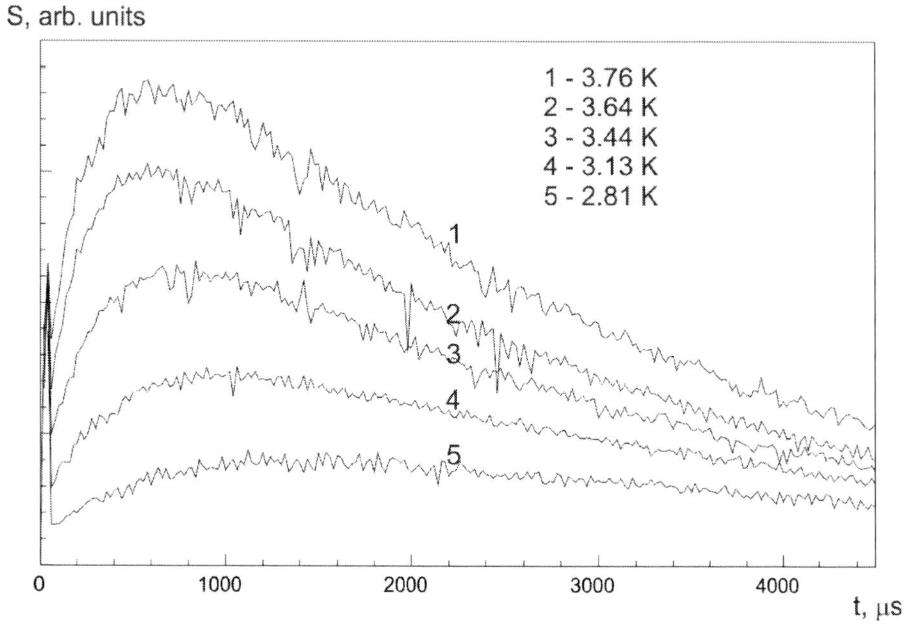

Figure 23. Phonon signals for sample no. 911 at different thermostat temperatures T = (*1*) 3.76, (*2*) 3.64, (*3*) 3.44, (*4*) 3.13, and (*5*) 2.81 K.

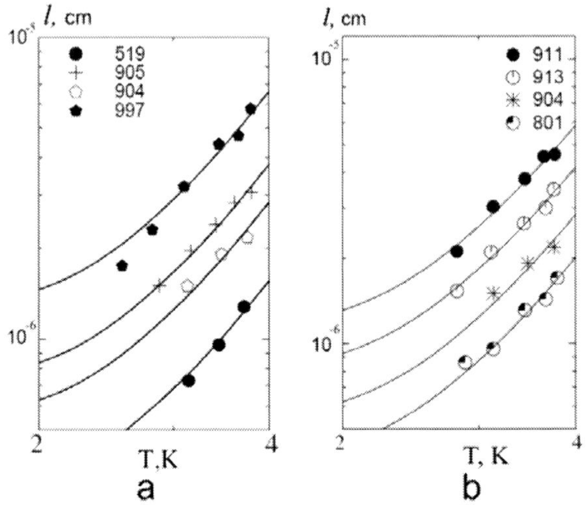

Figure 24. Temperature dependences of the mean free path of thermal phonons in samples from the table 4. a – the impurity contents x=0.2; b - impurity contents $x = 0.091$ (sample 911), 0.126 (sample 913), 0.20 (sample 904), and 0.30 (sample 801).

The model [53,54] is the phenomenological generalization of the microscopic model [64] and is applicable for trapping centers of any nature. By analogy to (10) we have for TLS:

$$D = \frac{D_0}{1 + n \cdot \frac{c_{tls}}{c_{ph}}} < D_0 \tag{12}$$

where c_{tls} and c_{ph} are the heat capacities of the two-level subsystems and phonons, respectively; n is the number of two-level subsystems; and D_0 is the diffusion coefficient in the absence of two-level subsystems. The relationship means that, at a high heat capacity of the two-level subsystem, the phonon in the course of the diffusive propagation through the sample with the diffusion coefficient $D_0=1/3l_0$v can be repeatedly trapped by two-level subsystems so that its delay time becomes fairly long. The phonon heat capacity per mole of the material:

$$c_{ph} = 12 R_G \pi^4 \frac{\left(T/\Theta\right)^3}{5} \tag{13}$$

where R_G is the universal gas constant and Θ is the Debye temperature.

The heat capacity of two-level systems (TLS) can be expressed as follows:

$$c_{tls} = k_B \left(\frac{\Delta}{T}\right)^2 \cdot \frac{\exp\left(-\Delta/T\right)}{\left(1 + \exp\left(-\Delta/T\right)\right)^2} \tag{14}$$

The other parameters (Δ, n, and l_0) can be fitted. In Figure 24, the solid lines are the theoretical dependences calculated from expression (12). The Debye temperatures for the samples with different impurity concentrations were calculated according to [65]. It was established that the value of Δ = 8.53 K makes it possible to achieve the best agreement between the theoretical dependences and the experimental data for all the samples under investigation. In this case, for all samples, we have $l_0 \cong 10^{-4}$ cm and the concentration of two-level systems is proportional to the concentration of gadolinium atoms but is approximately one order of magnitude lower, i.e., n =0.08x, which seems to be quite reasonable with due regard for the presence of cerium ions in different oxidation states in the system under consideration. The obtained values of l_0 are presented in the table 5. Thus the two-level systems Δ = 8.53K in the anion sublattice of the $Ce_{1-x}Gd_xO_{2-y}$ solid solution are the reason of NP transport peculiarities. The results obtained can be used for calculating of the full energy of $Ce_{1-x}Gd_xO_{2-y}$ system.

10. NP Transport Peculiarities in Ferroelectric Materials

The interest in studies of NP transport at helium temperatures in ferroelectric materials is stimulated by the existence of different phase states, domain structures, low-temperature anomalies of thermal conductivity and heat capacity [66, 67]. The temperature dependence behavior of non-equilibrium phonons in these materials was discussed in [68, 69]. As a rule the ceramics under investigation are polycrystals (figure 1d). Figure 25 shows phonon signals at several different temperatures in PZTL ferroelectric ceramic ($Pb_{1-1,5x}Zr_{0,65}Ti_{0,35}La_x$), x=0,08, where the ion disorder is considerable and translational symmetry is broken due to La doping [66]. It is an example of ceramic with diffusive (broadened) phase transition. Meanwhile $t_m \sim T^5$. The similar temperature dependence was observed in fused quartz [10]. It was shown in [70] that the dependence $t_m \sim T^5$ can be caused by redundant (in comparison with Debye) density of states in the plateau region of glassy materials where $K = c_v{\cdot}D$ = const. The inset to Figure 25 shows the signal of non-equilibrium phonons in pure PZT ceramic not doped with La. The propagation time in the pure material is considerably shorter, only a few microseconds.

This case can be regarded as quasi-ballistic propagation. The phase transition in this ceramic is narrow, and its kinetic properties are crystal-type [69]. Phonon scattering by low-energy vibrational excitations typical of dielectric glasses has been detected in samples of $PbMg_{1/3}Nb_{2/3}O_3$- $PbSc_{1/2}Nb_{1/2}O_3$ (PMN-PSN) ceramic solid solutions. For samples of all compositions $t_m \sim T^5$. The values of thermal conductivity K and heat capacity c_v at helium temperatures are well-known for a number of materials investigated. This allowed us to calculate the phonon diffusion coefficient by the standard formula $D = K/c_v$ and compare it with experimental results. It turned out that the discrepancy in the single crystals and $BaTiO_3$ ceramic was within 10% [66, 67]. In PMN and PZTL ceramics this discrepancy is notably higher, by a factor of up to two [66,72-74]. We do not think that this discrepancy is of major importance, since processes used to synthesize ceramic materials may vary. All in all, the analysis of data on ferroelectrics with glass-like kinetic properties leads us to conclude that the thermal pulse technique unambiguously identifies such properties, since the peak position of the NP signal at experimental temperature follows the law $t_m \sim T^5$.

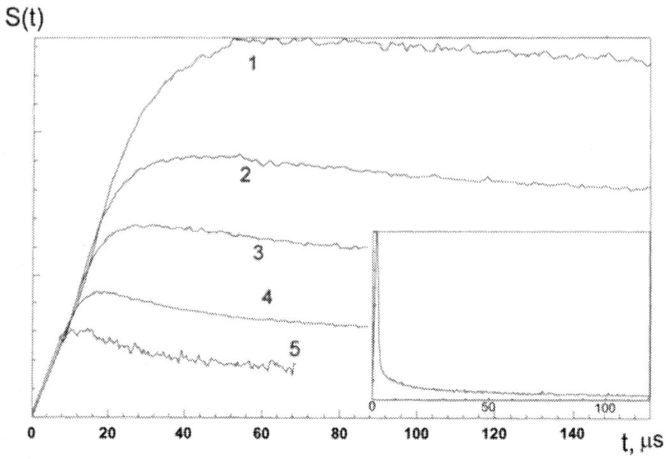

Figure 25. Phonon signals in a PZTL sample from [71]. L = 0.08 cm at (1) T=3.81 K; (2) 3.43 K; *(3)* 3.17 K; *(4)* 2.8 K; (5) 2.49 K. The inset shows the phonon signal in a PZT ceramic sample with L = 0.1 cm at T= 3.82 K.

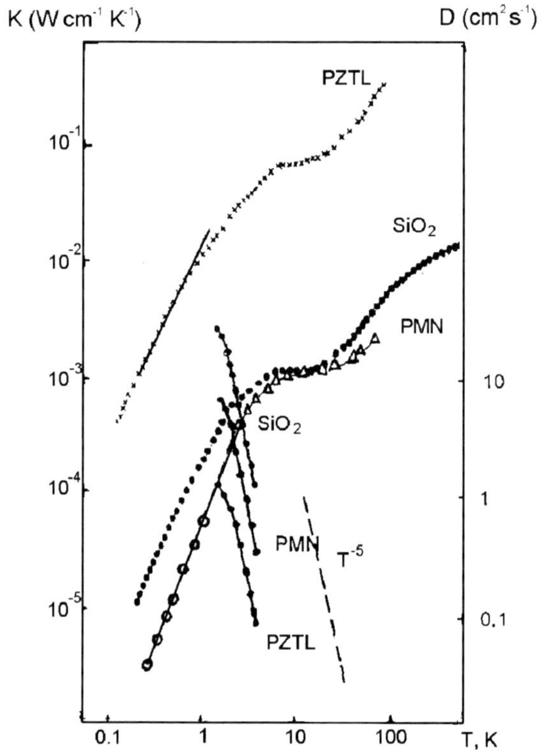

Figure 26. Temperature dependence of thermal conductivity of ceramics and glasses (left scale) and those of diffusion coefficients $D_{rff} = L^2/t_m$ from [71] (right scale).

Figure 26 shows curves of thermal conductivity of PZTL and PMN ceramics and the $Si0_2$ glass [66] versus temperature with the characteristic flat sections (plateau) near 10 K. The graph also plots our measurements of effective diffusion coefficients for the tested materials [71].

The temperature intervals corresponding to the flat sections of the thermal conductivity and to the diffusion coefficients described by the formula $D \sim T^{-5}$ are fairly close to each other taking into account that the maximum of diffusion signal is formed by the maximum of spectral density ($\hbar\omega \cong 3k_BT$ for Planck distribution). The assumption that scattering on grain boundaries is small and does not affect properties of glass-like ceramics in the low-temperature range is supported by direct measurements of the NP diffusion coefficient in a PMN single crystal. The diffusion coefficients D are the same for both single-crystal and ceramic samples to within the accuracy of the measurements.

Now let us consider curves of the phonon signal for ferroelectrics with a narrow phase transition. As was noted above, propagation of a thermal pulse in $PbZn_{0,65}Ti_{0,35}O_3$ not doped with La is quasi-ballistic (Figure 25). Measurements of the typical ferroelectric $BaTiO_3$ ceramic and single crystal (inset) are shown in Figure 27. The main set of the curves was obtained in a ceramic sample with thickness $L = 1$ mm. The main result, $\partial t_M/\partial T<0$, is typical for scattering on GB ($q_2d > 1$). The inset to Figure 27 shows curves of the non-equilibrium phonon signal for a $BaTiO_3$ single crystal, which implies $t_m \sim T^2$. In single crystals the effective diffusion coefficient for phonons is a factor of 10 higher, but, what is more important, it has different temperature dependence. The micrograph of domain structure in $BaTiO_3$ single crystal is shown in Figure 28.

A fairly regular pattern of a-type domains is observed. The boundary between the domains is a region where the order parameter is inhomogeneous, and its width r_c [75] can be treated as a domain wall boundary thickness. There is good reason to assume that r_c equals several lattice constants, i.e., in $BaTiO_3$ $r_c \cong$ 20-30A. The phonon wavelength is $\lambda >> r_c$. In real crystals a domain wall is pinned by defects, crystal inhomogeneities on crystal boundaries and in the wall region, i.e., elastic stress is present in the layer of thickness $\sim r_c$. At room temperature the mean domain size is $\delta \cong 3 \cdot 10^{-5}$ cm (Figure 28). We suppose that the δ value is almost constant when temperature decreases to helium.

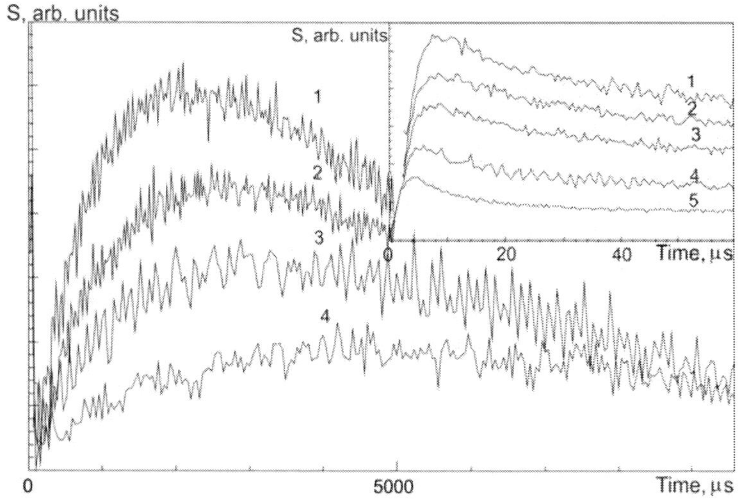

Figure 27. Non-equilibrium phonon signals in a $BaTiO_3$ ceramic sample with $L = 0.1$ cm. The inset shows non-equilibrium phonon signals in a $BaTiO_3$ single-crystal sample with $L = 0.043$ cm [71].

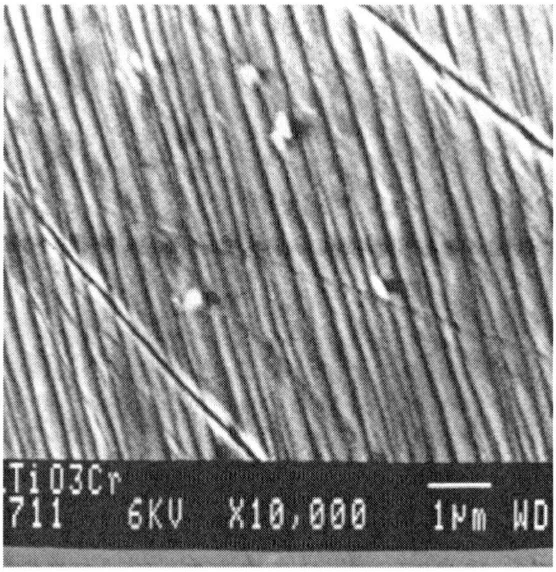

Figure 28. Electron micrographs of strip domains in the *xy* plane of a $BaTiO_3$ single crystal.

The results obtained unambiguously indicate that the contribution of GB to phonon scattering is decisive in the $BaTiO_3$ ceramics at helium temperatures. In the $BaTiO_3$ single crystal the phonon scattering is controlled by domain wall boundaries. In the model with diffusion of non-equilibrium phonons t_m value is related to the effective phonon free path l by the simple formula

$$t_m = \frac{3}{2} \cdot \frac{L^2}{l \cdot \mathrm{v}} \tag{15}$$

where L is the sample length, and v is the polarization-averaged velocity of sound.

According to estimates, in single crystals $l|_{3,80K} \cong 10^{-3}$cm and increases with decreasing temperature. Two neighboring domains in a ferroelectric are regions of the same crystal with different orientations related to one another by one or more symmetry operations. A phonon flux can undergo reflection on their boundary. According to (8), where $l_0 = \delta$ (domain thickness) and $s = 1\text{-}f_\omega$, the reflection coefficient s for NP is estimated to be $s|_{3,8K} \cong 10^{-2}$. This value is comparable with estimated reflection coefficient for NP flux on GB determined by the anisotropy of neighboring grains in alumina ceramics [22].

The processes occurred inside ceramic during sintering (grain growth, interface regions formation,...) could be demonstrated for the $BaTiO_3$-$WTiO_3$ ceramic as an example [76]. The structure and phase composition during sintering, in turn, depends on compacting conditions. The powders of $BaTiO_3$-$WTiO_3$ composition with an average particle size of 50 nm obtained by the sol-gel method [77] were compacted by uniaxial dry pressing under ultrasound action (US). The compacting parameters (compacting pressure, P, and US power, W) and phase composition (X-ray diffraction data) are listed in table 6.

Table 6. compacting parameters and phase composition of $BaTiO_3$-$WTiO_3$ samples

Sample	*P*, MPa	*W*, kW	*ρ*, g/cm^3	$BaTi_4O_9$, wt%	$Ba_2Ti_9O_{20}$, wt%	$BaWO_4$, wt%	WO_3, wt%	$BaTi_2O_5$, wt%
1	113	0	4.43	28	69	2	1	-
2	525	0	4.48	28	63	8	1	-
3	700.3	0	4.48	51	39	10	-	-
4	1050.4	0	4.45	39	49	10	2	-
5 (US)	1050.4	3	4.41	22	63	7	2	6
6 (US)	874.4	1.5	4.52	46	44	9	1	-

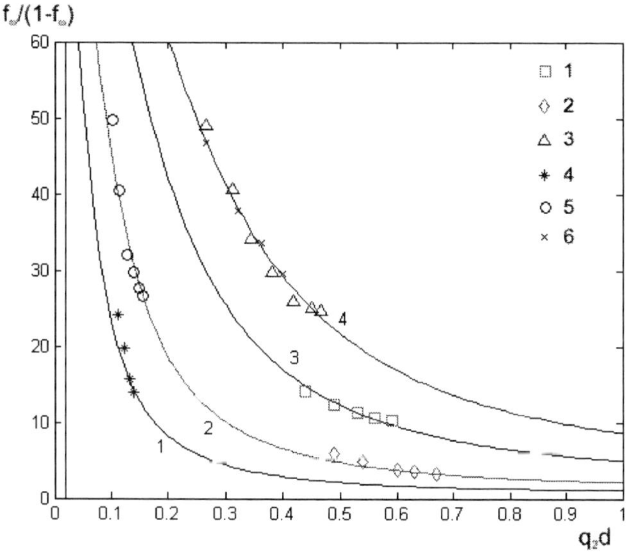

Figure 29. Theoretical dependences of $f_\omega/(1\text{-}f_\omega)$ value on acoustic thickness q_2d. 1 - ρ_2v_2/ρ_1v_1 =0.6; 2 – 0.7; 3 –0.8; 4 – 0.85. Points are experimental results from [76]: □ - sample.No.1; ◊- sample.No.2; Δ - sample No.3; * - sample No.4; O -sample No.5; x – sample No.6.

Table 7. The structure data for $BaTiO_3$-$WTiO_3$ samples

Sample	*P*, MPa	R_{ag}, nm	R_{gr}, nm	CSA (*r*), nm	ρ, g·cm^{-3}	*D* at 3.8K	*d*, nm	ρ_2v_2/ρ_1v_1
1	113.0	2010	250	86	4.48	39.0	1.50	0.80
2	525.0	1830	273	90	4.43	8.1	1.50	0.70
3	700.3	1747	232	96	4.48	52.0	1.28	0.85
4	1050.4	1414	236	150	4.45	29.5	0.27	0.60
5(US)	1050.4	2037	260	150	4.41	57.0	0.35	0.70
6(US)	874.4	1711	263	148	4.52	70.5	1.10	0.85

v=4.54 · 10^5 cm·s^{-1}.

The ceramics were sintered at T_s =1350°C for 3 hours. The fracture in the specimens always occur at the grain boundaries, therefore it was possible to define the composition of the ceramic interface regions. The samples have a complex grain boundary hierarchy. Large grains (agglomerates) R_{ag} consist of small grains R_g, which in turn consist of coherent scattering areas (CSA). The specimen structure data and experimental results according to (8) are presented in Figure 29 and table 7. The agglomerate boundaries assumed to consist of the same material as grain boundaries, i.e. $\rho_2 v_2(\text{ag}) = \rho_2 v_2(\text{gr})$. If the contribution of agglomerate boundaries to the NP scattering were comparable with grain boundaries, the experimental NP scattering dependences would lie on the wings of the theoretical curves and exhibit a weaker dependence of $l(T)/l_0$ on $q_2 d$ (temperature). Such a situation was observed for specimens 1 and 2 compacted at low pressure. The temperature dependence of NP scattering for the specimens compacted at pressure P > 700MPa is determined by the scattering on grains inside agglomerate.

In the case of weak temperature dependence which is observed for samples 1 and 2 (Low compacting pressure) the d value substantially corresponds to the agglomerate boundaries. The samples 4 and 5 compacted at P = 1040,5 MPa, have the thinnest GB and maximum of WO_3 phase in the grain interface areas (the lattice parameter a = 0.3837 is typical for film phase). The samples 2 and 5 have almost equal acoustic impedance ratios, which indicate the same interface regions structure. This fact was corroborated by X-ray investigation, revealing the existence of perovskite phase in the interface regions. Difference in temperature dependences between the samples indicates the reduction of interface layer thickness and hardening of sample 5 because of its higher WO_3 content. It was noted in [76] that the composition of interface regions is quite different from that of grain bulk because of tungsten atoms diffusion to grain boundaries during sintering. Ultrasound action leads to formation of complex structure of fragments and boundaries between them as in the case of YSZ ceramics [24]. It also activates the thermostimulated processes during the formation of interface regions composition.

CONCLUSION

Thus, the investigation of the NP kinetics at helium temperatures is the sensitive method of the comparative estimation of the degree of stabilization of GB in ceramic materials depending on the conditions for compacting and synthesis. In the range $qR >> 1$, at helium temperatures, the thermophysical properties of the dense ceramics based on the oxides are determined by the size and structure of crystallites (grains) and the GB thickness. The processes of the plastic deformation by the twinning in the grain structure can lead to the improvement of the thermophysical and optical characteristics of ceramics. It is demonstrated that the acoustic transparency of the YAG oxide ceramics with the elements of twinning in the grain (R=1-2μm) and GB structure in the terahertz frequency range is comparable with the acoustic transparency of the single crystals of the YAG:Re solid solutions. The ratio of phonon mean free path to mean grain size l/R is found to be several hundreds. This indicates the natural crystallographic character of GB and the absence of strains and defects in the grain structure.

The conditions of the gap formation in the phonon spectrum of ceramics at qR~1 was investigated. It was found that the presence or absence of the gap is determined by elastic properties of GB. The presence of inclusions (pores, other phases) with characteristic sizes smaller than that of grains of the main ceramic material shifts the gap to high frequencies. The nanodispersed polycrystal iron-containing corundum-based cermets were investigated. The insertion of 5% Fe effectively (by a factor of a hundred) suppress the sub-terahertz phonon transport. It was shown that the electron-phonon interaction and the metal fragments filter the high-frequency branch of phonon spectrum. The two-level systems (Δ = 8.53K) determined by formation of balanced vacancies configurations in the anion sublattice with respect to cation impurities (Gd) were revealed in ceramics on the base of $Ce_{1-x}Gd_xO_{2-y}$ solid solutions. The NP transport properties in ferroelectric ceramics and single crystals manifested diffusive (broadened) phase transition reflecting the anomalies of glass-like behavior, which is typical for low-temperature heat capacity and thermal conductivity. The contribution of domain structure to phonon scattering was estimated for $BaTiO_3$ single crystal. The structure, phase composition of ceramic grain boundaries and their dependence on synthesis conditions were analyzed for $BaTiO_3$-$WTiO_3$ solid solutions as example.

Acknowledgments

The study was supported by the Russian Foundation for Basic Research, and the program of the Presidium of the Russian Academy of Sciences.

References

[1] Nanotechnology Research Directions: Vision for Nanotechnology in the Next Decade. / Edited by M.C. Roco, S. Williams, P. Alivisatos Springer, Dordrecht, 2000.

[2] R.W.Siegel. *Nanostructured Materials*, 4, 121 (1994).

[3] A.I.Gusev. *Phys. Usp.*, 41, 49 (1998).

[4] D. R. Clarke. *J.Amer.Ceramic Soc.,* 70 (1), 15 (1987).

[5] P. Ernst, O. Kienzle, M. Ruhle. *J.Europ. Ceramic Soc.*,19 (6-7), 665 (1999).

[6] I.B. Levinson. *JETP*, 52 (4), 704 (1980).

[7] R. J. von Gutfeld, Jr. A. H. Nethercot. *Phys.Rev.Lett.*, 12 (23), 641 (1964).

[8] S. N. Ivanov, E. N. Khazanov, T. Paszkiewicz. *Z. Phys. B.,* 99 (4), 535 (1996).

[9] H. Yagi, T. Yanagitani, T. Numazawa, K. Ueda.*Ceramics International*. 33 (5), 711 (2007).

[10] S.N. Ivanov, A.G. Kozorezov, A.V. Taranov, E.N. Khazanov. *JETP*, 75 (2), 319 (1992).

[11] Ju. N. Barabanenkov, V. V. Ivanov, S. N. Ivanov. *JETP*, 102 (1), 114 (2006).

[12] Ju. N. Barabanenkov, V. V. Ivanov, S. N. Ivanov. *JETP*, 92 (3), 474 (2001).

[13] V. V. Ivanov, S. N. Ivanov, A. S. Kaigorodov, A. V. Taranov, E. N. Khazanov, V. R. Khrustov. *Inorganic Materials* , 43 (12), 1365 (2007).

[14] V. V. Ivanov, S. N. Ivanov, O. V. Karban, A. V. Taranov, E. N. Khazanov, V. R. Khrustov. *Inorganic Materials*, 40 (11), 1233 (2004).

[15] Ju. N. Barabanenkov, S. N. Ivanov, A. V. Taranov. *JETP Letters*, 79 (7), 342 (2004).
[16] S. N. Ivanov, L. M. Zhukova, Ya. M. Soifer, E. N. Khazanov, A. V. Taranov. *Inorganic Materials*, 36 (5), 504 (2000).
[17] W. A. Little. *Can .J. Phys* , 37, 334 (1959).
[18] L. M. Brekhovskih, O. A. Godin. Acoustics in continuum , Nauka, Moscow 1989 (In Russian).
[19] O. A. Khasanov, Ju. F. Ivanov, N. A. Popova. *Perspektivnye Materialy*, 5, 52 (1999).
[20] V. V. Ivanov, V. R. Khrustov. *Inorganic Materials,* 34 (4), 401 (1998).
[21] V. D. Kagan, A. V. Suslov. *Physics of the Solid State*, 36, 2672 (1994).
[22] A. A. Kaplyanskii, M. B. Mel'nikov, S. P. Feofilov. *Physics of the Solid State*, 38, 1434 (1996).
[23] F. F. Lange, G. L. Dunlap, B. I. Devis. *J. Am. Ceram. Soc.,* 69 (3), 237 (1986).
[24] O. V. Karban', E. N. Khazanov, O. L. Khasanov. *Perspektivnye Materialy*, 6, 76 (2010).
[25] A. V. Taranov, E. N. Khazanov. *JETP*, 107 (3), 509 (2008).
[26] J. Lu, M. Prabhu, J. Xu. *Appl. Phys. Lett.,* 77 (23), 3707 (2000).
[27] A. A. Kaminskii, S. N. Bagayev, K. Ueda. *Laser Phys. Lett.*, 3 (8), 375 (2006).
[28] T. Yanagitani, H. Yagi. *Jap.Pat.*No.10-101333; No.10-1014011, 1998.
[29] K. Ueda, J. Lu. K. Takaichi. *The Review of Laser Engineering*, 31, 465 (2003).
[30] M. Sh. Akchurin, R. V. Ganutdinov, R. M. Zakalyukin, A. A. Kaminski. *Doklady-Physics*, 52 (7), 373 (2007).
[31] P. P. Fedorov, V. V. Osiko, T. T. Basiev. *Nanotechnologies in Russia,* 2 (5-6) 95 (2007).
[32] T. T. Basiev, M. E. Doroshenko, P. P. Fedorov, V. A. Konyushkin, S. V. Kuznetsov, V. V. Osiko, M. Sh. Akchurin. *Optical Letters,* 33(5), 521 (2008).
[33] T. T. Basiev, V. V. Voronov, V. A. Konyushkin. *Doklady-Physics*, 52 (12), 677 (2007).
[34] T. T. Basiev, M. E. Doroshenko, V. A. Konushkinl. *Quantum Electronics*, 37 (11), 989 (2007).
[35] V. I. Al'shits, V. L. Indenbom .*Soviet Physics-Uspekhi*, 18 (1), 1 (1975).
[36] V. I. Al'shits, S. N. Ivanov, Ya. M. Soifer. *Soviet Physics Solid State*, 31 (11), 63 (1989).
[37] B. Reppich, G. Streb. *Phys. Stat. Sol. A,* 15, 77 (1973).
[38] G. Streb, B. Reppich B. *Phys. Stat. Sol. A*, 16, 493 (1973).
[39] E. N. Khazanov, A. V. Taranov, R. V. Gainutdinov. *JETP*, 110 (6), 983 (2010).
[40] E. N. Khazanov, A. V. Taranov, P. P. Fedorov. *Doklady-Physics,* 54 (1), 14 (2009).
[41] M. Sh. Akchurin, R. V. Gainutdinov, A. A. Kaminskii, A. V. Taranov and E. N. Khazanov. *JETP*, 108 (1), 83 (2009).
[42] A A Kaminskii, A V Taranov, E N Khazanov, M Sh Akchurin, *Quantum Electron*, 42 (10), 880 (2012).
[43] A. E. Solov'eva. *Inorganic Materials*, 21 (5), 808 (1985).
[44] E. I. Salamatov. *Phys.Stat.Sol. C,* 1, 2971 (2004).
[45] E. I. Salamatov. *Phys.Stat.Sol.B*, 244(6), 1895 (2007).
[46] V. V. Ivanov, E. I. Salamatov, A. V. Taranov. *JETP*, 106 (2), 288 (2008).
[47] V. V. Ivanov, S. N. Paranin, V. R. Khrustov. *The Physics of Metals and Metallography*, 94, suppl.1, S98 (2002).
[48] S. Schicker, T. Erny, D. E. Garcia. *J.Eur.Ceram.Soc.* 19, 2455 (1999).

[49] Y. Ji, J. A. Yeomans. *J.Eur.Ceram.Soc.* 22, 1927 (2002).
[50] D. -M. Liu, W. H. Tuan, Ch. -Ch. Chiu. *Meter. Science and Enginreering B*, 31, 287 (1995).
[51] A. B. Pippard. *Phill.Mag.* 46, 1104 (1955).
[52] O. V. Karban', E. I. Salamatov, A. V. Taranov, E. N. Khazanov, O. L. Khasanov. *JETP*, 108, 661 (2009).
[53] E. I. Salamatov. *Physics of the Solid State*, 44 (5), 978 (2002).
[54] E. I. Salamatov. *Physics of the Solid State*, 45 (4), 725 (2003).
[55] S. N. Ivanov, A. V. Taranov, E. N. Khazanov, V. A. Atsarkin, V. V. Demidov. *JETP*, 67, 1021 (1988).
[56] S. N. Ivanov, A. G. Kozorezov, A. V. Taranov, E. N. Khazanov. *JETP*, 73, 880 (1991).
[57] S. Ostanin, A. J. Craven, D. W. McComb. *Phys.Rev. B*, 65, 224109 (2002).
[58] S. Ostanin, E. Salamatov. *Phys.Rev.B*, 68, 172106 (2003).
[59] T. Tojo, T. Atake, T. Mori, H. Yamamura. *J. Therm Anal. Calorim.*, 57, 447 (1999).
[60] V. V. Ivanov, Yu. A. Kotov, V. P. Gorelov, V. N. Borisov, A. M. Murzakaev. *Russian Journal of Electrochemistry*, 41 (6), 612 (2005).
[61] T. Suzuki, I. Kosacki , H. U. Anderson. *Solid State Ionics,* 151, 111 (2002).
[62] V. V. Ivanov, E. I. Salamatov, A. V. Taranov, E. N. Khazanov. *JETP*, 110 (1), 34 (2010).
[63] A. Atkinson, A. Selcuk. *Sol.St.Ionics.*, 134, 59 (2000).
[64] A. G. Kozorezov. *JETP*, 73 (5), 872 (1991).
[65] Y. Wang, K. Duncan, E. D. Wachsman, F. Ebrahimi. *Solid State Ionics*, 178, 53 (2007).
[66] J. J. De Yoreo, R. O. Pohl. *Phys. Rev. B*, 32, 5780 (1985).
[67] W. N. Lawless. *Phys. Rev. B*, 14, 134 (1976).
[68] V. I. Kozub, A. M. Rudin, H. Schober. *Phys. Rev. B*, 50, 6032 (1994).
[69] В. И. Козуб, А. М. Рудин, ФТТ 38, 337 (1996).
[70] Yu. M. Galperin, V. G. Karpov, V. V. Kozub V.V. *Adv. in Phys*, 38, 669 (1989).
[71] S. N. Ivanov, E. P. Smirnova, A. V. Taranov, E. N. Khazanov. *JETP*, 88 (2), 342 (1999).
[72] E. Fisher, W. Hassler, E. Hegenbarth, V. I. Fritsberg. *Phys. St. Sol. A,* 66, K169 (1981).
[73] I. Henning, P. Frach, E. Hegenbarth, V. I. Fritsberg. *Phys. St. Sol. A,* 70, K7 (1982).
[74] D. A. Ackerman, D. May, R. C. Potter, A. C. Anderson. *Phys. Rev. B*, 23, 3886 (1981).
[75] B. A. Strukov, A. P. Levanyuk. *Physical Principles of Segnetoelectric Phenomena in Crystals Nauka*, 1983 (In Russian).
[76] O. V. Karban', O. M. Kanunnikova, E. N. Khazanov. *Chemical Physics and Mesoscopy*, 12(1), 102 (2010).
[77] O. Khasanov, E. Dvilis, V. Sokolov. *Proc. Nanotechnology Conf. and Trade show Boston*, MA, 2006 p.23-26.

In: Recent Advances in Ceramic Materials Research
Editors: Joan J. Roa Rovira and Merce Segarra Rubi
ISBN: 978-1-62417-729-3

Chapter 6

OPTICAL CERAMICS FOR FAST SCINTILLATOR MATERIALS

M. Nikl[1*], T. Yanagida[2], H. Yagi[3], T. Yanagitani[3], E. Mihokova[1] and A. Yoshikawa[4,5]

[1]Institute of Physics As CR, Prague, Czech Republic
[2] Kyushu Institute of Technology, Wakamatsu-ku, Kitakyushu, Fukuoka, Japan
[3]Konoshima Chemical Co., Ltd., Kouda, Takuma, Mitoyo, Kagawa Japan
[4]Institute for Materials Research (IMR), Tohoku University, Katahira, Aoba-ku, Sendai, Japan
[5]New Industry Creation Hatchery Center (NICHe), Tohoku University, Aoba, Aramaki, Aoba-ku, Sendai, Japan

ABSTRACT

Optical ceramics (OC) are transparent or translucent materials constituted by tight aggregating crystallite micro- or nanograins, each randomly oriented with respect to their neighbours. They have been under development as an alternative to single crystal materials to provide bulk optical elements in the case where single crystals cannot form or when ceramic materials show superior optical and luminescence properties. Development of OC for the scintillator applications was triggered by the needs of Computer Tomography (CT) medical imaging in the 90's of last century. The manufacturing and characterization of $(Y,Gd)_2O_3$:Eu, Gd_2O_2S:Pr,Ce,F and $Gd_3Ga_5O_{12}$:Cr,Ce and later on also the Eu^{3+} or Tb^{3+}-doped Lu_2O_3 slow ceramic scintillators were reported. More recently, fast optical ceramics with scintillation response in the tens-hundreds ns time scale and based on the Ce^{3+}-doped $Y_3Al_5O_{12}$ (YAG) was reported in late 1990's and its higher density analog based on LuAG host doped with Ce^{3+} or Pr^{3+} was reported only several years ago for the first time. This chapter will review the latest studies in the field of fast scintillation optical ceramics in the last decade focusing on the interconnection among the manufacturing technology, scintillation performance and defect occurence and nature in the discussed material systems. The most studied group of materials are the cubic structure aluminum or multicomponent garnets, but interesting results have been achieved also in sesquioxide,

* Corresponding Author's e-mail address: nikl@fzu.cz (M. Nikl).

silicate, hafnate, complex perovskite or rare earth halide compounds. Both basic physics and application aspects will be considered in the description of the above mentioned systems.

1. INTRODUCTION

In the introductory part of this chapter, we briefly summarize the history of scintillator development focused on single crystal and optical ceramic materials. We explain fundamental aspects of X (gamma)-ray transformation to ultraviolet/visible photons and list practical scintillation parameters, which are used to quantitatively evaluate new materials. Furthermore, we discuss the preparation technology of ceramic scintillators. Finally, we mention experimental techniques suitable for correlated characterization of physical phenomena in scintillators.

1.1. Historical Aspects

In November 1895 Wilhelm Conrad Roentgen noticed the glow of a barium platino-cyanide screen placed next to his operating discharge tube. This new, invisible and penetrating radiation [1] was named X-ray in English and "Roentgen radiation" in some other languages. In the X-ray registration, simple photographic film was found rather inefficient and the search for materials able to convert X-to-visible started immediately in order to couple with sensitive photographic film-based detectors. In 1896 Pupin introduced the $CaWO_4$ powder used for this purpose for more than 75 years. Together with ZnS-based powders introduced later by Crookes and Regener, these are the phosphor materials in longest use employed for the detection of X-rays.

The discovery of natural radioactivity (spontaneously released high energy radiation) of uranium by Henri Becquerel in early 1896 [2] was followed by pioneering works of Marie Curie Sklodowska and her husband Pierre Curie leading to the discovery of strongly radiactive elements polonium and radium [3,4]. This kind of radiation is commonly named gamma rays (γ–rays) and its energy is higher than the above mentioned X-rays. Among different methods for the detection of γ radiation, scintillator-based detectors have been widely employed. It is also interesting to note that in 1897 Braun invented the Cathode Ray Tube, in which the energy of an accelerated electron beam is converted into visible light (called cathodoluminescence) by a phosphor material.

The mechanism of this conversion is quite similar [5] to that of the X-ray conversion. Thus, at the end of the 19th century, two energetic (photon and particle) radiations were available together with γ rays generated by natural radiactive elements. Their enormous application potential was realized especially during the World War I, and stimulated the development of phosphor and scintillator materials necessary for their practical exploitation. The history of single crystal scintillators begins in the late 1940s with the introduction of NaI:Tl and CsI:Tl scintillators by Robert Hofstadter [6,7]. Since that time a number of material systems have been reported, see Figure 1 and [8] for an overview.

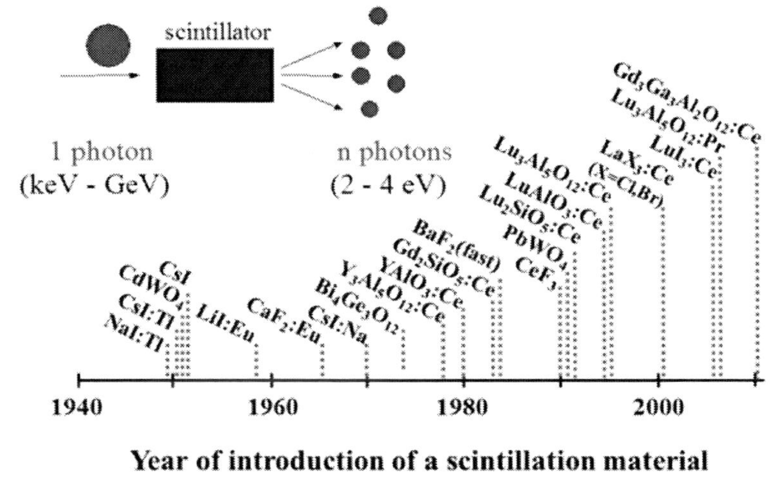

Figure 1. Sketch of scintillator principle and historical overview of single crystal scintillators.

The above two materials and $Bi_4Ge_3O_{12}$ (BGO) introduced by Weber and Monchamp [9] became widespread scintillators and are often used as "standard samples" to evaluate newly developed materials. Within the last 15 years there has been a considerable activity in this field triggered mainly by the need of high energy physics and advanced imaging applications in science, medicine and industry. This has led to a considerable number of articles in the literature mostly dealing with Ce^{3+}-doped and Pr^{3+}-doped materials, due to the fast decay time (typically 10-60 ns) and high quantum efficiency of their 5d-4f radiative transitions at room temperature (RT). For more information see [10, 11]. Despite the best figure-of-merit of the single crystal scintillators, not nearly all intrinsically efficient compounds can be grown in the form of bulk single crystals with sufficiently big dimensions and manufacturing price affordable for practical applications. Too high melting temperatures, the presence of phase transitions between the melting point and room temperature or stoichiometry problems resulting from different volatility of binary constituents from the high temperature melt of complex compounds, are just few examples of troubles which can prevent a single crystal preparation. Optical ceramics have been under development as an alternative to single crystal materials to provide bulk optical elements in the case where single crystals cannot form or when ceramic materials show superior properties, for example in the achievable concentration and homogeneity of the dopant. They are transparent or translucent materials constituted by tight aggregating crystallite micro-grains, each randomly oriented with respect to their neighbors.

Their technology has greatly developed mainly within the last two decades due to application of these materials in the field of solid state lasers [12]. In the case of cubic material the technological progress allowed to obtain bulk elements visually indistinguishable from single crystals and in some parameters (doping profile) clearly even superior. However, the application demands are higher in the case of scintillator with respect to laser optical ceramics. In the latter case the scattering losses due to pores and grain boundaries limit the material performance. The already achieved reduction of the pore volume down to a few ppm and grain interface thickness to a few nm reported in [13,14] is sufficient to make them fully comparable to their single crystal analogs. However, in the case of scintillators, even the point defects and structural irregularities at the atomic scale can seriously limit the material performance due to the introduction of trapping levels in the material forbidden gap. If we consider the grain size of the order of a few tens of microns and thickness of the interface layer 1-2 nm, then expected concentration of the interface-related (trapping) states may easily reach several hundreds of ppm. Such concentration can noticeably influence the characteristics of the transport stage due to retrapping of the migrating carriers or even by the introduction of nonradiative traps, which can seriously degrade the scintillation response and/or overall efficiency parameters. Development of optical ceramics for the scintillator applications was triggered by the needs of Computed Tomography (CT) medical imaging and started in the 1980s [15, 16].

The review of the results mainly achieved in the manufacturing and characterization of $(Y,Gd)_2O_3$:Eu, Gd_2O_2S:Pr,Ce,F and $Gd_3Ga_5O_{12}$:Cr,Ce ceramics was reported by Greskovich and Duclos [17]. In the struggle to reduce the afterglow of these materials, various co-doping strategies or annealing in a controlled atmosphere were applied to either diminish or passivate trapping states in the forbidden gap related to the grain interfaces. It is worth mentioning that scintillation response of all these ceramics is slow, in μs-to-ms time scale due to forbidden character of radiative transitions of the emission centers used (Eu^{3+}, Pr^{3+}, Cr^{3+}, Tb^{3+}). More recently, fast optical ceramics based firstly on the Ce-doped YAG have been reported in the literature [18]. In the last decade the R&D in the field of such fast scintillation ceramics became a hot topic in the hunt for new scintillation materials and will be described in more detail in *Section 2*. Due to the high practical importance and relatively long history of scintillators, a large amount of information was published on the topic. Let us mention the superb survey of luminescent materials written by Blasse and Grabmaier [19], the overview of methodology used in the radiation measurement authored by Knoll [20], the monograph devoted to scintillator materials from Rodnyi [21] and Lecoq *et al.* [22] and the useful phosphor handbook written by Shionoya and Yen [23]. Numerous featured review papers or book chapters on phosphor and scintillator materials and their applications exist in the scientific literature as well, for instance [24-31].

1.2. Fundamentals

Wide band-gap materials are employed for transformation of the X (γ)-ray to ultraviolet/visible photons, see the sketch in Figure 1. Consistent phenomenological description of the scintillation conversion process, efficiency criteria, etc.; were already developed in the 1970s [32] and later further refined [33]. Scintillation conversion is a relatively complicated issue, which can be divided into three consecutive sub-processes: *conversion, transport and luminescence* – see Figure 2. During the initial conversion a multi-step interaction of a high-energy photon with the lattice of the scintillator material occurs through the photoelectric effect. Compton scattering effect and pair production: for photon energies below 100 keV the first one is of major importance. Many electron-hole pairs are created and thermalized in the conduction and valence bands, respectively. This first stage is concluded within less than 1 ps. More detailed considerations about the conversion processes have been published [34,35]. In the transport process the electrons and holes (*possibly created excitons*) migrate through the material, they may be repeatedly trapped at defects, or lose their energy due to nonradiative recombinations. Considerable delay in the migration towards luminescent center can be introduced due to the mentioned charge carrier recapture at trapping levels in the material forbidden gap. This stage is the least predictable as point defects, flaws, surfaces and interfaces can introduce energy levels into the forbidden gap and strongly modify/degrade otherwise high intrinsic scintillation performance. These phenomena are strongly dependent upon manufacturing technology [36].

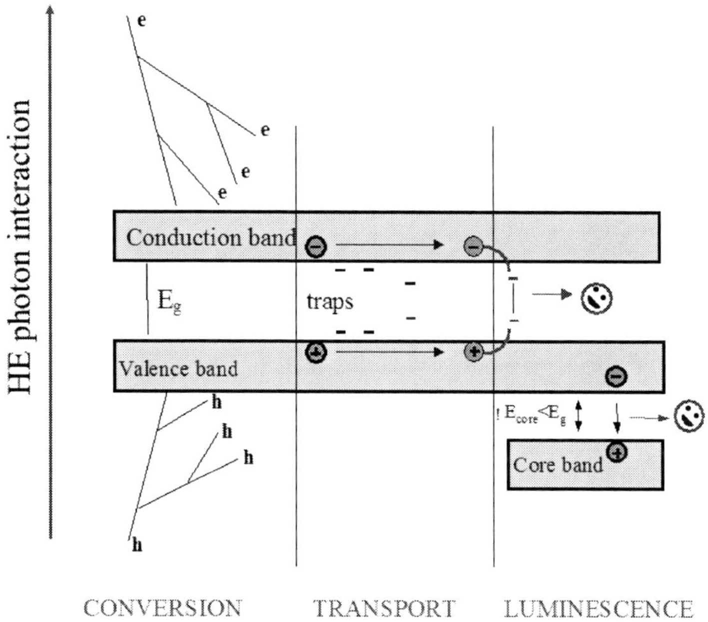

Figure 2. Sketch of scintillator mechanism.

During the final stage, luminescence consists of consecutive trapping of the electron and hole at the luminescent center and their radiactive recombination. Non-radiactive energy transfer towards the emission center via exciton-based state is also possible. In a particular group of so called cross-luminescence materials, see the review [21], the light generation occurs in radiactive transitions between the valence and first core bands – sketched in Figure 2. Cross-luminescence mechanism enables very fast, even subnanosecond, scintillation response, which is, however, usually accompanied by much slower exciton-related luminescence. This phenomenon is reported in the literature mainly for BaF_2 and other halide single crystals [37]. The physics of luminescent centers is usually well understood due to advanced experimental methods available for their selective and time-resolved study, see e.g. [19, 20].

As mentioned above, X-rays, in their early days, were registered exclusively by phosphor powders in the form of thin screens coupled to a photographic film [24]. Later, due to the need to detect and monitor higher energy X- or γ-rays, scintillator materials in the form of single crystals were introduced, see Figure 1. Despite the identical underlying physics, scientific communities working on phosphors and scintillators have been partially separated, mainly due to the different demands of related applications and different preparation technologies employed [24-27]. In general, materials are called phosphors when applied in the photon integrating (steady-state) detection mode, while scintillators are employed in the

(X- or γ-ray) photon counting regime. At present, the separation between phosphor (powders) and scintillator (bulk) materials is somewhat suppressed as some materials are used in both detection modes, in powder, bulk or other forms, depending on the application [30].

In the case of scintillators, X (γ)-ray photon counting consists of accumulating the generated light arriving soon after the initial conversion stage is accomplished (Figure 2), since the scintillator works as a high-energy photon counter. Significantly delayed light such as that due to retrapping processes mentioned above cannot be technically exploited in the counting mode. Furthermore, in some applications one should be able to distinguish X (γ)-ray photons of different energy. The most important characteristics of scintillation materials for detectors are following:

1) Light yield (LY)
2) X(γ)-ray stopping power
3) Scintillation response – decay time
4) Spectral matching between the scintillator emission spectrum and photo-detector
5) Chemical stability and radiation resistance
6) Linearity of light response with the incident X(γ)-ray photon energy – energy resolution

The overall scintillation efficiency of X (γ)-ray-to-light conversion is determined both by intrinsic and extrinsic material characteristics. The number of UV/visible photons, N_{ph}, produced in the scintillation conversion per energy E of incoming X (γ)-ray photon can be expressed as [32-34]:

$$N_{ph} = \frac{E}{\beta E_g} SQ \tag{1}$$

where E_g represents the forbidden gap of the material, S and Q are quantum efficiencies of the transport and luminescence stages respectively, and β is a phenomenological parameter which is typically found between 2 and 3 for most materials. The relative efficiency can then be obtained as:

$$\eta = \frac{E_{vis} N_{ph}}{E} \tag{2}$$

where E_{vis} is the energy of generated UV/vis photons. The most efficient material among the phosphors and scintillators today is ZnS:Ag with $\eta \sim 0.2$. If we consider that the bandgap of the here discussed scintillator materials is higher than ZnS (~ 3 eV), then even more efficient materials could still be found within those with the narrower bandgap.

The LY of a scintillator is always a value inferior to that given by eq. (1) since it represents only a fraction of generated visible photons, namely those arriving to the photo-detector within a certain time gate after the high energy photon absorption. Typical values of such a time gate are usually set between 100 ns and 10 μs. The X (γ)-ray stopping power (attenuation coefficient) of a material of given thickness depends on its density, ρ, and its effective atomic number, Z_{eff} (for the calculation of Z_{eff}, see e.g. [17]). Considering only the interaction associated with the photoelectric effect, the stopping power is proportional to $\rho Z_{eff}^{3\text{-}4}$ [25, 29].

The kinetics of the light response of a scintillator is governed by the characteristics of the transport and luminescence stages in Figure 2, as they are far slower than the initial conversion. The decay rate of the luminescence center itself is defined by its transition dipole moment from the excited-to-ground state and can be further enhanced by additional nonradiative quenching or energy transfer processes from the excited state. Such quenching or energy transfer, however, result in the decrease of parameter Q in eq. (1) and the overall conversion efficiency is also decreased. In the most simple case of an exponential decay, the emission intensity $I(t)$ is:

$$I(t) \approx e^{-t/\tau} \tag{3}$$

where τ is the decay time. While decay times of the parity and/or spin forbidden transitions for most of the rare earth ions are typically on the order of several tens of μs up to ms, in the case of allowed 5d-4f transitions of Ce^{3+} and Pr^{3+} the values scale down to tens of ns and fully allowed singlet-singlet transition in organic molecules is still about ten times faster [38]. The fastest emission transitions are offered by the radiative decay of Wannier excitons in direct gap semiconductors, where subnanosecond values have been reported for compounds such as ZnO, CuX, $CsPbX_3$ (X=Cl,Br), PbI_2 and HgI_2 [39]. The emission rate is enabled by the coherent nature of the exciton state spread over a (large) number of elementary cells [40]. However, a serious disadvantage of the latter group of materials consists in the low binding energy of their excitonic state, which results in partial ionization and consequently quenching of the exciton related emission at *RT*. In bulk materials, further efficiency decrease follows from the small Stokes shift, which leads to noticeable reabsorption losses [41].

Due to the afore-mentioned retrapping processes during the transport stage sketched in Figure 2, the light emission response of a material under high energy excitation is often further complicated by slow non-exponential components. These processes are currently quantified by the afterglow (sometimes called a persistence), which is defined as a residual light intensity at a certain time (from a few ms to minutes) after excitation cut-off, see [36]. The delayed radiative recombination will contribute to slow scintillator response, lowered *LY* values and deteriorated signal-to-background ratio [42]. In some studies, a so-called parameter alpha has been introduced to quantify the occurrence of delayed radiative recombination within tens to hundreds of μs.

Due to repetitive nature of the measurement, the delayed recombination is reflected by the elevated signal level before the rising edge of the decay curve [42,43], which is illustrated in Figure 3. In practical evaluation for both phosphors and scintillators, either simple 1/e or 1/10 decay times are sometimes provided, defined as the time when the light intensity decreases to 1/e or 1/10 of its initial value after an excitation flash, or so called mean decay time can be calculated [27]. Spectral matching between the scintillator emission band and the photodetector spectral sensitivity dependence is an obvious requirement. Classical criteria were defined some time ago as near UV-blue emission is an optimum for a photomultiplier detector, while for a photodiode the green-red spectral region was considered the best. In recent years there has been an enormous development of semiconductor photodetectors and the latest generation of the back illuminated "charge coupled device" CCD photo detector shows an enhanced sensitivity down to 200 nm. Chemical stability mainly concerns the hygroscopicity of materials, which in some cases severely limits their long-time operation in the open air (NaI:Tl, CsI:Na, LaX_3:Ce, X=Cl,Br).

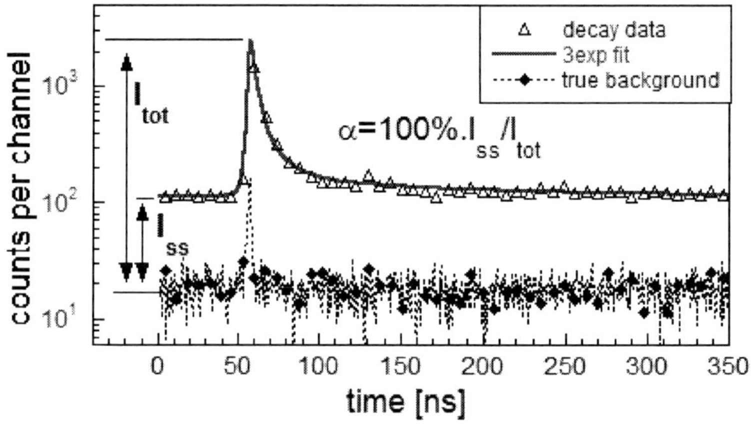

Figure 3. Spectrally unresolved scintillation decay of undoped PbWO4, fit by 3-exp. function *I(t)= 0.521exp[-t/3.0ns]+0.072exp[-t/12.1ns]+0.00 8exp[-t/102ns]+ 110* convoluted with the instrumental response (solid line). True background level is displayed as well. Background enhancement in the repetitive decay measurement is reflected in I_{SS}: superslow decay component amplitude which comes from preceding excitations and is evaluated using the coefficient alpha.

Radiation resistance of materials mainly regards the performance changes and instabilities due to the induced absorption resulting from the material irradiation and color centers creation. It is a matter of concern mainly in the bulk scintillation materials [27]. In the case of the induced absorption overlap with the emission spectrum, the re-absorption losses occur with resulting loss of the overall efficiency/*LY*. An example of this behavior is given in Figure 4.

While this parameter has been mainly considered in the research of scintillators for high energy physics, it should be noted that it is also important in several medical imaging techniques [29] as well as in the case of industrial flaw detection or synchrotron beam diagnostics applications.

The linearity of light response with incident *X (γ)*-ray photon energy in scintillators is the dependence of *LY* on the incident photon energy. The value for *LY* is energy dependent: it is partly due to abrupt changes of the attenuation coefficient around the *K* and *L* edges of the elements constituting the compound, but it also follows from the non-equal conversion efficiency of the photoelectric effect and Compton scattering, which become progressively more important with the increasing energy of incoming *X (γ)*-rays. As a result, the energy resolution of a scintillating material is degraded with respect to the intrinsic limits based purely on statistical grounds [44]. Nonproportionality of scintillating materials became systematically studied in recent years [45] and first comprehensive models and explanations have appeared recently [46].

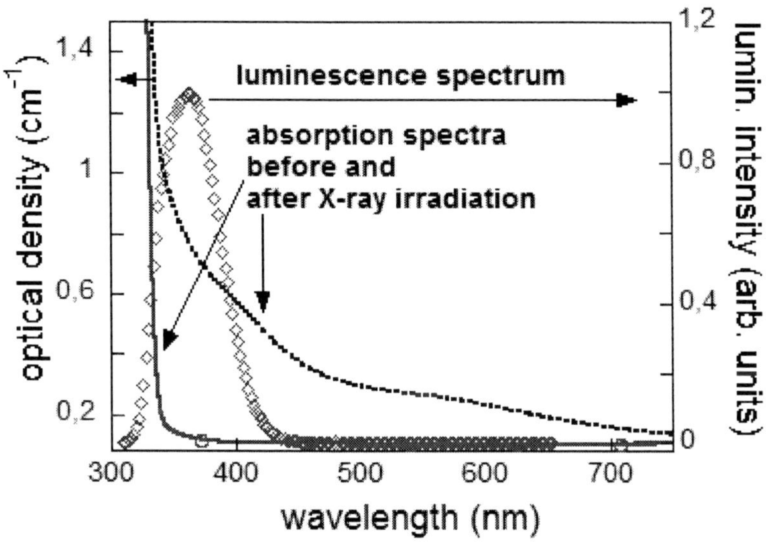

Figure 4. Absorption spectra of 2 cm thick $YAlO_3$:Ce single crystal before and after γ-ray irradiation (^{60}Co, 500 Gy dose) at RT. Overlap of the irradiation induced absorption with luminescence spectrum of the material is demonstrated.

1.3. Preparation Technology

As already mentioned, the optical ceramic materials are ensembles of tightly aggregated and randomly oriented nano-micro crystals, see Figure 5. The first technical trial to obtain transparent ceramic was accomplished with α-Al_2O_3 half a century ago [47]. It was revealed that ceramic α-Al_2O_3 could be made translucent by reducing pores and controlling fine structure of grains. In order to obtain transparent ceramics, numerous light scattering loss processes have to be suppressed [12, 48, 49]. In Figure 6, numerous sites for light scattering are schematically drawn. Unlike single crystals, there are many grain boundaries in ceramics and sometimes the grain boundary becomes a scattering center (i). Inclusions or pores in each grain can also become scattering centers (ii). Second phases on the grain boundaries (iii), and double refraction from birefringent materials (iv), are also candidates for the scattering centers. In addition to these internal phenomena, the surface roughness (v) could cause light scattering as well. In recent technical trends, optical transparent ceramic materials, including those for scintillator and laser applications, can be synthesized mainly in two ways, namely so called non-reactive [50] or reactive [12] sintering. The main difference between them consists in the preparation of precursors.

Figure 7 (a) illustrates the manufacturing procedure of the former in the case of Nd-doped $Y_3Al_5O_{12}$ for the solid state laser [51]. In preparation of the ceramics, aqueous solutions of aluminum, yttrium, and neodymium chlorides are mixed together. The mixed aqueous solution is added dropwise and mixed with aqueous solution of ammonium hydrogen carbonate.

Figure 5. Scaning Electron Microscopy (SEM) micrograph of fractured surface of Ce:LuAG ceramic.

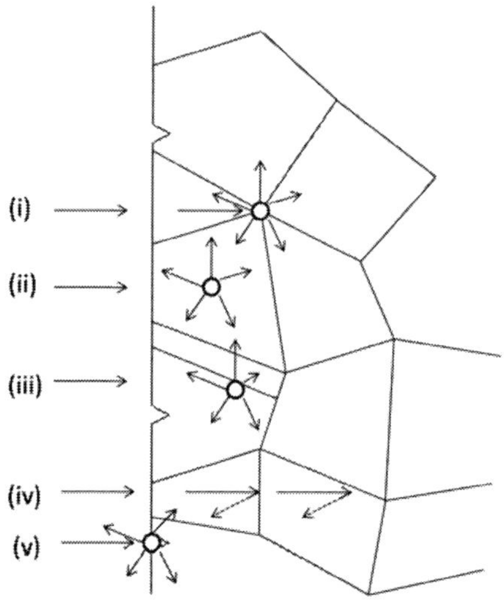

Figure 6. Schematic illustration of photon scattering in ceramics.

Then, steps of filtration and washing with water were repeated several times, and the obtained powder material was dried for two days at 120 °C in an oven. The obtained precursor consists of yttrium and aluminum hydrates ($Y(OH)_3$, $Al(OH)_3$) and carbonates ($Y(CO_3)_3$, $Al(CO_3)_3$). This precursor is calcinated at approximately 1200 °C to produce raw YAG oxide powder with an average particle diameter of about 200 nm.

The powder was ball-milled with the solvent, binder and dispersion medium, and mixed for 24 hours. The milled slurry is put into a gypsum mold and dried to obtain a desired form. By the slip casting forming method, a high density and uniform green body was obtained. After removing the organic components by calcinations, the materials were sintered in vacuum 10^{-6} torr at 1750 °C for 10 h. After annealing, highly transparent Nd:YAG ceramic samples are obtained.More details regarding the preparation technology can be found in [52,53]. In Figure 7 (b) the scheme of the reactive manufacturing procedure is displayed. To fabricate ceramic samples, starting powders of α-Al_2O_3, Y_2O_3 and Nd_2O_3 are weighed for the desired Nd concentration.

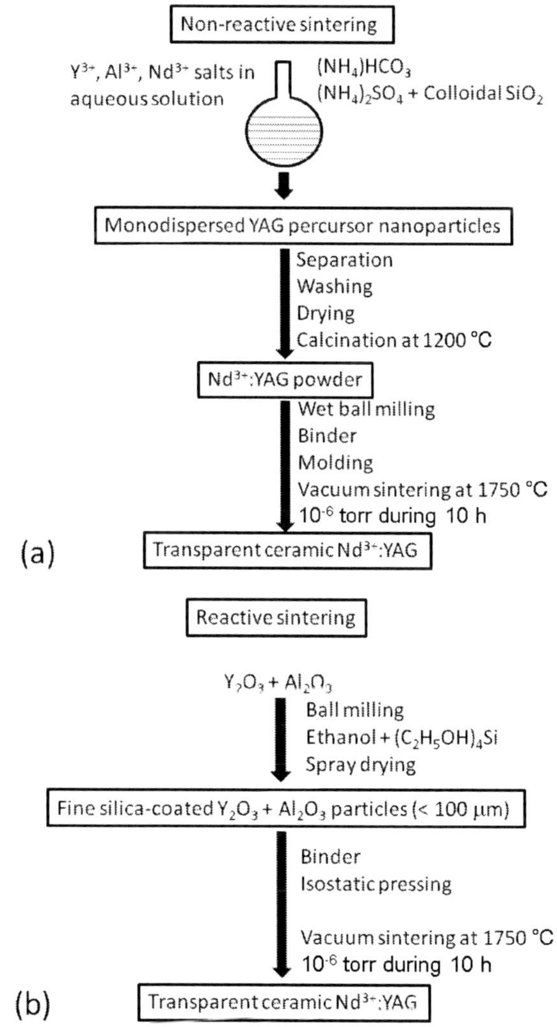

Figure 7. Schematic drawing of the fabrication procedure of transparent ceramic based on the non-reactive (a) and reactive (b) sintering.

After these powders are mixed with tetraethyl orthosilicate as a sintering aid, the mixture is milled for 12 h with the high-purity α-Al_2O_3 balls. The milled slurry including a dispersant and a binder is dried using a spray drier. The granulated, spherical powder grains with a diameter less than 100 μm, are isostatically pressed at 140 MPa into disks with a diameter of several mm.

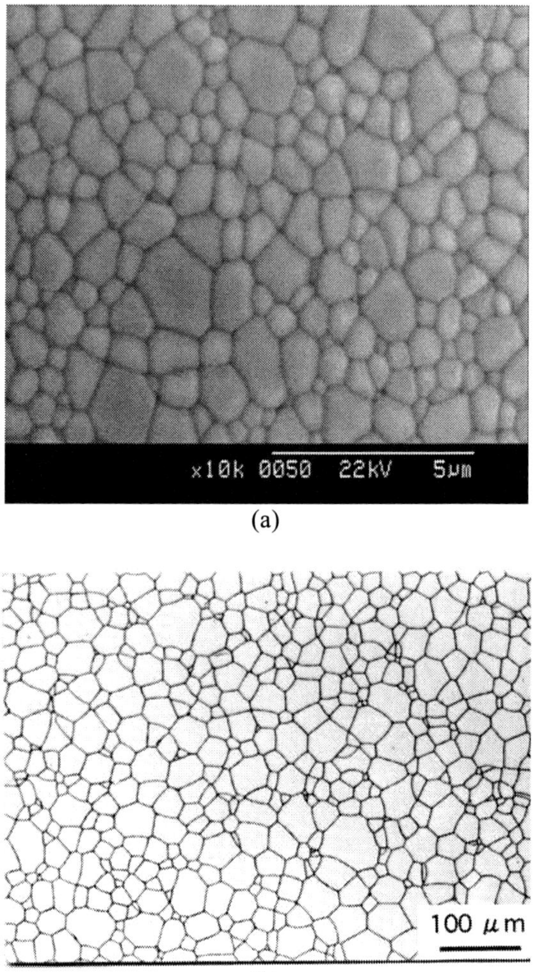

(a)

(b)

Figure 8. SEM images of ceramic YAG (a) fabricated by the non-reactive sintering [50] and Nd-doped YAG (b) by the reactive sintering [56].

The powder compacts are sintered at 1750 °C for 10-20 h in vacuum using a high-vacuum furnace with 10^{-5}-10^{-6} torr. Detailed description of the procedure can be found in [54-57]. Figure 8 compares the SEM images of YAG manufactured by the non-reactive sintering [50] and Nd-doped YAG synthesized by the reactive sintering [56]. Generally, the grain size depends on the sintering temperature and larger grains can be obtained at higher temperature.

However, in these ceramic YAGs, the difference in preparation of precursors resulted in different grain sizes although the sintering temperature is the same. Furthermore, the different conditions of raw material powders for the green body fabrication may affect the ceramics micro structure as well. The mean grain size obtained by the former method is a few microns while that of the latter is several tens of microns. Although typical sizes of the grains are very different, the ceramics synthesized by both procedures exhibit a high optical transparency. Photographs of the Ce:YAG ceramics fabricated by Konoshima is shown in Figure 9.

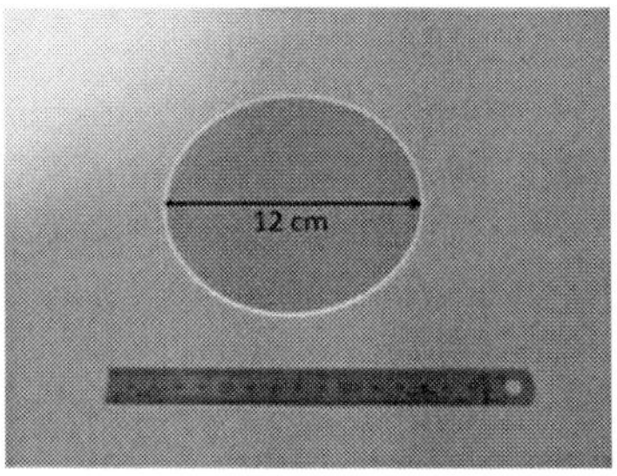

Figure 9. Photograph of ceramic Ce:YAG (Ø 120 mm) manufactured by Konoshima.

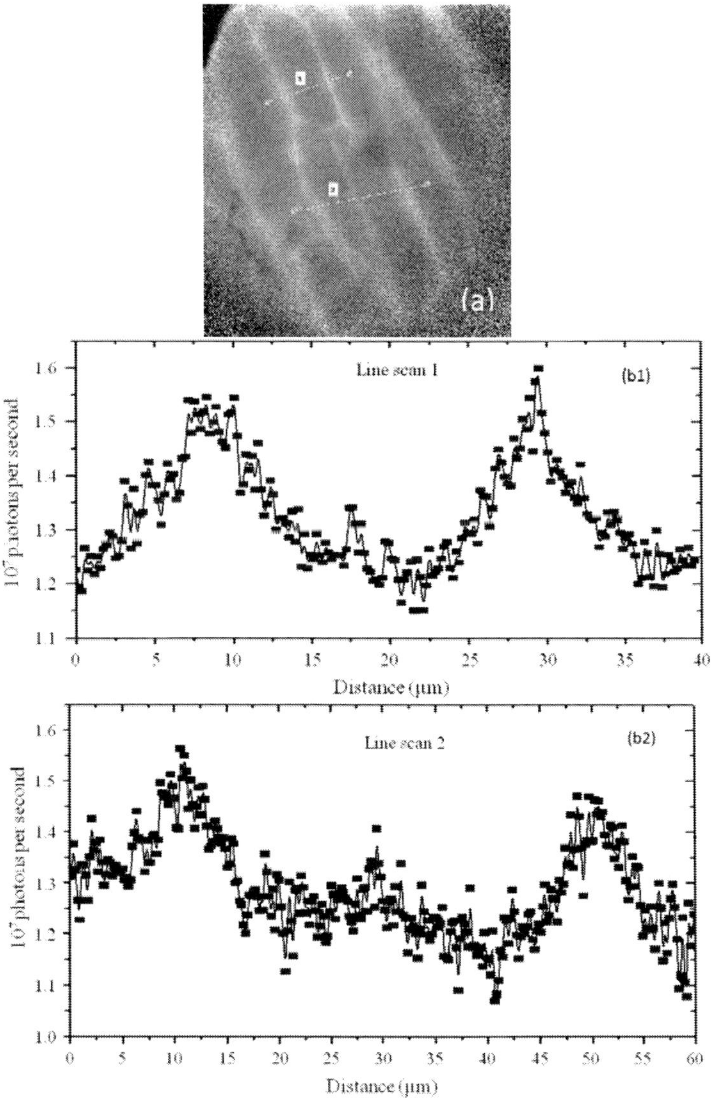

Figure 10. Imaging Confocal Microscopy measurements on the 0.1%Ce^{3+}-doped YAG ceramic sample. (a) Micro-luminescence image (standard mode). (b1,b2) Luminescence intensity measurement in confocal mode for two lines marked by (1) and (2) in (a).

At present, such a big (Ø 120 mm) fully transparent optical element is available. In addition to the above mentioned conventional manufacturing procedures, the spark plasma sintering (SPS) method can also be a candidate for fabrication of transparent ceramic materials. In SPS, a pulse continuous current is supplied to the specimen and highly densified ceramic can be obtained with relatively shorter sintering time than those of conventional procedures. Using this novel method, transparent ceramic of undoped Y_2O_3 [58], Lu_2O_3 [59], Lu_3NbO_7 [60], and $Lu_2Ti_2O_7$ [61] have been recently developed. Despite common belief that the doping profile is much more homogeneous in ceramic with respect to a single crystal, in a few recent years it has been revealed that in transparent ceramic the dopant segregation does occur at the grain boundaries [62-64], especially when the dopant is substituting for the host cation of much different ionic radius. These reports provide a clear evidence of the inhomogeneous dopant distribution across the grain and grain boundaries in optical ceramics, see Figure 10.

Considering ceramic formation as independent growth of number of individual crystals, we may apply common knowledge on segregation phenomenon that is generally well studied for the crystals grown by conventional melt/flux growth techniques. Therefore, the grain boundaries should be enriched with doping cation when segregation coefficient is $K << 1$, and oppositely the concentration of the dopant on the boundaries will be decreased as compared with bodies of the grains when $K >> 1$. Moreover, if the segregation coefficient of the dopant is close to unity, then the dopant distributes uniformly along all the ceramic material including both the grain bodies and the grain boundaries. As an example, no increased concentration of Yb^{3+} was detected on the grain boundaries of Yb^{3+}-doped Y_2O_3 transparent ceramic. This corresponds well to observed segregation coefficient of about $K = 1$ for the Yb^{3+}-doped Y_2O_3 crystals grown from the melt by micro-pulling-down technique [65, 66]. Thus, prediction of grain boundary segregation can be performed at least qualitatively based on following statement: the lower is the segregation coefficient of the dopant in the melt growth the greater amount of dopand will be collected at the grain boundaries of the ceramic material.

1.4. Characterization Methods

To understand the relation between material defects and the occurrence of traps in the material forbidden gap, their involvement in the energy capture and storage and interconnection with the production technology, one has to employ a set of specific characterization methods closely coordinated with the manufacturing technology. Correlated experiments using the time-resolved luminescence spectroscopy, wavelength-resolved thermally stimulated luminescence and electron spin resonance in an extended temperature interval (10-300 K at least) are very powerful tools for the above purpose. Furthermore, special "one-purpose" setups are used to measure practically important scintillation characteristics, especially light yield or scintillation decay, see also [20]. The former is measured with the help of so called pulse-height spectra using a multichannel analyser and fast photomultiplier or a hybrid photomultiplier in the current regime [67], while the latter is obtained by the time-correlated single photon counting method after the excitation by ^{22}Na isotope (511 keV photons) [68].

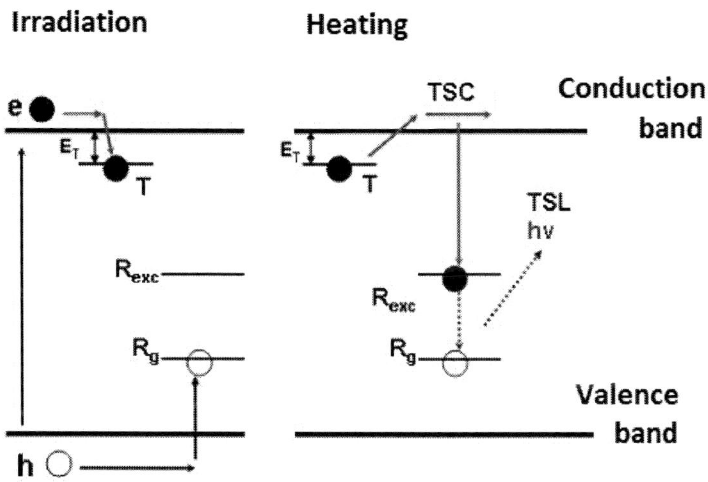

Figure 11. Synthetic sketch of the TSL process, see also [31].

Nowadays, the scintillation decay is also routinely measured using the detection part based on a fast digital oscilloscope and fast photomultiplier in the current regime. Time-resolved UV-visible emission spectroscopy enables detailed understanding of the luminescence center itself, its intracenter transitions and possible nonradiative quenching or excited state ionization phenomena [11, 19]. At the same time its interconnection with the host lattice environment via energy transfer processes can be monitored. Selective pulse excitation in the VUV-UV-visible spectral region completed with X-ray or single photon gamma-ray excitations is used for this purpose. Thermally Stimulated Luminescence (TSL) or thermoluminescence allows the study of point defects giving rise to localized energy levels (traps) in a material forbidden gap. The TSL measurement is a two-stage process, involving at least one electron trap and one hole trap. A very simplified scheme is proposed in Figure 11, see [69] for a systematic description. During irradiation at a given temperature (RT or cryogenic) free carriers are trapped at electron or hole centres. A subsequent heating cycle allows the carriers in shallow traps to be released and recombine radiatively with carriers of the opposite charge stably localized at recombination centres. In the example scheme above, electron traps are emptied while the hole traps act as recombination centres. The emitted light as a function of temperature (glow curve) displays a peak at a characteristic temperature related to the trap structure. If, before recombination, electrons are transferred to the conduction band, a Thermally Stimulated Conductivity (TSC) signal can also be measured. In the simplest first order kinetics model [69] the intensity $I^{TSL}(T)$ of emitted photons is:

$$I^{TSL}(T) = sn_0 \exp(-E_T / kT)x \\ \exp\left[-\frac{s}{\beta}\int_{T_0}^{T}\exp\left(-E_T / kT' dT'\right)\right] \tag{4}$$

where n_0 (cm^{-3}) is the number of traps filled by irradiation, s (s^{-1}) is a « frequency factor », β is the heating rate and E_T (eV) is the trap depth. In the same simplified model, if instead of being heated the sample is held at a constant temperature after irradiation, the mean time spent by carriers in the trap before recombination is:

$$\tau = \frac{1}{s} e^{\left(\frac{E_T}{kT}\right)} \qquad (5)$$

This time governs the phosphorescence which can be spontaneously emitted at constant temperature. According to trap parameters, τ can vary in a very wide interval from ns-μs to hundreds or thousands of years. In suitable cases a detailed analysis of eq. *(4)* or of more general ones by numerical methods [69] results in the determination of the trap parameters (E_T, s). The mean time τ at RT can thus be evaluated, allowing a direct comparison with slow components in time resolved scintillation measurements or correlation with *LY*. Spatial correlation between traps and recombination centres can also be sometimes elucidated. Moreover, more detailed information can be obtained by measuring the wavelength spectrum of the emitted light (see an example in Figure 11). Altogether, while temperature resolution can separate different trapping processes, the measurement of the TSL emission spectrum can evidence different recombination processes when several centres act simultaneously. The nature and local structure of the defects, traps and recombination centers mentioned above can be studied in a detailed manner using electron spin resonance (ESR) and related methods, such as Electron Nuclear Double Resonance (ENDOR) and Optically Detected Magnetic Resonance (ODMR) [70, 71]. All of these methods belong to so-called local methods and exploit the interaction of the magnetic moment of the unpaired electron(s) and nuclei in a defect with each other and with external magnetic fields. In a magnetic resonance experiment, the position of the resonance lines reflects both local symmetry and a degree of lattice distortion around paramagnetic particle.

In addition, the electron-nuclear interaction can give direct evidence for the nuclear (thus also the chemical) species coupled to the paramagnetic particle. This can be illustrated in Figure 12 with an example related to the Pb^+ center in a $PbWO_4$ crystal [72] where an electron is localized at the Pb^+ p_z orbital around an oxygen vacancy. With application of an external magnetic field B_0, the lowest energy level of Pb^+ splits nearly linearly with the field strength. The microwave energy is absorbed by the sample when the frequency of the irradiation is in resonance with the energy splitting. For lead ions with non-zero nuclear spin there is an additional splitting of the energy levels (hyperfine splitting) due to the interaction of electron and nuclear magnetic moments, that gives rise to small intensity satellites around the main resonance. Detection of hyperfine satellites in many cases allows unambiguous determination of the local structure of defects. In order to obtain the maximum amount of information from the ESR experiment it should be performed on a single crystal.

However a great deal of information can still be obtained from experiment on powdered solids or on bulk ceramics. The essential feature of these experiments is that we are now dealing not with selectively oriented single crystal but with a large number of randomly oriented crystallites which all contribute to the spectrum.

This essentially reduces the spectrum resolution of a powder or ceramics. Spectral resolution enhancement for such spatially disordered systems can be achieved by using high frequency microwaves, e.g. W-band (95 GHz) and even mm-wave 263 GHz.

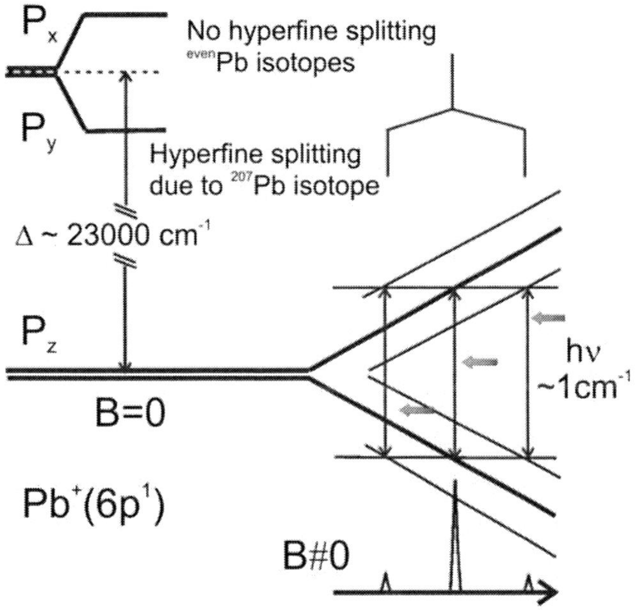

Figure 12. Sketch of an ESR measurement principle, see also [31].

Further improvement of the informativity of ESR experiment is provided by application of advanced pulsed ESR techniques, such as spin echo envelope modulation (ESEEM) spectroscopy and hyperfine sublevel correlation (HYSCORE) spectroscopy [73,74].

In favorable cases one can thus obtain the following information from ESR experiments:

1) symmetry of a defect,
2) concentration of defects,
3) identification of nuclei (atoms) constituting a defect,
4) charge state of paramagnetic ion(s) involved,
5) change of defect charge state under external influences (irradiation, temperature etc.), and
6) dynamic properties of a defect, e.g. decay or growth of the concentration of defects after the irradiation is switched on (off). This last item allows the direct determination of the energy of the local electronic level of a defect created by irradiation and thus can be correlated with TSL characteristics.

Further refinement of the electronic structure of absorption or luminescence centers can be obtained using the technique of optically detected ESR via the magnetic circular dichroism (MCD) [71]. This technique provides a direct correlation between ESR and MCD by the "tagged MCD" method, when one keeps the magnetic field constant on one of the resonance positions and changes the wavelength at which the optical absorption or luminescence is detected.

Therefore, ODMR offers the most compelling interrelation between atomic scale defects and the related optical absorption or luminescence bands.

2. MATERIALS

In Section 2 of this chapter we summarize the results so far reported on scintillating optical ceramics. These results include their preparation and characterization in several material groups, namely the aluminum and multicomponent garnets, silicates, sesquioxides, hafnates, halides, nanocomposites and others. An emphasis is put on the interconnection among the preparation technology, material performance and occurrence of typical defects and their role in scintillation mechanism.

2.1. Aluminum Garnets

The chemical composition of aluminum garnets is $X_3Al_5O_{12}$(XAG), where X is either one or combination of these trivalent ions: Y, Dy, Ho, Er, Tm, Yb and Lu. Determined by the average ionic radius, aluminum garnets can partially consist of RE^{3+}elements which have a higher atomic number than Dy^{3+} (e.g. Tb^{3+} or Gd^{3+}). Single crystal YAG was among the first oxides grown by the conventional Czochralski technique in the 1960s [75]. Single crystal Nd^{3+}-doped YAG manifesting laser oscillations at room temperature was developed in 1964 [76] and good scintillation properties of Ce^{3+}-doped YAG were reported in the 1970s [77]. After these inventions, continuous effort was paid to overcome technological and economic problems in the single crystal growth of laser materials, and the development of transparent ceramics based on YAG appeared to be promising R&D direction. In the 1980s, translucent YAG ceramics were developed and tested [78-80]. Unfortunately, poor laser oscillation was obtained because the material was of low optical grade. Then in the 1990s, the first laser oscillation of ceramic Nd:YAG was reported [81]. In that stage, transparent optical ceramics of aluminum garnet became available for optical applications.

Ceramic Ce^{3+}-doped YAG for fast scintillator applications was reported for the first time in the late 1990s [18, 82]. Unlike laser media, fast scintillators require the decay time of emission centers to be deep in the sub-microsecond range. That is why the Ce^{3+} (Pr^{3+}) activators are the ones mostly used. In those studies, the materials were not transparent but translucent so that the scintillation light output remained about 50% of that of their single crystalline counterpart. In the early aughts, a transparent ceramic Ce:YAG scintillator was developed by Konoshima Chemical, and interestingly, its light output exceeded that of a single crystal [83, 84]. The Ce:YAG emission peaking at about 530 nm allows efficient coupling to the Si photodiode detector. When coupled to the Si avalanche photodiode, the energy resolution at 662 keV was approximately 6% at -20 °C [81,82] and 6.6% at -15 °C [85]. In 2007, a detailed study of optical, luminescence and scintillation properties of ceramic Ce:YAG was reported: it revealed that theY_{Al} antisite defect-related emission around 300 nm and dominant thermoluminescence peak at 92 K associated with a shallow electron trap disappeared; see Figure 13 [86]. It was suggested that the lower preparation temperature of ceramics compared to that of single crystals grown from the melt results in suppression of antisite defects in ceramics.

Theoretical calculations also showed that among defect configurations that can form in the garnet lattice with two isovalent cations the antisite defects require the lowest formation energy [87-89]. Consequently, they become an intrinsic feature in the crystals prepared from the melt. With respect to practical applications, suppression of shallow traps associated with antisite defects would be quite important because related slow emission and energy trapping processes may cause a pileup in applications that have a high counting rate, such as Positron Emission Tomography, and in general deteriorate the scintillator performance [42].

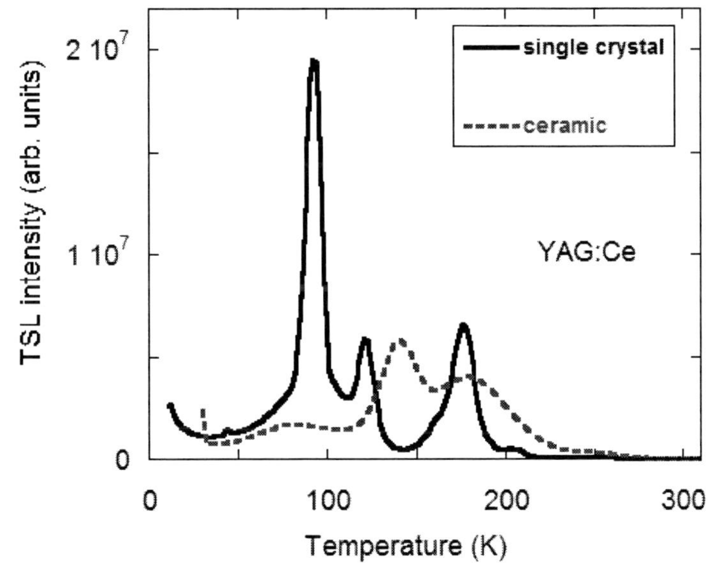

Figure 13. X-ray excited TSL glow curves of single crystal (solid line) and optical ceramic (dashed line), measured after X-ray irradiation at 10 K, see also [86].

Due to low density and Z_{eff} (ρ=4.56 g·cm^{-3}, Z_{eff} = 32.6), YAG cannot be used for detection of high energy photons above several tens of keV because of low stopping power. To overcome this problem, Lu or Gd ions were admixed into YAG. $(Gd_{0.5}Y_{0.5})_3Al_5O_{12}$ doped with Ce^{3+} was developed to acquire higher stopping power than YAG [84, 90]. Photoelectric fraction increased several times with respect to that of YAG [90]. However, fully transparent ceramic sample could not be obtained. Figure 14, shows photographs of Ce:YAG and GYAG manufactured by Konoshima Chemicals [90] and of Ce:GYAG by Lawrence Livermore [91]. When compared with transparent ceramic Ce:YAG (left piece in (a)), other Gd-substituted ceramics are opaque. When coupled to Si avalanche photodiode, the energy resolution at 662 keV was around 8% in GYAG [90]. Based on these results, R and D of the Gd-admixed garnet moved to multicomponent garnet compositions (see Section 2.2). The other way to achieve higher stopping power is to switch the host lattice into LuAG. The density of LuAG is 6.7 g·cm^{-3} and its Z_{eff} = 63. In the single crystal growth from the melt, due to higher melting temperature and smaller difference in ionic radii of Lu^{3+} and Al^{3+}, the problem of Lu_{Al} antisite defects and related energy trapping phenomena are even more serious than in the YAG host [92], see Figure 15. Therefore, ceramic form of this material appeared very attractive to avoid the above mentioned degradation of scintillation performance. In 2005-2008, transparent ceramic Ce-doped LuAG was reported by several groups [91, 93-96]. Similarly to the ceramic Ce:YAG, the emission at 300-350 nm due to antisite defects and shallow electron trap concentration responsible for dominant TSL peaks within 140-190 K in single crystals [97] was comparatively much smaller than in a single crystal Ce:LuAG [95]. However, the pulse height measurements showed that the absolute light yield value of the ceramic Ce:LuAG was noticeably inferior with respect to that of the single crystal Ce:LuAG [91, 95].

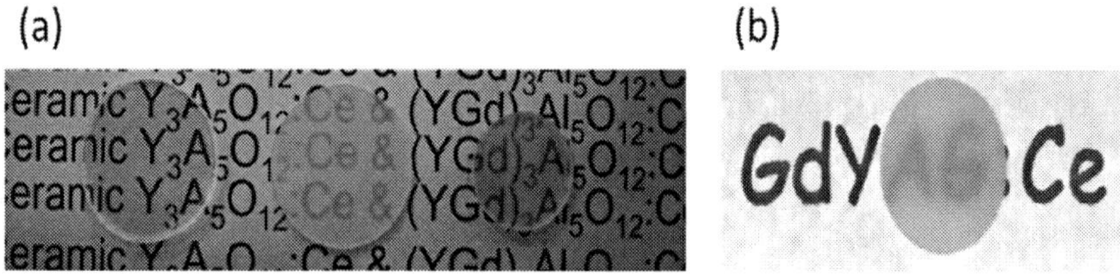

Figure 14. (a) From left to right, Ce:YAG, Ce:GYAG (former) [84, 90], and Ce:GYAG (improved) fabricated by Konoshima. (b) Ce:GYAG synthesized by Lawrence Livermore, see also [91].

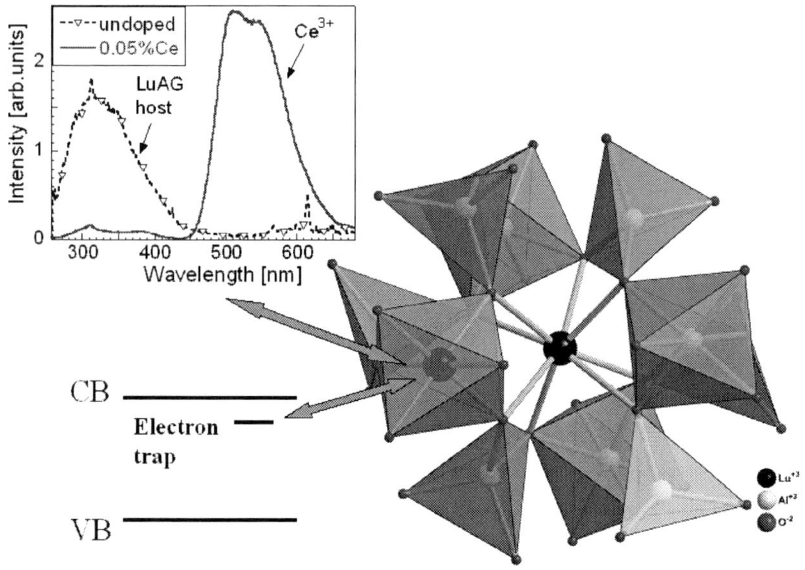

Figure 15. The Lu_{Al} antisite defect in the LuAG structure. Resulting electron trap in the material forbidden gap is sketched on the left. Emission band within 300-350 nm due to antisite defect and its competition with that of the Ce^{3+} center can be derived from radioluminescence spectra at RT - upper left. Emission lines around 312 nm and 615 nm in the undoped sample are due to Gd^{3+} and Eu^{3+} accidental impurities, respectively, see also [92].

In fact, TSL characteristics between 200 and 300 K and enhanced intensity of very slow components in scintillation decay in LuAG:Ce point to the occurence of deep traps situated most probably at the surface and interface of grains in ceramics [95] which are responsible for ligh yield deterioration. After these pioneering studies, the recently synthetized Ce 0.5% doped transparent ceramic Ce:LuAG by Konoshima did show superior light yield compared to that of the Czochralski grown high quality single crystal Ce:LuAG under ^{137}Cs excitation when coupled to Si avalanche photodiode [98]. In the LuAG host, Pr^{3+} doping has also been studied. Due to strong crystal field at dodecahedral Lu site of LuAG, the $5d_1$ level of Pr^{3+} is situated noticeably below the $^1S_0$ 4f one and a fast $5d_1$-4f emission appears. In 2005, a single crystal Pr:LuAG scintillator was discovered and showed a fast (~ 20 ns) and intense $5d_1$-4f emission peaking at 310 nm in the near ultraviolet region [99]. Following this discovery, the

Czochralski growth technique was applied to obtain high quality and larger size crystals [100]. Two-inch diameter crystals became soon commercially available from Furukawa Co. [101]. The most important advantage of Pr:LuAG is its high energy resolution of approximately 4-4.5% at 662 keV [102] which is the best value among heavy scintillators with $Z_{eff} \geq 60$. First report about a transparent ceramic Pr:LuAG was published in 2009 [103] and its light yield was approximately half of that of the single crystal counterpart. Following this study, several groups intensely worked on the development of Pr:LuAG transparent ceramics to achieve higher scintillation light yield with respect to that of a single crystal [104-107], however without any apparent success. An interesting study regarding the final annealing treatment of LuAG:Pr ceramics to restore oxygen stochiometry after the Hot Isostatic Pressing (HIP) cycle has been recently published. Too high annealing temperature apparently causes migration of nanometer pores and their aggregation into bigger ones which remarkably deteriorate transparency of the material [108]. Only very recently, the Pr 0.25 mol% doped ceramic LuAG prepared by Konoshima Chemicals exceeded the light yield value of a high quality single crystal by about 20% [109], as shown in Figure 16. In this ceramic sample, the antisite defect related emission completely disappeared in the radioluminescence spectrum similarly to Ce:YAG and Ce:LuAG ceramics discussed above. In this section, we reviewed R&D results regarding transparent ceramics of aluminum garnets including Ce:YAG, Ce:GYAG, Ce:LuAG and Pr:LuAG.

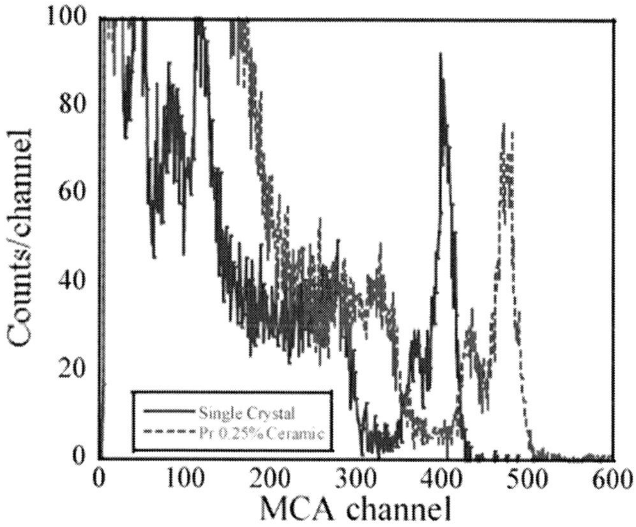

Figure 16.^{137}Cs excited pulse height spectra of single crystal and transparent ceramic Pr:LuAG [109].

Compared to the single crystal counterparts these ceramic materials do not show the host emission due to antisite defects and degrading phenomena based on the associated shallow electron traps. Transparent ceramics are generally sintered at lower temperature with respect to the melting point of each material. It seems to be a critical step to suppress the traps related to antisite defects. The superior light yield with respect to single crystal counterparts, however, also requires the perfection and suppression of trapping states at the grain surfaces and interfaces and that is why only few manufacturers have, so far, been able to achieve this goal. The most recent reports do convincingly show that transparent ceramics of Ce and Pr-doped aluminum garnet scintillators are potentially even superior to their single crystal counterparts in terms of timing characteristics and light yield. The next goal is to reach the industrial stage production.

2.2. Multicomponent Garnets

Multicomponent garnets of general chemical formula $RE_3(Al,Me)_5O_{12}$, where RE= Gd,Y,Lu or their mixture and Me= Ga, Sc, differ from aluminum garnets mainly in the occupancy of the octahedral or possibly tetrahedral sites where Al is partially substituted by Ga or Sc. In general, due to this substitution a forbidden gap shrinks and a composition of the bottom of the conduction band profoundly changes [110-112], although the detailed and systematical band structure calculations have not been published yet. In the period of R&D of slow scintillation ceramics the Cr-doped $Gd_3Ga_5O_{12}$ (GGO) has been already mentioned in *Section 1* [17]. It is an effective host for hard X-ray detection due to elevated density of 7.1 g/cm^3 and high Z_{eff} = 53.4. However, little attention has been paid in that time to Ce or Pr-doped materials of this kind with fully Ga-substituted Al site, because it was reported a long time ago that Ce^{3+} does not show any luminescence in $Y_3Ga_5O_{12}$ due to positioning of its 5d level in the host conduction band [113]. Scintillation of powder solid solutions $Y_3(Al_{1-x}Ga_x)_5O_{12}$ doped with Ce^{3+} and Pr^{3+} was studied under X-ray or accelerated electron excitations and the emission intensities were found still reasonably high for x = 2-3 for both dopants [114]. $(Gd,Y)_3(Ga,Al)_5O_{12}$:Ce powder phosphors have recently been extensively studied for the application as white LED phosphors. The color in such a white LED source can be tuned by manipulating the chemical composition that shifts the Ce^{3+} 5d-4f band position [115,116]. The crucial point in the study of the Ga-admixed Pr:$Lu_3(Ga_xAl_{1-x})_5O_{12}$ scintillator was reported in 2006: it was found that shallow electron traps in single crystals grown by the micro-pulling down method are substantially suppressed when the content of Ga is within x = 2-3. Consequenly, their scintillation response manifested significantly lower intensity of slower decay components [117]. Later, a refined study completed with DFT calculations of the band structure [112] has shown that also the thermal depth of deep electron traps monitored in TSL above RT in such material decreases with the Ga content even in quantitative agreement with the lowering of the conduction band edge. Combinatorial study of the family of $(Gd,Y)_3(Ga,Al)_5O_{12}$:Ce and $(Gd,Lu)_3(Ga,Al)_5O_{12}$:Ce single crystals prepared by the same technique [118, 119] then lead to a discovery of a new ultraefficient scintillator $Gd_3Al_2Ga_3O_{12}$:Ce. The red shift of $5d_1$ level of Ce^{3+} due to the Gd-induced stronger crystalline field at the dodecahedral site ensured both the sufficiently high energy barrier of an unwanted Ce^{3+} excited state ionization and efficient and fast energy transfer toward Ce^{3+} centers provided the Ga content was sufficiently high. This composition proved feasible also for the Czochralski growth. Two inch diameter crystal with the light yield of about 46 000 $ph{\cdot}MeV^{-1}$ and energy resolution of 4.9%@662 keV was already grown[120].

Analogous R&D activities concerning the Ga substitution in Ce:$(Gd,Y)_3(Al,Ga)_5O_{12}$ (Ce:GYGAG) optical ceramics have been performed by the Lawrence Livermore group [121]. Transparent ceramic Ce:GYGAG scintillators with high scintillation performance, namely the light yield of 50000 $ph{\cdot}MeV^{-1}$ and energy resolution of 4.5-4.8% at 662 keV, were reported [122,123]. In contrast to other Ce-doped ceramic garnets which in general manifested the principal scintillation decay time of several tens of ns, in the reported

Ce:GYGAG scintillator its value was 250 ns. This phenomenon was interpreted as due to energy migration over the Gd^{3+} sublattice followed by an energy transfer to Ce^{3+} [123]. In fact, also single crystals of $Ce:Gd_3Al_4Ga_1O_{12}$ and $Ce:Gd_3Al_3Ga_2O_{12}$ compositions showed scintillation decay time of 316 ns and 221 ns, respectively, [118] which is fully consistent with the decay time of ceramics. The value does point to somewhat slower energy migration towards the Ce^{3+} emission center whose photoluminescence decay time in these hosts is about 50-55 ns [118,119]. Apparently, this somewhat slower scintillation response can be improved by fine tuning of the chemical composition. An interesting observation was also reported for $Gd_3(Al,Ga)_5O_{12}$:Ce regarding the relation between the afterglow and stoichiometry of the host [124]. In particular, the afterglow substantially diminished for a small excess of Gd cation with respect to a stoichiometric composition. Formation of isoelectronic traps was discussed based on the stoichiometry of the material and difference in the electronegativity of the dopant and host ions. Regarding transparent ceramics, the $Ce:(Gd,Y)_3(Al,Sc)_5O_{12}$ composition family with various Sc content (0-40 mol% with respect to Al) was also prepared and characterized [125]. However, the light yield was only about a half of that of Ce:GYAG ceramics described in *Section 2.1* and manufactured in the same laboratory. Apparently, transparent ceramic and single crystal multicomponent garnets show similar luminescence and scintillation characteristics and are very promising for X- and gamma-ray detectors due to their high scintillation efficiency and especially light yield values. However, further investigations are required to profoundly understand the scintillation mechanism and underlying energy transfer processes and to further optimize these complex material systems.

2.3. Silicates

Scintillation characteristics of the Ce-doped rare-earth (RE) oxyorthosilicate were reported for the first time almost 30 years ago. After the first study revealing promising properties of Gd_2SiO_5:Ce (GSO) [126], other studies of high density orthosilicate Lu_2SiO_5:Ce (LSO) [127] and later also pyrosilicate $Lu_2Si_2O_7$:Ce (LPS) [128] materials appeared. The Ce-doped mixed LYSO [129], LGSO [130] and GYSO [131] compositions were studied as well. Due to their intended applications, e.g. in medical imaging, single crystals were predominantly grown: the Czochralski technique was applied using an iridium crucible, which is necessary due to very high melting point of these compounds (ranging between ~1950-2150 ^{0}C) [132]. Optimization of all the manufacturing process enabled the growth of large single crystals (approximately 70-80 mm diameter and 200 mm long) with very good and reproducible scintillation performance [133]. GSO possesses the monoclinic *P2₁/c* crystal structure, while LSO and YSO possess the monoclinic *C2/c* structure [134]. The LSO structure is depicted in Figure 17. In particular, there are two Lu sites with seven (Lu1) and six (Lu2) oxygen neighbors, the structure is built by SiO_4 tetrahedra and O anions unbound to Si surrounded by four Lu cations in a distorted tetrahedron. Consequently, in Ce-doped samples there are two emission bands belonging to Ce^{3+} at Lu1 and Lu2 sites [127]. A relative occupancy of Ce^{3+} at Lu1 site is about 90% [135] because this site is somewhat larger. It was determined that unbound oxygen sites are the most likely to give rise to oxygen vacancies [136] which become deep electron traps causing unwanted afterglow effects in LSO:Ce single crystal scintillator [137].

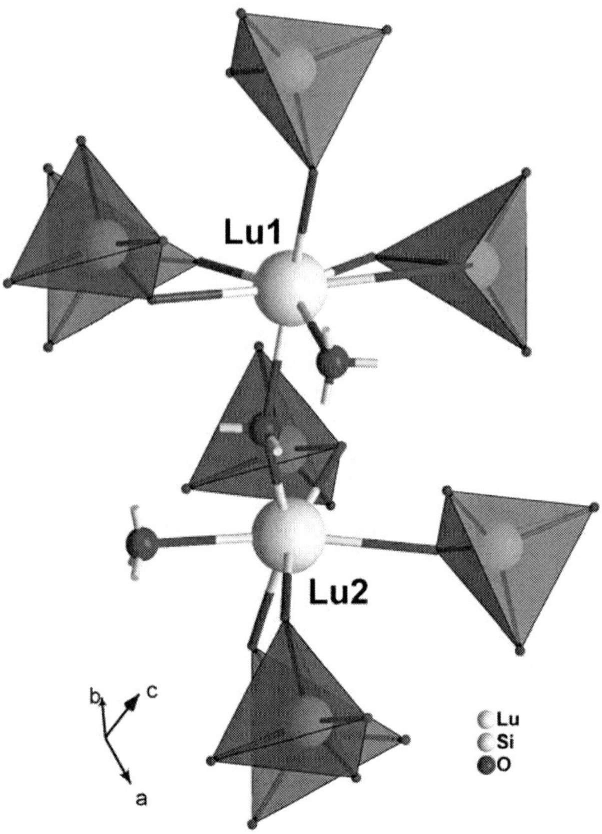

Figure 17. The structure of LSO with two Lu sites and surrounding SiO_4 tetrahedra. Three oxygen ions with cut bonds form OLu_4 tetrahedra and do not participate in Si-O bonds.

Despite the distinctly non-cubic, low symmetry structure of LSO several groups tried to prepare a ceramic version of LSO:Ce scintillator because the single crystal technology is very demanding and expensive. The first reports were published in 2007-2008 period [138,139]. Consolidation of ceramics was achieved by hot pressing at temperatures up to 1700 °C and pressure of 8000 psi. Hot pressing causes a loss of oxygen resulting in a strong coloration of ceramics, which had to be removed by annealing in air at approximately 1100 °C. The final specimens were colorless and highly translucent, but the anisotropic nature of the crystal structure precluded achievement of full transparency [139]. In [138] the scintillation light yield of a LSO:Ce ceramic reached about 86 % of that of a good LSO:Ce single crystal, its decay time was even faster, and the long term afterglow lower than in LSO single crystals. Combining nanotechnology with a sinter-HIP approach [140] it was found that the densities of the LSO ceramics increased with increasing sintering temperature, which corresponds to a systematic decrease of porosity as confirmed by a SEM examination. After sintering, the residual pores were found to segregate at grain boundaries and were essentially removed by subsequent HIP procedure.

The process increased the density to practically the value characteristic of a single crystal and considerably improved transparency. Under excitation by a ^{22}Na source the light output as high as 30,100 ph·MeV^{-1}, energy resolution of 15% (FWHM) at 662 keV (^{137}Cs source) and principal decay time of 40 ns were obtained. Also the spark plasma sintering (SPS) technique was employed to manufacture the cerium-doped lutetium orthosilicate polycrystalline scintillation ceramics starting from nanosized Ce:LSO powders synthesized by the sol-gel processing [141]. Fully-densified polycrystalline ceramics with fine grains were fabricated under optimal sintering conditions at 1350°C for 5 min under pressure of 50 MPa. Although some of reported results regarding scintillation parameters of LSO:Ce ceramics are encouraging, apparently, the insufficient transparency due to optical anisotropy of the material structure is a serious problem which inhibits its usage in the form of bulk optical elements. Further research is clearly needed to still optimize technological procedure and to study the energy transfer and trapping at grain surfaces and interfaces in more detail.

2.4. Sesquioxides

Sesquioxides have a chemical composition RE_2O_3, where RE stands for rare earth elements and includes also yttrium and scandium. Similarly to other transparent ceramics, pioneering research was done for a cubic Y_2O_3 for laser media [142] to avoid a phase transition at 2280 °C in a single crystal (melting point ~ 2400 ±C). For scintillator applications, Re = Sc, Y, La, Gd, Lu and their mixtures of cubic structure, further suitably doped, have so far been reported. Unlike other hosts, Ce^{3+} or Pr^{3+} doping are not attractive for sesquioxides, because their 5d excited state is situated in the conduction band. In such a case, the 5d electron becomes delocalized in the conduction band and the 5d-4f transition based luminescence is fully quenched even at low temperatures [143]. Due to this photoionization effect, Ce- or Pr-doped sesquioxides do not show the 5d-4f emission (e.g., $Ce:La_2O_3$ [144]). As mentioned in the introductory part the ceramic version of sesquioxide, namely the Eu^{3+}-doped $(Y, Gd)_2O_3$ was developed and commercialized by General Electric [145] for X-ray computed tomography (X-ray CT). Unlike Positron Emission Tomography (PET), X-ray CT accumulates signals with several tens of ms integration time (integrated-mode), so that a scintillation response even of several ms, which is the case of Eu^{3+}-doped $(Y, Gd)_2O_3$, is fully sufficient. The emission maximum of $Eu:(Y, Gd)_2O_3$ is at 610 nm which is well suited to the photodiode readout. In CT, the afterglow is a matter of concern and an optimization of this parameter in $Eu:(Y, Gd)_2O_3$ was achieved by the Pr^{3+} co-doping [146]. In $Eu:(Y, Gd)_2O_3$, the Eu^{3+} activator acts as an electron trap and the afterglow of this material is attributed to the trapping of mobile holes at negatively charged defects such as oxygen interstitials which causes the delay in their delivery to the Eu emission centers [147]. Pr^{3+} codopant competes for the excess mobile hole capture and a nonradiative recombination in Pr^{4+}–Eu^{2+} pairs brings the material back to the ground state without the delayed radiative recombination effects. Pr co-doped $Eu:(Y, Gd)_2O_3$ is called HiLight™ (GE) and is commonly used in X-ray CT machines.

The Tb^{3+} codoping to $Eu:(Y, Gd)_2O_3$ was reported to cause similar effect as Pr^{3+} [17]. Following the success of $Eu:(Y, Gd)_2O_3$,the $Eu:Lu_2O_3$ has been recently studied due to its higher X-ray stopping power: its density 9.4 g/cm^3 and Z_{eff} = 69 makes it an excellent candidate for this purpose. A ceramic powder $Eu:Lu_2O_3$ was first introduced in 2002 and similar emission properties to those of $Eu:(Y, Gd)_2O_3$ were observed [148]. Transparent ceramic $Eu:Lu_2O_3$ was fabricated by Radiation Monitoring Devices, Inc (RMD) and the imaging capability coupled with the charge coupled device (CCD) was examined [149,150]: the X-ray transmission images of Eu:Lu_2O_3 transparent ceramics showed superior 2D-spatial resolution with respect to Pr,Ce,F:Gd_2O_2S translucent ceramics. Transparent ceramic Eu:Lu_2O_3 was also prepared from nano-powder and characterized [151]. Similarly to Eu:(Y, Gd)$_2$O$_3$ the afterglow effect deteriorated performance of Eu:Lu_2O_3 and its mechanism was studied [152,153]. In an attempt to improve optical efficiency and decrease the afterglow, the effect of aliovalent doping was theoretically studied in sesquioxide structures [154]. Atomistic scale computer simulation has been used to predict the defect processes associated with the solution of extrinsic divalent and tetravalent ions. These calculations provide a mechanistic framework with the help of which one can identify how specific doping schemes modify the populations of defects affecting a scintillator performance. Very recently, Eu^{3+}-doped (Lu,Gd)$_2$O$_3$ was synthesized and its scintillation responses were compared with Eu:Lu_2O_3 [155]. Partially substituting Gd for Lu, the afterglow was dramatically suppressed. Therefore, Eu^{3+}-doped (LuGd)$_2$O$_3$ might become a preferred candidate in CT application, because of its superior density and effective atomic number.

In the family of transparent ceramic sesquioxide, the Yb^{3+} doped Re_2O_3 have recently been discovered as ultrafast scintillators [156] and Yb:Lu_2O_3 might become a suitable material for the Time-of-flight (TOF) PET application. The Yb^{3+}center in Re_2O_3 manifests the charge transfer (CT) luminescence in the near UV-blue spectral region [157, 158]. This emission is intrinsically fast because it is inhibited neither by parity nor spin selection rules. CT luminescence was systematically studied in Yb doped aluminum garnets and perovskites [159, 160]: its radiative lifetime at 10 K is about 100 ns, but at RT is severely quenched. So far no detectable scintillation light yield has been reported.

Temperature dependence of both the Yb^{3+} CT luminescence intensity and decay time in a Lu_2O_3 ceramic host are presented in Figure 18. Radiative lifetime is about 86 ns (T = 8 K) and a shortening of the decay time as well as emission intensity reduction with increasing temperature is a bit less pronounced when compared to an Yb-doped YAP [161, 162]. Scintillation responses of transparent ceramic of Yb doped Sc_2O_3, Y_2O_3, Lu_2O_3, and Yb_2O_3 were satisfactorily measureable. Under ^{137}Cs (662 keV) excitation, all of them except Yb_2O_3 exhibited a clear photo-absorption peak [156]. Figure 19 displays the ultra-fast decay curve of Yb:Lu_2O_3 under the ps X-ray pulse excitation [163].

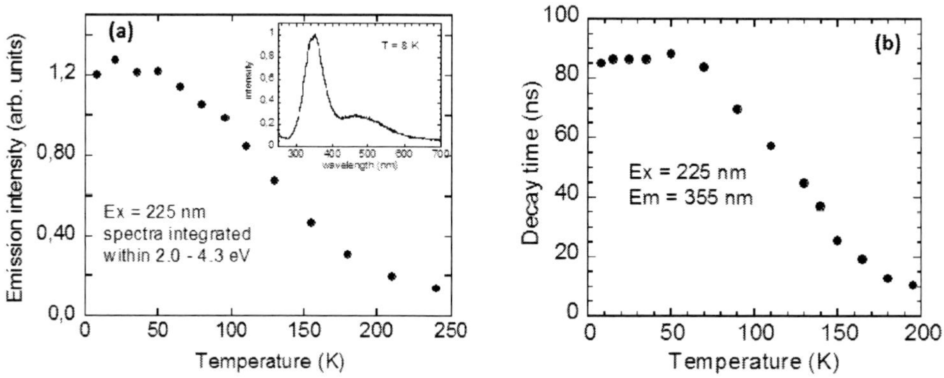

Figure 18. Temperature dependence of (a) emission intensities (spectrum integrals) and (b) decay times for Lu_2O_3:Yb 0.3% ceramics. In the inset of (b) the PL spectrum is shown at 8 K.

Corresponding decay time is about 1 ns. In the high count rate applications like PET, not only the decay time but also a time resolution is an important parameter. Figure 20 shows the time resolution of $Yb:Lu_2O_3$ compared with the conventional $Ce:Lu_2SiO_5$ (Ce:LSO) single crystal scintillator [156] which is nowadays widely used in PET. The time resolution of $Yb:Lu_2O_3$ was two times better than that of Ce:LSO. At present stage, the light yield of this material is not sufficiently high, nevertheless there is a considerable space for optimization of both the fabrication process itself and Yb^{3+} concentration.

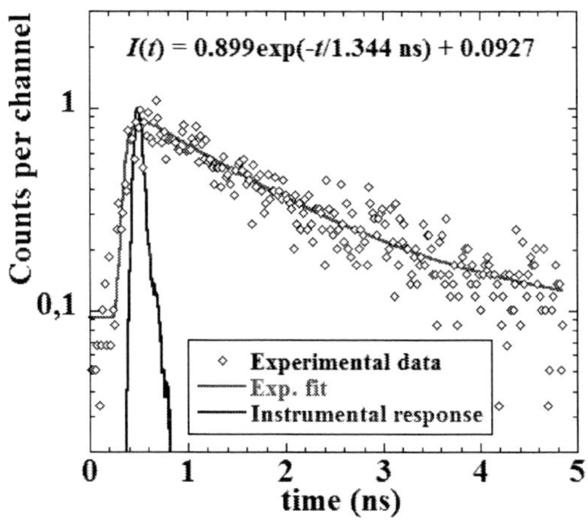

Figure 19. Scintillation decay of $Yb:Lu_2O_3$ ceramics [156]. Excitation by ps X-rays. Solid line is the convolution of instrumental response and the function *I(t)* in the figure.

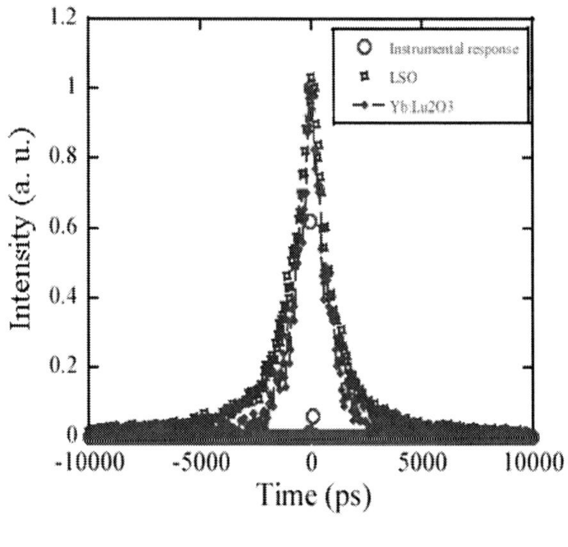

Figure 20. Time resolution of $Yb:Lu_2O_3$ and Ce:LSO.

Undoped sesquioxides are also of interest. Unlike some reports about undoped sesquioxide single crystals in a recent literature [164], there are only few investigations of transparent ceramics. In the early aughts, luminescence properties of ceramic Sc_2O_3 and Y_2O_3

were studied [165]. The intrinsic emission mechanism of a single crystal Y_2O_3 was determined as due to the self-trapped exciton (STE) but in Sc_2O_3 the interpretation was unclear [166]. In 2010, scintillation properties of a transparent ceramic Y_2O_3 were evaluated and interestingly, the emission wavelength blue-shifted (350 nm) with respect to a single crystal (370-385 nm) [167]. Further studies are required to clarify such a difference. The light yield of an undoped Y_2O_3 ceramic was relatively high, 9300 ph·MeV^{-1},and the decay was fast, with dominant decay times of 34 and 107 ns.

Apparently, regarding a scintillation response of transparent ceramic undoped sesquioxides there is a large space for further study and understanding of scintillation mechanism, especially in Sc_2O_3 and Lu_2O_3.

In this section, we provided an outline of the characteristics of doped and undoped transparent ceramic sesquioxide scintillators. Eu^{3+}-doped sesquioxides have reached the industrial stage and the Yb^{3+}-doped ones started to be studied for the high counting rate applications. To clarify the nature of the luminescence and scintillation mechanism of undoped sesquioxides, further research is necessary. From the industrial point of view, only ceramic sesquioxides are suitable for mass production because single crystal materials are difficult to manufacture due to the very high melting point of 2400-2500 °C. By contrast, transparent ceramics are made at lower temperatures and a fabrication cost can decrease to reasonable levels.

2.5. Hafnates

Alkaline earth hafnate perovskites $AHfO_3$ have been investigated for more than 70 years [168]. At room temperature (RT), barium hafnate has ideal cubic perovskite structure (space group Pm3m), while $CaHfO_3$ and $SrHfO_3$ have orthorhombic structures. The structure of $SrHfO_3$ has been refined at RT as orthorhombic Pnma [169,170]. Raman spectra of $BaHfO_3$, $CaHfO_3$ and $SrHfO_3$ powders are reported in [171]. Research interest in hafnate materials is based on a number of potential applications. Their high dielectric permitivity (high k) makes them candidates for replacing conventional SiO_2–based gate dielectrics in silicon-based metal-oxide semiconductor (MOS) electron devices. This feature would suppress undesired electron tunnelling [172,173]. The hafnate $SrHfO_3$ has thermal and mechanical properties that suggest that it may be useful for thermal barrier coatings [174]. Yet another potential application concerns alkaline earth hafnates for microwave ceramic resonators. These are important components in wireless communication devices and were considered in [175]. High density of $BaHfO_3$ (8.5 g·cm^{-3}) and $SrHfO_3$ (7.56 g·cm^{-3}) as well as their high effective atomic number make them promising hosts for scintillating materials when activated with rare earths. Blue emitting phosphor $SrHfO_3$: Tm^{3+} is reported in [176].

$SrHfO_3$:Ce^{3+}, discovered as an efficient phosphor [177], is at present being intensely investigated. Its Ce^{3+} luminescence peaks around 400 nm. Compared to BGO it has twice the light output, and its scintillation response is five times faster at a similar cost. Because of its high melting point (>2400 °C) it is not available as a single crystal. Possible alternatives include transparent ceramics or nanocrystals embedded in an inert host. Stability of Ce^{3+} activators, an important feature when converting hafnate powders to transparent ceramics, as well as codopant effects in $MHfO_3$ (M=Ba, Sr), are reported in [178]. Scintillation properties of Ce-doped $SrHfO_3$ and $BaHfO_3$ ceramics obtained from powders prepared by solid-state synthesis were studied in [19]. However, only a few fully transparent samples were obtained and the significantly reduced light yield compared to radioluminescence (RL) yield was probably due to large fraction of slow components. Ceramics of $BaHfO_3$:Ce were sintered from powders prepared by a modified Pechini technique [180]. The powders were of complex morphology with a broad distribution of crystallite sizes and high agglomeration.

Consequently, the vacuum sintering produced practically totally opaque pellets. Ce-doped $SrHfO_3$ ceramics can be made transparent by carefully controlling the stoichiometry of the precursor powders [181], see Figure 21. Ceramics close to stoichiometry (Sr/Hf ~1) appear to show good transparency and a reasonable light output several times that of calibrated BGO crystals. The 'contact' and 'distance' transparencies of ceramics hot pressed at about1450 °C are very good, but deteriorate at increasingly higher hotpress temperatures.

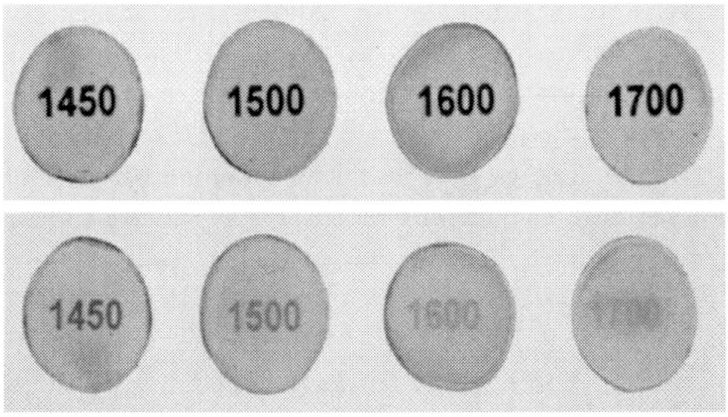

Figure 21. Transparency of disks hot pressed at indicated temperatures (°C). In the upper picture the disks are in contact with the back-lit text, while in the lower they are 1 cm above the text. Note that image degradation is evident only at a distance, illustrating the inadequacy of contact transparency, reprinted with permission from [181].

Combustion synthesis to prepare nanocrystals of $SrHfO_3$ was developed in [182, 183]. The emission intensity of Ce doped samples was increased by Li codoping [182]. Ce-doped $SrHfO_3$ elaborated via the sol-gel route is presented in [184,185]. Optical, structural and morphological characteristics of $SrHfO_3$ powder samples prepared by acetate and citrate combustion and solid state reaction were reported in [186,187] and [188], respectively. In addition to Ce^{3+}-doped $SrHfO_3$, a novel potentially interesting scintillator, Pb^{2+}-doped $SrHfO_3$ was also introduced. Its RL efficiency was slightly higher than that of $SrHfO_3{:}Ce^{3+}$ while photoluminescence (PL) decay time at RT was about 190 ns, comparing to the remarkable 15 ns decay time of $SrHfO_3{:}Ce^{3+}$. Such short Ce^{3+} decay time was soon understood as due to thermally induced ionization of the Ce^{3+} excited state in the $SrHfO_3$ host evidenced in [189]. The onset of ionization appears somewhat below room temperature and shifts with increasing Ce concentration to higher temperatures. A fast efficient luminescence at 334 nm from a defect center was reported in a nonstoichiometric Sr deficient $SrHfO_3$ sample [190]. This new phosphor shows remarkable scintillation efficiency of about ten times of that of BGO, decay time of about 170 ns at RT and onset of thermal quenching at about 400 K. To better understand the underlying scintillation mechanisms in the $SrHfO_3$ host, the RL emission spectra of undoped $SrHfO_3$ together with corresponding PL emission and excitation spectra were thoroughly analyzed and the observed RL bands were tentatively associated with excitonic or defect related emissions. The similar analysis performed for $SrHfO_3$ doped by Ce^{3+} or Pb^{2+} luminescent activators demonstrates the competition between activators and intrinsic or defect centers in the carrier capture [191]. Some theoretical work was also done. Density functional theory was used to study electronic properties of the cubic structure of strontium hafnate [192] and barium hafnate [193] as well as structural, elastic and mechanical properties of orthorhombic strontium hafnate under pressure [194]. Rare earth hafnates can crystallize in two different structures. $RE_2Hf_2O_7$ compounds crystallize in the disordered fluorite structure [195], with the space group Fm3m. They can also appear as interesting scintillating materials, in particular, due to their significantly high density, approximately 9.4 $g \cdot cm^{-3}$ in the case of $Lu_2Hf_2O_7$. The synthesis of $Y_2Hf_2O_7$ and $Lu_2Hf_2O_7$ powders by nitrate solution combustion is described in [196], while Tb^{3+} -doped $Lu_2Hf_2O_7$ ultrafine powders prepared by a solution combustion process are reported in [197]. Transparent ceramics of $Lu_2Hf_2O_7$:Ce were prepared in [198]. The RT radioluminescence spectrum showed a broad emission band due to Ce^{3+} emission peaking at 400 nm. The light yield was low, about 1000 $ph \cdot MeV^{-1}$. The principal scintillation decay time was about 20 ns.

There is disagreement in the literature concerning $La_2Hf_2O_7$:Ce as it has been reported to show both bright [199] as well as no Ce-activated luminescence [200,201]. Recent first-principles electronic structure calculations of $RE_2M_2O_7$ (RE=Y, La; M=Ti, Zr, Hf) doped with Ce using the pseudopotential method based on the generalized gradient approximation in density functional theory [202] determined the positions of the Ce 4f states relative to the top of the valence band and the position of the Ce 5d states relative to the bottom of the conduction band of the host material. For all these compounds, it was found that the Ce 5d state lies above the bottom of the conduction band which would prevent any luminescence from the Ce site. An alternative structural phase $RE_4Hf_3O_{12}$ has a rhombohedral structure with the space group R-3. Optical properties of undoped and Eu^{3+},Tb^{3+}, Ce^{3+}, Pr^{3+} and Bi^{3+} –doped $Lu_4Hf_3O_{12}$ prepared by the solid state reaction method were studied in [203]. 5d-4f emissions of Ce^{3+} and Pr^{3+} were absent which is probably also due to localization of the 5d states in the conduction band of the host. Nanosized $Lu_4Hf_3O_{12}$ powder doped with the same RE^{3+} ions as in [203] was also prepared by acetate-citrate gel combustion [204]. Although recent results, both experimental and theoretical indicate that rare earth hafnates do not represent the most suitable route for developing fast scintillating materials, alkaline earth hafnates, especially the mentioned undoped nonstoichiometric compositions, remain the class of materials worth pursuing further research effort.

2.6. Halides

In this section, halide transparent ceramic scintillators are introduced. Materials based on fluorides, chlorides, bromides, iodides and their mixtures will be reviewed. While quite some of them have a cubic structure, even those with lower symmetry often manifest low optical anisotropy [205] being of great advantage as far as their preparation in ceramic form is considered. Within fluoride scintillators, the cubic BaF_2 is one of the well-known materials due to its fast core-valence luminescence [37]. Ceramic BaF_2 powder was examined in the 1990s and the near UV emission, assigned to the self trapped exciton, was observed [206].

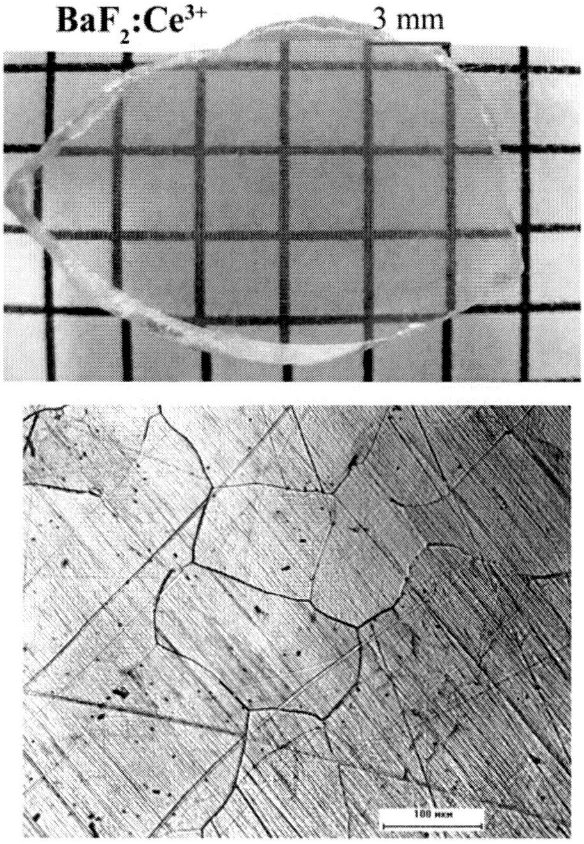

Figure 22. (a) Exterior view and (b) microstructure of BaF_2:Ce^{3+}($Ba_{0.99}Ce_{0.01}F_{2.01}$) ceramics, reprinted with permission from [207].

Recently, the Ce^{3+} doped transparent ceramic BaF_2 bulk material was successfully obtained and its luminescence properties were compared with CeF_3 single crystal [207]. Figure 22 shows a photograph of Ce:BaF_2 [207]. Then in 2010, a transparent ceramic pure BaF_2 and Ce:BaF_2 were fabricated and their scintillation properties were studied [208]. Fast core valence luminescence and slow self trapped exciton luminescence were clearly observed as well as the Ce^{3+} 5d-4f emission in Ce-doped ceramics. Recently, BaF_2 crystals and ceramics with essentially higher light yield in the fast VUV-UV component of emission spectra have been reported [209]. The light yield in the fast component increased by a factor of 2 to 2.5 in crystals both annealed in the CF_4 atmosphere and doped by scandium at a concentration from 0.5 to 2.0 mol%. In ceramic BaF_2 doped by Sc, an increase was by a factor of 1.7 to 2.5. The light yield of the 0.9 ns decay component can be obtained between 3000 to 4000 ph·MeV^{-1}. In addition to BaF_2, CaF_2 is also a candidate for transparent ceramic scintillator because it has a cubic structure and shows a high light yield when doped with Eu^{2+} [210]. In the late aughts, transparent ceramic CaF_2 became frequently reported. The Yb^{3+}-doped CaF_2 was successfully synthesized for solid state laser [211]. In the same year, nanocrystallized glass ceramic of Eu^{2+}-doped CaF_2 was also developed and it manifested 30% higher light yield than that of the single crystal counterpart [212]. Optical properties of transparent ceramic of Eu:CaF_2 were also studied in [213]: both the Eu^{2+} 5d-4f emission at

420 nm and Eu^{3+} 4f-4f emission around 500-600 nm were observed. Undoped transparent ceramic CaF_2 is also intensely studied like a passive element for optical applications [214]. In 2005, Radiation Monitoring Devices developed $CaF_2:Eu^{2+}/^6LiF$ neutron sensitive composite scintillator [215]. In this material, neutrons are stopped within the ^6LiF grains due to $^6Li(n,\alpha)^3H$ reaction. Created alpha particles deposit their energy partly also in Eu:CaF_2 grains and as a result, a scintillation in Eu^{2+} emission centers appears. The phenomenon was proved by direct detection of thermal neutrons.

Together with fluorides, also chloride and bromide scintillators can be used in the γ-ray spectroscopy. $Ba_{1-x}La_xCl_{2+x}$ is one of the examples of transparent ceramic chloride scintillator. It was developed in 2008 and showed 4000 ph·MeV^{-1} light yield under γ-ray excitation with 100 ns decay time [216]. Very recently, R&D of transparent ceramic Ce:$LaBr_3$ has been reported [217,218]. Single crystal Ce:$LaBr_3$ shows a high energy resolution below 3% at 662 keV [219] with the disadvantage of both the high hygroscopicity and cost of the fabrication process. Thus, a transparent ceramic counterpart is expected to reduce the fabricating costs. Transparent ceramic Ce:$LaBr_3$ has been synthesized by the hot wall evaporation (HWE) method which is a well known vapor deposition technique [217,218]. It showed inferior energy resolution (8.5% at 662 keV) but higher LY (86000 ph·MeV^{-1}) with respect to the high quality commercial single crystal [218]. Therefore, transparent ceramic Ce:$LaBr_3$ has a high potential as a scintillator and requires further research. New cubic materials from elpasolite family (A_2BLnX_6; A,B-alkali metal, Ln-lanthanide and X-halogen) hold promise for scintillator materials due to their high light output, proportionality, and toughness as well as a possibility to detect thermal neutrons when B=Li [220]. The isotropic nature of the cubic elpasolites leads to minimal thermomechanical stresses during the single crystal solidification, and eliminates the problematic light scattering at the grain boundaries. Therefore, it may be possible to produce these materials in large sizes as either single crystals or transparent ceramics with both the high production yield and reduced costs. Synthesis and the structural variations of four different compounds, including $Cs_2NaLaBr_6$, $Cs_2LiLaBr_6$, Cs_2NaLaI_6, and Cs_2LiLaI_6 were studied and the attempts to produce a large-area detector by a hot forging technique were explored [221]. In iodides, cubic Tl-doped CsI is one of the first single crystal scintillators which became used in the late 1940s. It is explored till today due to its high light yield of 61000 ph·MeV^{-1} [222] and low price, despite its medium decay time of 1 μs and low density of 4.51 g/cm^3. Transparent ceramic (polycrystal) Tl:CsI was studied in the 1990s, fabricated by Oak Ridge National Laboratory in United States. They synthesized ceramic Tl:CsI from a powder that was compressed uniaxially at 100 MPa at a temperature of 150 °C (hot pressing technique) [223-225]. In these papers, the radioluminescence and transmittance in infrared wavelength range were investigated.

In scintillator studies including both single crystal and ceramic halides, the most recent activity is focused on Eu^{2+}-doped SrI_2. The material itself was first patented in 1968 [226], but it was not further developed until very recently. Numerous reports about single crystal Eu:SrI_2 were published starting from 2008 [227,228]. As reported, it showed an excellent light yield of 100000-120000 ph·MeV^{-1} with high energy resolution below 3% at 662 keV. The energy resolution was similar to that of Ce:$LaBr_3$ mentioned above. Therefore, this material attracts a lot of interest as an alternative to classical Tl:NaI scintillator though its principal decay time is somewhat longer, about 1-1.5 μs and it suffers from re-absorption effects in bulk elements due to the small Stokes shift. It has a cubic structure and transparent ceramic Eu:SrI_2 became investigated recently [229].

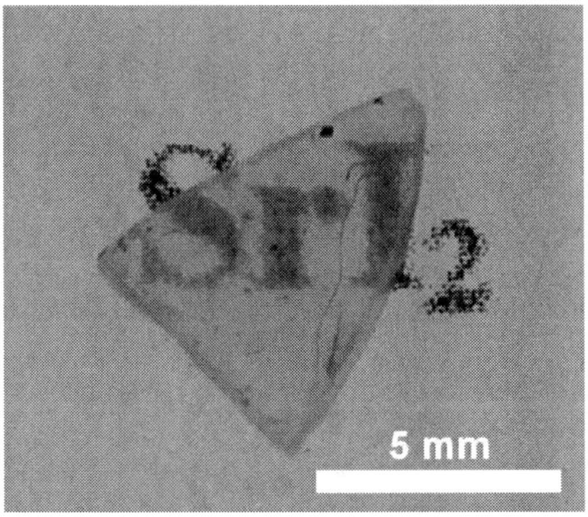

Figure 23. Photograph of ceramic Eu:SrI_2, reprinted with permission from [227].

Eu-doped and undoped SrI_2 was fabricated by hot-pressing and translucent samples were obtained. Figure 23 shows a photograph of ceramic SrI_2 [229]. At this very early stage, the light yield of the ceramic Eu:SrI_2 is around 20000 ph·MeV^{-1}which is about a fifth of that of the single crystal [229]. However, scintillation properties can certainly be improved by an optimization of fabrication technique.

In this section, a survey of activity in the field of transparent ceramic halide scintillators has been provided. Activities in this field have started just a few years ago. Ceramic technology apparently offers a great possibility for manufacturing scintillators for practical applications related especially to the homeland security-oriented sector.

2.7. Nanocomposite Materials

Alternative approach in developing new scintillation materials consists in embedding the dispersed scintillator nanophase into an optically isotropic host which gives rise to so called nanocomposite scintillators. Since for sufficiently small nanocrystals (less than 20–30 nm) dispersed in an optically isotropic host the classical laws of light scattering are no longer valid it should be possible to obtain transparent bulk systems even with non-cubic scintillator nanophase [49, 228]. Some trials to create such "crystal-in-crystal" systems have been reported in literature [231, 232], but the creation of sufficiently high concentration of nanocrystals in a bulk crystalline matrix does not appear too feasible. Furthermore, nanocrystals of direct-gap semiconductors (so called quantum dots) were also embedded both in organic or inorganic host for this purpose [233, 234]. In the latter, the interest lies in a possibility to obtain superfast scintillation response thanks to the Wannier exciton-based luminescence response. Promising energy resolution was also found in [234]. The most frequently reported material systems in question are based on silica glass, in which the $YAlO_3$:Ce [235], ZnO [236], GaN [237], $NaYF_4$:RE [238] and especially LPS:Ce and

YPS:Ce [239, 240] luminescence- and scintillation-active nanophase was embedded, Figure 24. Some attempts have also been done with fluoride hosts [230, 241].

The Eu-doped hexagonal $BaCl_2$ nanocrystals in the glass matrix have shown surprisingly high scintillation efficiency and can possibly compete with $CdWO_4$ single crystals in imaging applications [242]. There is an interesting question whether such a "scintillator-dense" nanomorphological composite system cannot provide higher intrinsic scintillation efficiency with respect to its single crystal analog. Some intrinsic reasons for obtaining higher luminescence efficiency in nanocrystals have been discussed [230, 241,243]. One of the issues is that in scintillation nanocrystals the localization of electrons and holes may become spatially correlated because of small nanocrystal dimensions. Consequently, mechanisms of the radiative recombination within the forbidden gap (without involvement of conduction or valence bands) can be reinforced. Therefore, the fraction of electron-hole pairs born in scintillation conversion stage that is nonradiatively lost can be reduced hence increasing both the scintillation efficiency and speed of the response. Such effects were noticed in literature, even if their explanation was rather vague [244].

It is worth noting that in several Ce-doped single crystals of complex oxide scintillators such mechanisms of radiative recombination based on tunneling were observed [90]. However, the major problem in all nanocomposite systems will be the surface/interface formation between the nanocrystal and host phases and related creation of defects. Enhanced surface/volume ratio in the nanocrystal scintillator phase results in the elevated presence of surface/interface traps or nonradiative recombination centers which can remarkably diminish the figure-of-merit of these materials [245]. On the other hand, fabrication of nanocomposite materials might exploit experience and also profit from new technological tools consisting in manipulating the surface/interface states by chemical reaction and bonding with the host phase which is apparently not available in classical optical ceramics discussed above.

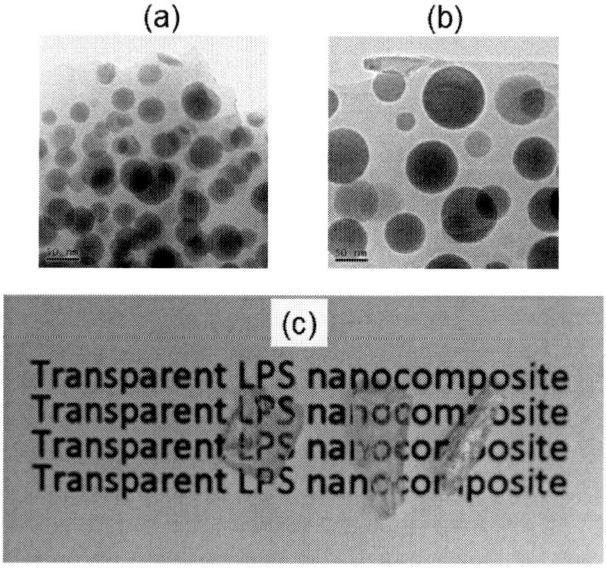

Figure 24. HRTEM of $SiO_2/Lu_2Si_2O_7$ nanocomposite heat-treated at 1100 °C (a), 1200 °C (b) [240] and (c) photo of the 1100°C heat-treated non-stoichiometric sample (TEOS : (Ce+Lu) = 100:20) [239].

2.8. Other Materials

In this last part of Section 2 we will mention some other materials recently appeared in literature. Ceramic form of ZnO scintillator is studied. ZnO is a direct bandgap semiconductor manifesting subnanosecond luminescence decay at room temperature and good scintillation intensity when doped with the n-type dopants such as gallium or indium and treated in hydrogen atmosphere [246]. Due to its sub-nanosecond decay times, medium density (5.6 $g \cdot cm^{-3}$), and potentially high light output, after further optimization it could be used in scintillation detectors in the homeland security, nuclear non-proliferation, and high energy physics fields. Recently, ceramic based Zn:ZnO scintillators were reported [247]. Zn:ZnO is zinc oxide containing excess zinc in the lattice interstitials and is known as a blue-green phosphor. Because ZnO does not have a cubic structure, a fully transparent ceramic is difficult to obtain. Transparency of the ceramic Zn:ZnO resulted to be 30% and the X-ray induced radioluminescence spectra featured both free exciton (390 nm) and defect-related (500 nm) luminescence peaks [247]. X-ray excited decay measurement revealed slower free exciton emission of ZnO ceramic with a decay time around 10 ns. An increase in the decay time of the ZnO:Zn ceramic compared with crystalline ZnO was explained due to the presence of shallow traps, i.e. due to the delay in the transfer stage of scintillation mechanism. Such traps can be formed at the boundaries between ceramic grains. An interesting attempt is done by Murata Manufacturing. In previous studies, most of the efforts related to transparent ceramic scintillators were focused on garnet or sesquioxide type materials (see *Sections 2.1*, *2.2* and *2.4*). In addition to these conventional ceramic scintillators, complex perovskite structured transparent ceramics can also be candidates for scintillator applications. Murata Manufacturing developed the ceramic material for a dielectric resonator and during its R&D, Ba(Mg, Ta)O_3 translucent ceramic was developed [248, 249]. It is known that non-substituted Ba($Mg_{1/3}Ta_{2/3}$)O_3 has a hexagonal structure, and partial substitution of the B-site cations (Mg^{2+} and/or Ta^{5+}) for tetravalent cations (e.g., Zr^{4+}) increases disorder of the B site sub-lattice.

Increasing the Zr^{4+} content promotes a transformation of original structure into a cubic perovskite structure. Taking into consideration this structure transformation, a fabrication method of transparent ceramic Ba(M, Mg, Ta)O_3 (M = Ti^{4+}, Sn^{4+}, Zr^{4+}, or Hf^{4+}) was developed [250]. We have recently evaluated scintillation responses of transparent ceramic of pure and Yb-doped Ba(ZrMgTa)O_3 from Murata manufacturing and rather poor light yield of the host-only emission was found [251]. Another type of complex perovskite, such as (LaSr)(AlTa)$O_3$was also studied [251]. This material (single crystal) was independently developed for the substrate applications for the GaN epitaxial growth by solid state reaction as well as the conventional Czochralski method [252]. Then, numerical approach was applied to study the formability of cubic perovskites in complex oxide systems [253]. Because these materials contain Ta^{5+}, high stopping power to high energy photons was expected. Another attractive property of this perovskite structure is its ability to accept rare-earth doping cations at both A-sites (Ba^{2+}) and B-sites. We also measured scintillation responses of pure and Yb-doped transparent ceramic (LaSr)(AlTa)O_3 provided by Murata Manufacturing, and only intrinsic emission of the host was observed in radioluminescence spectra [251]. Unlike Ba(ZrMgTa)O_3 series, pulse height could be observed under ^{241}Am γ-ray excitation and the light yield resulted to be 500 ph/5.5 MeV. As an example, Figure 25 shows a photograph of fully transparent Ba(ZrMgTa)$O_3$and (LaSr)(AlTa)O_3 ceramic.

Figure 25. Photograph of transparent complex perovskite scintillators made by Murata manufacturing [251].

Furthermore, $Ba_2YZrO_{6-\delta}$ was also examined by an Indian group [254]. It showed 60-80% transparency in a visible spectral range. No luminescence characteristics were reported. A study of transparent ceramic complex perovskite scintillators has just started and further investigations are required.

3. APPLICATIONS

In most cases the scintillation ceramics are competing with single crystals. As an example, in Computern Tomography (CT) medical imaging $CdWO_4$ single crystal has been gradually replaced by slow scintillation ceramics based on $(Y,Gd)_2O_3$:Eu or Gd_2O_2S:Pr compounds [15-17] . R&D of scintillating materials in last two decades is triggered by modern medical imaging methods using X-ray sources or radioisotopes. A thorough overview of medical imaging principles, techniques and devices can be found in Refs. [29, 255, 256]. Applications in medical imaging can be divided into *(i)* static and *(ii)* dynamic imaging. Both general (chest, bone fractures, etc.) and dental radiography together with mammography are the most frequently mentioned applications included in static imaging. Photographic film coupled to a Gd_2O_2S:Tb phosphor screen is gradually being replaced by the modern flat panel detectors [255, 256]. General radiography requires a large image format (up to 43×43 cm^2 flat panels are commercially available) and pixel size is typically 100-150 μm^2. In dental imaging an image size of about 3×4 cm^2 is typically used with the pixel size of about 20-40 μm^2. In mammography a special attention is paid to the detection of both very low contrast tumors in soft tissue and dense microcalcifications of a few hundred μm dimensions [258]. While higher X-ray energy between 60-120 keV is needed in general radiography, in the latter two application areas usual values are between 20-30 keV. With respect to classical film-screen cassettes, the modern flat panel detectors offer at the same 2D-resolution the better sensitivity, higher dynamical range and comfort of all-electronic image processing. In practice, such features of flat panel detectors result in an improved image quality, easier and faster data handling, reduced running costs and also lower irradiation doses of patients. In dynamical imaging it is namely fluoroscopy and Computer Tomography, which uses the speedy data aquisition in real time and stringent requirements as for the afterglow are put on the phosphor/scintillator material. Fluoroscopy is an analog of the above static imaging, where a progression-in-time of an image of a body part is to be followed, e.g. during interventional surgery, angiography examinations or other more general situations in which a distribution of a contrast agent in time is followed to detect a lesion. Typical frame rates (number of images taken per second - fps) are between 15-30 fps and the above mentioned flat panel technology was explored for this purpose as well [259].

Computer Tomography (CT) scanning became a widely used technique to obtain transversal images of human body [29]. An X-ray fan beam with a 1D-position sensitive detector on a fixed construction rotates around the patient's body in the center with a typical frequency of 0.5-1 Hz and a large number of attenuation profiles are registered during the one full (360 degree) rotation. From these data the cross-sectional images of the body are reconstructed. Applying simultaneously the slow gradual move of the patient through the CT machine (so called multi-slice spiral scan mode) even the volume 3D-reconstruction image can be obtained. X-ray energies up to 140 keV are applied, which led to the need to employ dense ceramic materials with suppressed afterglow which was the main challenge in their development [15-17, 145-147].

Positron emission tomography (PET) is the most sophisticated imaging technique used at present. PET allows visualization of intriguing processes, organ functioning or malformed tissue in human or animal body which can be labeled by positron emitting radioisotopes [256, 260]. When time-of-flight (TOF) information from the coincidence detection events is evaluated in so called TOF-PET, all the measurement sequence can be made shorter or isotope dose smaller which is highly benefitial so that this principle and its technical application is nowadays of great interest [261, 262].

The field of homeland security applications is rapidly developing after the events of September 11, 2001. Namely, there is the demand for so called radiation portal monitor (RPM) systems to be applied e.g. at country borders, sea ports etc. They must be able to rapidly detect localized sources of radiation with a very high detection probability and low false-alarm rate, while screening all of the traffic without impeding the flow of commerce [263]. Revealing illicit traffic of nuclear weapons, special nuclear material, and radiation dispersal device materials is of the utmost importance. Due to the need of isotope identification excellent energy resolution is indispensable. Under certain circumstances such screening procedures can be simultaneously used for the search for drugs and explosives especially when combination of X-ray and portable neutron source irradiation is applied. In this respect, various novel principles of scintillator-based neutron detection are studied [264, 265]. Scintillator detectors are also used in radio (X-ray) astronomy. The first detection of X-ray emission from an extra-solar object was reported in 1962 [266] and from that time monitoring radiation in this energy window became one of the drivers of modern astronomy. The hard X-ray energy range (~20-600 keV), so far, has not been explored much as X-rays are more difficult to focus and available scintillation detectors offer less advantageous spatial and energy resolution. NaI:Tl-based scintillation detectors were repeatedly used in recent satellite missions [267, 268]. As a pioneering challenge, Suzaku satellite launched in 2005 mounted Ce:GSO-BGO type phoswitch detector to observe hard X-rays [269] and succeeded to detect hard X-rays from the Black Hole [270]. In this application, mechanically hard scintillators are required to endure the shock at the launch so that ceramic scintillators are one of the candidates. As a new choice for these detectors also the new scintillators $LaBr_3$:Ce or $LaCl_3$:Ce are under study due to their unique energy resolution capability. A crucial condition for all scintillation materials to be used in such applications is an extremely low spontaneous background coming from radioactive isotope contamination due to extremely low level of registered signals.

4. Future Prospects

Acceleration of R&D in the field of scintillation materials over the last 20 years has been motivated mainly by new demanding applications in the medical, industrial, scientific and security imaging sectors. Further development and exploitation of powerful X-ray sources, e.g. synchrotron radiation, electron and particle beams, the availability of new radiopharmaceuticals significantly enhancing PET imaging capability in oncology [271] and the wider exploitation/monitoring of radioisotopes in different branches of human activity continue to contribute to the increasing interest in fast and efficient scintillators. Currently, there is surviving need for dense, fast and high-light-yield scintillators for PET. While LYSO:Ce and GSO:Ce were successfully introduced and replaced the previously used BGO, the search for a second material potentially used in combined PET detectors to correct for the depth-of-interaction error [272] or superfast scintillators for TOF detection [156, 209] continues. In PET detectors the sufficiently fast and bright Ce- or Pr-doped materials allow for the evaluation of the TOF information related to the coincidence detection of two 511 keV photons also from the rising edge of the scintillation decay [262]. Hot candidate could be e.g. the earlier mentioned LuAG:Pr scintillation ceramics [109] due to perfect combination of fast decay, high light yield and excellent energy resolution.

As the majority of applications are using X-ray sources with energies below 150 keV, there is significant market potential for very high light yield, medium density, fast scintillators, where the latest developments led to the discovery of the group of the Ce-doped rare earth halides and re-discovery of the Eu-doped alkali earth iodides mentioned in Section 2.6. Also in this case, the understanding of material point defects is at the very beginning and improvements in the crystal quality are expected in near future. Renewed interest in the scintillation characteristics of direct band-gap semiconductors based on ZnO or other binary compounds has arised recently, [39,246] due to their superfast emission with decay times below 1 ns. ZnO ceramic was mentioned in Section 2.9. Emission in these materials is due to Wannier excitons, free or localized at defects or dopants. In the latter case it shows a low energy shift and still faster decay times of tens of ps. Embedding the quantum dots of such materials in an optically inactive host has been mentioned in the section 2.8 on nanocomposite scintillators. Finally, particular combinations of organic-inorganic materials [273] can also bring unexpected material solutions in the field of fast scintillators.

A separated group of scintillation materials is under development for neutron detection. These materials must contain elements with high cross-sections for neutron capture (^{6}Li, ^{10}B, ^{155,157}Gd) and ideally should have a low effective Z-number providing insensitivity to gamma rays. These constraints substantially limit the number of candidate material compositions [274]. Such scintillation detectors are needed in mentioned homeland security applications to be employed at the airports, seaports or country borders to defend civil sector against terrorist activities and to detect illicit trafficking or inadvertent circulation of nuclear materials or radioactive sources and chemical warfare agents. Several different material compositions in the group of Li-containing complex halide crystals are under development [275]. Practical use of such monitoring portals is conditioned by the availability of sufficiently intense portable neutron sources, which are under development as well.

CONCLUSION

In this chapter the overview of R&D in the field of scintillation ceramic materials has been introduced. The variety of materials considered or currently under development is due to inequivalent demands of the ever-growing number of practical applications and the need to tailor scintillation materials to every one of them. Technological progress in the field of fast scintillation ceramics convincingly shows that some of them may become serious competitors to fast single crystal scintillators in near future. Attention was paid not only to review the materials and their scintillation characteristics, but also to address the issue of specific material defects and their relationship with the manufacturing technology used. Material defects and irregularities often give rise to energy levels in the host forbidden gap. Such energy levels can form the trapping states, which serve as centres of nonradiative recombination or simply as elements delaying the process of energy transport to the emission centers. Consequently, both the scintillator efficiency and timing characteristics are degraded. Understanding the underlying physical mechanisms of energy transfer and storage and the role of particular material defects is of crucial importance for bringing the material performance close to the intrinsic limits. Necessity to use correlated experiments of time-resolved spectroscopy, thermoluminescence and magnetic resonance for the defect nature understanding has been also demonstrated by several examples.

ACKNOWLEDGMENTS

Financial support of Czech MSMT KONTAKT II LH12150 and GA AV KAN300100802 projects is gratefully acknowledged. Authors also thank to X-J. Liu, A. Ikesue, N. Cherepy, P. Fedorov and R.S. Feigelson for providing material for figures 5, 8b, 14b, 22 and 23, respectively. V.V. Laguta and E. Jurkova are acknowledged for useful expert opinions and technical help, respectively, in the course of manuscript preparation.

REFERENCES

[1] a) W. C. Röntgen. *Sitz. Ber. Phys. Med. Ges. Wuerzb.*, 9, 132 (1895). b) W. C. Röntgen. *Science.*, *3,* 227 (1896).
[2] H. Becquerel. *C R Acad. Sci.*, 122, 501 (1896).
[3] M. Curie. *C R Acad. Sci.*, 126, 1101 (1898).
[4] P. Curie, M. Curie. *C R Acad. Sci.*, 127, 175 (1898).
[5] L. Ozawa, M. Itoh. *Chem. Rev.*, 103, 3835 (2003).
[6] R. Hofstadter. *Phys. Rev.*, 75, 796 (1949).
[7] W. Van Sciver, R. Hofstadter. *Phys Rev.*, 84, 1062 (1951).

[8] M. J. Weber. *J. Lumin.*, 100, 35 (2002).
[9] M. J. Weber, R. R. Monchamp. *J. Appl. Phys.*, 44, 5495 (1973).
[10] C. W. E. van Eijk. *Nucl. Instrum. Methods Phys. Res. A*, 460, 1 (2001).
[11] J. Pejchal, M. Nikl, E. Mihóková, J. A. Mareš, A. Yoshikawa, H. Ogino, K. M. Schillemat, A. Krasnikov, A. Vedda, K. Nejezchleb, V. Múčka. *J. Phys. D Appl. Phys.*, 42, 055117 (2009).
[12] A. Ikesue, Y. N. Aung. *Nat Photonics.*, *2*, 721 (2008).
[13] J. Lu, K. Ueda, H. Yagi, T. Yanagitani, Y. Akiyama, A. A. Kaminskii. *J. Alloy Compd.*, 341, 220 (2002).
[14] V. Lupei, A. Lupei, A. Ikesue. *J. Alloy Compd.*, 380, 61 (2004).
[15] Y. Ito, H. Yamada, M. Yoshida, H. Fujii, G. Toda, H. Takeuchi, Y. Tsukuda. *Jpn. J. Appl. Phys.*, 27, L1371 (1988).
[16] B. C. Grabmaier, W. Rossner, J. Leppert. *Phys Status Solidi A.*, 130, K183 (1992).
[17] C. Greskovich, S. Duclos. *Annu. Rev. Mater. Sci.*, 27, 69 (1997).
[18] E. Zych, C. Brecher, A. J. Wojtowicz, H. Lingertat. *J. Lumin.*, 75, 193 (1997).
[19] G. Blasse, B. C. Grabmaier. *Luminescent Materials;* Springer: Berlin, GE, 1994.
[20] G. F. Knoll. *Radiation Detection and Measurement;* Wiley: New York, US, 2000.
[21] P. A. Rodnyi. *Physical Processes in Inorganic Scintillators;* CRC Press: Boca Raton, US, 1997.
[22] P. Lecoq, A. Annenkov, A. Gektin, M. Korzhik, C. Pedrini, C. *Inorganic Scintillators for Detector Systems;* Springer: Berlin, GE, 2006.
[23] S. Shionoya, W. M. Yen. Ed.; *Phosphor Handbook;* CRC Press: Boca Raton, US, 1998.
[24] L. H. Brixner. *Mater. Chem. Phys.*, 16, 253 (1987).
[25] M. Ishii, M. Kobayashi. *Prog. Cryst. Growth Charact. Mater.*, 23, 245 (1992).
[26] G.Blasse. *Chem. Mater.*, 6, 1465 (1994).
[27] M. Nikl. *Phys. Status Solidi A*, 178, 595 (2000).
[28] A. J. Wojtowicz. *Nucl. Instrum. Methods Phys. Res. A*, 486, 201 (2002).
[29] C. W. E. van Eijk. *Phys. Med. Biol.*, 47, R85 (2002).
[30] M. Nikl. *Meas. Sci. Technol.*, 17, R37 (2006).
[31] M. Nikl, A. Vedda, V. V. Laguta. Single-crystal scintillation materials. In *Springer Handbook of Crystal Growth*; Dhanaraj G.; Byrappa K.; Prasad V.; Dudley, M.; Ed.; Springer: Heidelberg, GE, 2010; 1663-1700.
[32] D. J. Robbins. *J. Electrochem. Soc.*, 127, 2694 (1980).
[33] A. Lempicki, A. J. Wojtowicz, E. Berman. *Nucl. Instrum. Methods Phys. Res. A*, 333, 304 (1993).
[34] P. A. Rodnyi, P. Dorenbos, C. W. E. van Eijk. *Phys. Status Solidi. B*, 187, 15 (1995).
[35] A. N. Vasil'ev. *Nucl. Instrum. Methods Phys. Res. B*, 107, 165 (1996).
[36] J. A. Shepherd, S. M. Gruner, M. W. Tate, M. Tecotzky. *Opt. Eng.*, 36, 3212 (1997).
[37] S. Kubota, J. Ruan, M. Itoh, S. Hashimoto, S. Sakuragi. *Nucl. Instrum. Methods Phys. Res. A*, 289, 253 (1990).
[38] M. Nikl, N. Solovieva, K. Apperson, D. J. S. Birch, A. Voloshinovskii. *Appl. Phys. Lett.*, *86*, 101914 (2005).
[39] S. E. Derenzo, M. J. Weber, M. K. Klintenberg. *Nucl. Instrum. Methods Phys. Res. A*, 486, 214 (2002).
[40] J. Wilkinson, K. B. Ucer, R. T. Williams. *Radiat. Meas.*, 38, 501 (2004).

[41] D. Ehrentraut, H. Sato, Y. Kagamitani, A. Yoshikawa, T. Fukuda, J. Pejchal, K. Polak, M. Nikl, H. Odaka, K. Hatanaka, H. Fukumura. *J. Mater Chem.*, 16, 3369 (2006).

[42] M. Nikl. *Phys. Status Solidi. A*, 202, 201 (2005).

[43] M. Nikl, K. Nitsch, K. Polak, E. Mihokova, I. Dafinei, E. Auffray, P. Lecoq, P. Reiche, R. Uecker. *Phys. Status Solidi. B*, 195, 311 (1996).

[44] P. Dorenbos, J. P. M. de Haas, C. W. E. van Eijk. *IEEE Trans Nucl. Sci.*, 2, 2190 (1995).

[45] M. W. Moses, S. A. Payne, W. –S. Choong, G. Hull, B. W. Reutter. *IEEE Trans Nucl. Sci.*, 55, 1049 (2008).

[46] G. Bizarri, W. W. Moses, J. Singh, A. N. Vasil'ev, R. T. Williams. *J. Appl. Phys.*, 105, 044507 (2009).

[47] R. L. Coble. *J. Appl. Phys.*, 32, 787 (1961).

[48] A. Ikesue, Y. L. Aung, T. Taira, T. Kamimura, K. Yoshida, G. L. Messing. *Annu. Rev. Mater Res.*, 36, 397 (2006).

[49] A. Krell, J. Klimke, T. Hutzler. *Opt. Mater.*, 31, 1144 (2009).

[50] H. Yagi, T. Yanagitani, T. Numazawa, K. Ueda. *Ceram Int.*, 33, 711 (2007).

[51] J. Lu, H. Yagi, K. Takaichi, K. Uematsu, J. F. Bisson, Y. Feng, A. Shirakawa, K. Ueda, T. Yanagitani, A. A. Kaminskii. *Appl. Phys. B*, 79, 25 (2004).

[52] H. Yagi, T. Yanagitani, K. Takaichi, K. Ueda, A. A. Kaminskii, A. A. *Opt. Mater.*, 29, 1258 (2007).

[53] T. Yanagida, K. Kamada, Y. Fujimoto, Y. Yokota, A. Yoshikawa, H. Yagi, T. Yanagitani. *Nucl. Instrum. Methods A*, 631, 54 (2011).

[54] A. Ikesue, I. Furusato, K. Kamata. *J. Am. Ceram. Soc.*, 78, 225 (1995).

[55] A. Ikesue, K. Kamata, K. Yoshida. *J. Am. Ceram. Soc.*, 78, 2545 (1995).

[56] A. Ikesue. *Opt. Mater.*, 19, 183 (2002).

[57] A. Ikesue, Y. L. Aung, T. Yoda, S. Nakayama, T. Kamimura. *Opt. Mater.*, 29, 1289 (2007).

[58] L. An, A. Ito, T. Goto. *J. Eur. Ceram. Soc.*, 32, 1035 (2012).

[59] L. An, A. Ito, T. Goto. *J. Eur. Ceram. Soc.*, 31, 1597 (2011).

[60] L. An, A. Ito, T. Goto. *Mater Lett.*, 65, 3167 (2011).

[61] L. An, A. Ito, T. Goto. *J. Eur. Ceram. Soc.*, 31, 237 (2011).

[62] S. Kochawattana, A. Stevenson, S. Lee, M. Ramirez, V. Gopalan, J. Dumm, V. K. Castillo, G. J. Quarles, G. L. Messing. *J. Eur. Ceram. Soc.*, 28, 1527 (2008).

[63] V. Chani, G. Boulon, W. Zhao, T. Yanagida, A. Yoshikawa. *Jpn. J. Appl. Phys.*, 49, 075601 (2010).

[64] W. Zhao, C. Mancini, D. Amans, G. Boulon, T. Epicier, Y. Min, H. Yagi, T. Yanagitani, T. Yanagida, A. Yoshikawa. *Jpn J Appl Phys.*, 49, 022602 (2010).

[65] J. H. Mun, A. Novoselov, A. Yoshikawa, G. Boulon, T. Fukuda. *Mat. Res. Bull.*, 40, 1235 (2005).

[66] A. Yoshikawa, V. I. Chani. *MRS Bul.*, 34, 266 (2009).

[67] J. Mares, C. D'Ambrosio. *Opt. Mater.*, 30, 22 (2007).

[68] L. M. Bollinger, G. E. Thomas. *Rev Sci Instrum.*, 32, 1044 (1961).

[69] R. Chen, S. W. S. McKeever. *Theory of Thermoluminescence and Related Phenomena;* World Scientific: SI, 1997.

[70] J. A. Weil, J. R. Bolton, J. E. Wertz. *Electron Paramagnetic Resonance: Elementary Theory and Practical Applications;* Wiley: New York, US, 1994.

[71] J. –M. Spaeth, J. R. Niklas, R. H. Bartram, R. H. *Structural Analysis of Point Defects in Solids: An Introduction to Multiple Magnetic Resonance Spectroscopy;* Springer: Berlin, GE, 1992.

[72] V. V. Laguta, M. Martini, A. Vedda, M. Nikl, E. Mihokova, P. Bohacek, J. Rosa, A. Hofstatter, B. K. Meyer, Y. Usuki. *Phys. Rev. B*, 64, 165102 (2001).

[73] S. A. Dikanov, Y. D. Tsvetkov. *Electron spin envelope modulation (ESEEM) spectroscopy;* CRC Press Inc.: Boca Raton, US, 1992.

[74] A. Schweiger, G. Jeschke. *Principles of pulse electron paramagnetic resonance;* Oxford University Press: New York, US, 2001.

[75] C. D. Brandle. *J. Cryst. Growth.*, 264, 593 (2004).

[76] E. Geusic, H. M. Marcos, L. G. van Uitert. *Appl. Phys. Lett.*, 4, 182 (1964).

[77] A. Autrata, P. Schauer, J. Kvapil, J. Kvapil. *J Phys E.*, 11, 707 (1978).

[78] G. de With, H. J. A. van Dijk. *Mater. Res. Bull.*, 19, 1669 (1984).

[79] C. A. R. Mulder, G. de With. *Solid State Ionics*, 16, 81 (1985).

[80] M. Sekita, H. Haneda, T. Yanagitani, S. Shirasaki. *J. Appl. Phys.*, 67, 453 (1990).

[81] A. Ikesue, T. Kinoshita, K. Kamata, K. Yoshida. *J. Am. Ceram. Soc.,* 78, 1033 (1995).

[82] E. Zych, C. Brecher, H. Lingertat. *Mater. Sci. Forum.*, 239-241, 257 (1996).

[83] T. Yanagida. Basic experiment for gamma-ray detector in astrophysics by using ceramic scintillator and avalanche photodiode. Master Thesis; University of Tokyo: Tokyo, JA, 2004 (in Japanese)

[84] T. Yanagida, H. Takahashi, T. Ito, D. Kasama, T. Enoto, M. Sato, S. Hirakuri, M. Kokubun, K. Makishima, T. Yanagitani, H. Yagi, T. Shigeta. *IEEE Trans Nucl. Sci.*, 52, 1836 (2005).

[85] N. J. Cherepy, J. D. Kuntz, J. J. Roberts, T. A. Hurst, O. B. Drury, R. D. Sanner, T. M. Tillotson, S. A. Payne. *Proc SPIE Int. Soc. Opt. Eng.*, 7090, 70790X (2008).

[86] E. Mihokova, M. Nikl, J. A. Mares, A. Beitlerova, A. Vedda, K. Nejezchleb, K. Blazek, C. D'Ambrosio. *J. Lumin.*, 126, 77 (2007).

[87] M. Kuklja. *J Phys Condens Matter.*, 12, 2953 (2000).

[88] C. R. Stanek, K. J. McClellan, M. R. Levy, R. W. Grimes. *Phys. Status Solidi. B*, 243, R75 (2006).

[89] C. R. Stanek, K. J. McClellan, M. R. Levy, C. Milanese, R. W. Grimes. *Nucl. Instrum. Methods Phys. Res. A*, 579, 27 (2007).

[90] T. Yanagida, T. Itoh, H. Takahashi, S. Hirakuri, M. Kokubun, K. Makishima, M. Sato, T. Enoto, T. Yanagitani, H. Yagi, T. Shigetad, T. Ito. *Nucl. Instrum. Methods Phys. Res. A*, 579, 23 (2007).

[91] G. Hull, J. J. Roberts, J. D. Kuntz, S. E. Fisher, R. D. Sanner, T. M. Tillotson, A. D. Drobshoff, S. A. Payne, N. J. Cherepy. *Proc. SPIE Int. Soc. Opt. Eng.*, 6706, 670617 (2007).

[92] M. Nikl, E. Mihokova, J. Pejchal, A. Vedda, M. Fasoli, I. Fontana, V. V. Laguta, V. Babin, K. Nejezchleb, A. Yoshikawa, H. Ogino, G. Ren, G. *IEEE Trans. Nucl. Sci.*, 55, 1035 (2008).

[93] H. –L. Li, X. –J. Liu, L.-P. Huang. *J. Am. Ceram. Soc.*, 88, 3226 (2005).

[94] N. J. Cherepy, J. D. Kuntz, T. M. Tillotson, D. T. Speaks, S. A. Payne, B. H. T. Chaib, Y. Porter-Chapman, S. E. Derenzo. *Nucl. Instrum. Methods Phys. Res. A*, 579, 38 (2007).

[95] M. Nikl, J. A. Mares, N. Solovieva, H. –L. Li, X. –J. Liu, L. –P. Huang, I. Fontana, M. Fasoli, A. Vedda, C. D'Ambrosio. *J. Appl. Phys.*, 101, 033515 (2007).
[96] H. –L. Li, X. -J. Liu, R. –J. Xie, G. –H. Zhou, N. Hirosaki, X. –P. Pu, L.-P. Huang. *Jpn. J. Appl. Phys.*, 47, 1657 (2008).
[97] M. Nikl, E. Mihokova, J. Pejchal, A. Vedda, Yu Zorenko, K. Nejezchleb. *Phys. Status Solidi. B*, 242, R119 (2005).
[98] T. Yanagida, Y. Fujimoto, Y. Yokota, K. Kamada, S. Yanagida, A. Yoshikawa, H. Yagi, T. Yanagitani. *Radiat. Meas.*, 46, 1503 (2011).
[99] M. Nikl, H. Ogino, A. Krasnikov, A. Beitlerova, A. Yoshikawa, T. Fukuda. *Phys. Status Solidi. A*, 202, R4 (2005).
[100] H. Ogino, A. Yoshikawa, M. Nikl, K. Kamada, T. Fukuda. *J. Cryst. Growth.*, 292, 239 (2006).
[101] K. Kamada, T. Yanagida, K. Tsutsumi, Y. Usuki, M. Sato, H. Ogino, A. Novoselov, A. Yoshikawa, M. Kobayashi, S. Sugimoto, F. Saito. *IEEE Trans. Nucl. Sci.*, 56, 570 (2009).
[102] W. Drozdowski, P. Dorenbos, J. T. M. de Haas, R. Drozdowska, A. Owens, K. Kamada, K. Tsutsumi, Y. Usuki, T. Yanagida, A.Yoshikawa. *IEEE Trans. Nucl. Sci.*, 55, 2420 (2008).
[103] T. Yanagida, A. Yoshikawa, A. Ikesue, K. Kamada, Y. Yokota. *IEEE Trans. Nucl. Sci.*, 56, 2955 (2009).
[104] T. Yanagida, A. Fukabori, Y. Fujimoto, A. Ikesue, K. Kamada, J. Kataoka, Y. Yokota, A. Yoshikawa, V. Chani. *Phys. Status Solidi. C*, 8, 140 (2011).
[105] J. Zhao, Z. Wang, C. Wang, X. Wei, Y. Li, W. Zhang, M. Yin. *J. Rare Earths*, 27, 376 (2009).
[106] Y. Shi, X. Q. Feng, Y. B. Pan, J. Zhou, Y. Huang, Z. H. Gao. *Radiat. Meas.*, 45, 457 (2010).
[107] Y. Shi, M. Nikl, X. Feng, J. A. Mares, Y. Shen, A. Beitlerova, R. Kucerkova, Y. Pan, Q. Liu. *J. Appl. Phys.*, 109, 013522 (2011).
[108] W. H. Rhodes, Y. Wang, C. Brecher, J. Gary Baldoni. *J. Am. Ceram. Soc.*, 94, 3655 (2011).
[109] T. Yanagida, Y. Fujimoto, K. Kamada, D. Totsuka, H. Yagi, T. Yanagitani, Y. Futami, S. Kurosawa, Y. Yokota, A. Yoshikawa, M. Nikl. presented at SCINT 2011 Conference, IEEE Trans. Nucl. Sci., accepted 2012
[110] Y. –N. Xu, W. Y. Ching, B. K. Brickeen. *Phys. Rev. B*, 61, 1817 (2000).
[111] A. B. Muñoz-García, L. Seijo. *Phys. Rev. B*, 82, 184118 (2010).
[112] M. Fasoli, A. Vedda, M. Nikl, C. Jiang, B. P. Uberuaga, D. A. Andersson, K. J. McClellan, C. R. Stanek. *Phys. Rev. B*, 84, 081102(R) (2011).
[113] G. Blasse, A. Bril. *J. Chem. Phys.*, 47, 5139 (1967).
[114] Yu. Zorenko. *Opt Spectrosc.*, 88, 551 (2000).
[115] J. L. Wu, G. Gundiah, A. K. Cheetham. *Chem. Phys. Lett.*, 441, 250 (2007).
[116] M. Kottaisamy, P. Thiyagarajan, J. Mishra, M. S. R. Rao. *Mater. Res. Bull.*, 43, 1657 (2008).
[117] M. Nikl, J. Pejchal, E. Mihokova, J. A. Mares, H. Ogino, A. Yoshikawa, T. Fukuda, A. Vedda, C. D'Ambrosio. *Appl. Phys. Lett.*, 88, 141916 (2006).
[118] K. Kamada, T. Endo, K. Tsutumi, T. Yanagida, Y. Fujimoto, A. Fukabori, A. Yoshikawa, J. Pejchal, M. Nikl. *Cryst. Growth Des.*, 11, 4484 (2011).

[119] K. Kamada, T. Yanagida, J. Pejchal, M. Nikl, T. Endo, K. Tsutumi, Y. Fujimoto, A. Fukabori, A. Yoshikawa. *J. Phys. D Appl. Phys.*, 44, 505104 (2011).

[120] K. Kamada, T. Yanagida, T. Endo, T. Tsutumi, Y. Usuki, M. Nikl, Y. Fujimoto, A. Fukabori, A. Yoshikawa. http://dx.doi.org/10.1016 /j.jcrysgro.2011.11.085 (2012).

[121] N. J. Cherepy, J. D. Kuntz, Z. M. Seeley, S. E. Fisher, O. B. Drury, B. W. Sturm, T. A. Hurst, R. D. Sanner, J. J. Roberts, S. A. Payne. *Proc. SPIE*, 7805, 78050I (2010).

[122] N. J. Cherepy, S. A. Payne, B. W. Sturm, J. D. Kuntz, Z. M. Seeley, B. L. Rupert, R. D. Sanner, O. B. Drury, T. A. Hurst, S. E. Fisher, M. Groza, L. Matei, A. Burger, R. Hawrami, K. S. Shah, L. A. Boatner. *IEEE NSS/MIC Conference Record.*, 1288 (2010).

[123] N. J. Cherepy, S. A. Payne, B. W. Sturm, S. P. O'Neal, Z. M. Seeley, O. B. Drury, L. K. Haselhorst, B. L. Rupert, R. D. Sanner, P. A. Thelin, S. E. Fisher, R. Hawrami, K. S. Shah, A. Burger, J. O. Ramey, L. A. Boatner. *Proc. SPIE*, 8142, 81420W (2011).

[124] T. Kanai, M. Satoh, I. Miura. *J. Am. Ceram. Soc.*, 91, 456 (2008).

[125] N. J. Cherepy, S. A. Payne, S. J. Asztalos, G. Hull, J. D. Kuntz, T. Niedermayr, S. Pimputkar, J. J. Roberts, R. D. Sanner, T. M. Tillotson, E. V. Van Loef, C. M. Wilson, K. S. Shah, U. N. Roy, R. Hawrami, A. Burger, L. A. Boatner, W. –S. Choong, W. W. Moses. *IEEE Trans. Nucl. Sci.*, 56, 873 (2009).

[126] K. Takagi, T. Fukazawa. *Appl. Phys. Lett.*, 42, 43 (1983).

[127] H. Suzuki, T. A. Tombrello, C. L. Melcher, J. S. Schweizer. *Nucl. Instrum. Methods Phys. Res. A*, 320, 263 (1992).

[128] L. Pidol, A. Kahn-Harari, B. Viana, B. Ferrand, P. Dorenbos, J. T. M. De Haas, C. W. E. Van Eijk, E. Virey. *J. Phys. Condens. Matter.*, 15, 2091 (2003).

[129] D. W. Cooke, K. J. McClellan, B. L. Bennett, J. M. Roper, M. T. Whittaker, R. E. Muenchausen, R. C. Sze. *J. Appl. Phys.*, 88, 7360 (2000).

[130] O. Sidletskiy, V. Bondar, B. Grinyov, D. Kurtsev, V. Baumer, K. Belikov, K. Katrunov, N. Starzhinsky, O. Tarasenko, V. Tarasov, O. Zelenskaya. *J. Cryst. Growth.*, 312, 601 (2010).

[131] M. Jie, G. Zhao, X. Zeng, L. Su, H. Pang, X. He, J. Xu. *J. Cryst. Growth.*, 277, 175 (2005).

[132] C. L. Melcher, R. A. Manente, C. A. Peterson, J. S. Schweizer. *J. Cryst. Growth.*, *128*, 1001 (1993).

[133] C. L. Melcher, M. A. Spurrier, L. Eriksson, M. Eriksson, M. Schmand, G. Givens, R. Terry, T. Homant, R. Nutt. *IEEE Trans. Nucl. Sci.*, 50, 762 (2003).

[134] T. Gustafsson, M. Klinterberg, S. E. Derenzo, M. J. Weber, J. O. Thomas. *Acta. Crystallogr. C*, 57, 668 (2001).

[135] L. Pidol, O. Guillot-Noël, A. Kahn-Harari, B. Viana, D. Pelenc, D. Gourier. *J. Phys. Chem. Solids*, 67, 643 (2006).

[136] B. Liu, Z. Qi, M. Gu, X. Liu, S. Huang, C. Ni. *J. Phys. Condens. Matter.*, 19, 436215 (2007).

[137] P. Dorenbos, C. W. W. van Eijk, A. J. J. Bos, C. L. Melcher. *J. Phys. Condens. Matter.*, 6, 4167 (1994).

[138] D. J. Wisniewski, L. A. Boatner, J. S. Neal, G. E. Jellison, J. O. Ramey, A. North, M. Wisniewska, A. Lempicki, C. Brecher, J. Glodo. *Proc. SPIE*, 6706, 670619 (2007).

[139] A. Lempicki, C. Brecher, H. Lingertat, S. R. Miller, J. Glodo, V. K. Sarin. *IEEE Trans. Nucl. Sci.*, 55, 1148 (2008).

[140] Y. Wang, W. H. Rhodes, G. Baldoni, E. Van Loef, J. Glodo, C. Brecher, L. Nguyen, K. Shah. *Proc. SPIE*, 7393, 73930H (2009).
[141] T. Lin, L. Fan, Z. Xu, Y. Shi, J. Xie, L. Deng, Y. Ren, Y. *Adv. Mater. Res.*, 295-297, 1300 (2011).
[142] C. D. Greskovich, K. N. Woods. *Am. Ceram. Soc. Bull.*, 52, 473 (1973).
[143] W. M. Yen. *Phys. Status Solidi. A*, 202, 177 (2005).
[144] G. Blasse, W. Schipper, J. J. Hamelink. *Inorg. Chim. Acta.*, 189, 77 (1991).
[145] C. D. Greskovich, D. Cusano, D. Hoffman, R. J. Riedner. *Am. Ceram. Soc. Bull.*, 71, 1120 (1992).
[146] W. Kostler, A. Winnacker, W. Rossner, B. C. Grabmaier. *J. Phys. Chem. Solids*, 56, 907 (1995).
[147] S. J. Duclos, C. D. Greskovich, R. J. Lyons, J. S. Vartuli, D. Hoffman, R. J. Riedner, M. J. Lynch. *Nucl. Instrum. Methods Phys. Res. A*, 505, 68 (2003).
[148] E. Zych, D. Hreniak, W. Strek. *Mater. Sci.*, 20, 111 (2002).
[149] A. Lempicki, C. Brecher, P. Szupryczynski, H. Lingertat, V. V. Nagarkar, S. V. Tipnis, S. R. Miller. *Nucl. Instrum. Methods Phys. Res. A*, 488, 579 (2002).
[150] V. V. Nagarkar, S. V. Tipnis, S. R. Miller, A. Lempicki, C. Brecher, P. Szupryczynski, H. Lingertat. *IEEE Trans. Nucl. Sci.*, 50, 297 (2003).
[151] Y. Shi, Q. W. Chen, J. L. Shi. *Opt. Mater.*, 31, 729 (2009).
[152] C. Brecher, R. H. Bartram, A. Lempicki. *J. Lumin.*, 106, 159 (2004).
[153] R. H. Bartram, A. Lempicki, L. A. Kappers, D. S. Hamilton. *J. Lumin.*, 106, 169 (2004).
[154] M. R. Levy, C. R. Stanek, A. Chroneos, R. W. Grimes. *Solid. State Sci.*, 9, 588 (2007).
[155] H. Rétot, S. Blahuta, A. Bessière, B. Viana, B. LaCourse, E. Mattmann. *J. Phys. D Appl. Phys.*, 44, 235101 (2011).
[156] T. Yanagida, Y. Fujimoto, S. Kurosawa, K. Watanabe, H. Yagi, T. Yanagitani, V. Jary, Y. Futami, Y. Yokota, A. Yoshikawa, A. Uritani, T. Iguchi, M. Nikl. *Appl. Phys. Express.*, 4, 126402 (2011).
[157] J. Legendziewicz, J. Sokolnicki. *J. Alloy Compd.*, 451, 600 (2008).
[158] D. Amans, A. Belsky, C. Dujardin, A. Hovsepyan, I. Kamenskikh, A. Kotlov, G. Ledoux, N. Fedorov, C. Pedrini, A. N. Vasil'ev. *IEEE Trans. Nucl. Sci.*, 57, 1355 (2010).
[159] L. van Pieterson, M. Heeroma, E. de Heer, A. Meijerink. *J. Lumin.*, 91, 177 (2000).
[160] M. Nikl, A. Yoshikawa, T. Fukuda. *Opt. Mater.*, 26, 545 (2004).
[161] M. Nikl, N. Solovieva, J. Pejchal, J. B. Shim, A. Yoshikawa, T. Fukuda, A. Vedda, M. Martini, D. H. Yoon. *Appl. Phys. Letters*, 84, 882 (2004).
[162] J. B. Shim, A. Yoshikawa, T. Fukuda, J. Pejchal, M. Nikl, N. Sarukura, D. H. Yoon. *J. Appl. Phys.*, 95, 3063 (2004).
[163] T. Yanagida, Y. Fujimoto, A. Yoshikawa, Y. Yokota, K. Kamada, J. Pejchal, N. Kawaguchi, S. Ishizu, K. Fukuda, T. Suyama, K. Uchiyama, K. Mori, K. Kitano, M. Nikl. *Appl. Phys. Express.*, 3, 056202 (2010).
[164] A. Fukabori, V. Chani, K. Kamada, T. Yanagida, Y. Yokota, F. Moretti, N. Kawaguchi, A. Yoshikawa. *J. Cryst. Growth.*, 318, 823 (2011).
[165] O. M. Bordun. *J. Appl. Spectrosc.*, 68, 304 (2001).

[166] A. Lushchik, M. Kirm, Ch. Lushchik, I. Martinson, G. Zimmerer. *J. Lumin.*, 87-89, 232 (2000).

[167] A. Fukabori, T. Yanagida, Y. Yokota, J. Pejchal, T. Ikegami, A. Yoshikawa. *J. Appl. Phys.*, 107, 073501(2010).

[168] A. Hoffmann. *Z Phys. Chem. B.*, 28, 65 (1935).

[169] J. A. Guevara, S. L. Cuffini, Y. P. Mascarenhas, R. E. Carbonio, J. A. Alonso, M .T. Fernandez, P. De la Presa, A. Ayala, A. L. Garcia. *Mater. Sci. Forum*, 278-281, 720 (1998).

[170] B. J. Kennedy, C. J.; Howard, B. C. Chakoumakos. *Phys. Rev. B*, 60, 2972 (1999).

[171] C. I. Park, R. A. Condrate, R. L. Snyder. *Appl. Spectr.*, 30, 352 (1976).

[172] V. V. Afanas'ev, A. Stesmans. *J. Appl. Phys.*, 95, 2518 (2004).

[173] C. Rossel, M. Sousa, C. Marchiori, J. Fompeyrine, D. Webb, D. Caimi, B. Mereu, A. Ispas, J. P. Locquet, H. Siegwart, R. Germann, A. Taponnier, K. Babich. *Microelectron. Eng.*, 84, 1869 (2007).

[174] S. Yamanaka, T. Maekawa, H. Muta, T. Matsuda, S. Kobayashi, K. Kurosaki. *J. Alloys Comp.*, 381, 295 (2004).

[175] A. Feteira, D. C. Sinclair, K. Z. Rajab, M. T. Lanagan. *J. Am. Ceram. Soc.*, 91, 893 (2008).

[176] H. Yamamoto, M. Mikami, Y. Shimomura, Y. Oguri. *J. Lumin.*, 87-89, 1079 (2000).

[177] S. L. Dole, S. Venkataramani U. S. Patent No. 5, 124,072 (23 June 1992).

[178] S. M. Loureiro, Y. Gao, V. Venkataramani. *J. Am. Ceram. Soc.*, 88, 219 (2005).

[179] E. V. van Loef, W. M. Higgins, J. Glodo, C. Brecher, A. Lempicki, V. Venkataramani, W. W. Moses, S. E. Derenzo, K. S. Shah. *IEEE Trans. Nucl. Sci.*, 54, 741 (2007).

[180] A. Grezer, E. Zych, L. Kepinski. *Radiat. Meas.*, 45, 386 (2010).

[181] E. V. Van Loef, Y. Wang, S. R. Miller, C. Brecher, W. H. Rhodes, G. Baldoni, S. Topping, H. Lingertat, V. K. Sarin, K. S. Shah. *Opt. Mater.*, 33, 84 (2010).

[182] Y. M. Ji, D. Y. Jiang, L .S. Qin, J. J. Chen, T. Feng, Y. K. Liao, Y. P. Xu, J. L. Shi. *J. Cryst. Growth.* 280, 93 (2005).

[183] H. Retot, A. Bessiere, A. Kahn-Harari, B. Viana. *Opt. Mater.*, 30, 1109 (2008).

[184] M. Villanueva-Ibanez, C. Le Luyer, S. Parola, O. Marty, J. Mugnier. *J. Sol.-Gel. Sci. Technol.*, 31, 277 (2004).

[185] M. Villanueva-Ibanez, C. Le Luyer, S. Parola, C. Dujardin, J. Mugnier. *Opt. Mater.*, 27, 1541 (2005).

[186] E. Mihóková, F. Moretti, N. Chiodini, A. Lauria, M. Fasoli, A. Vedda, A. Nale, M. Nikl, P. Boháček. *IEEE Trans. Nucl. Sci.*, 57, 1245 (2010).

[187] A. Lauria, N. Chiodini, E. Mihóková, F. Moretti, A. Nale, M. Nikl, A. Vedda. *Opt. Mater.*, 32, 1356 (2010).

[188] V. Jarý, M. Nikl, E. Mihóková, P. Boháček, B. Trunda, K. Polák, V. Studnička, V. Múčka. *Radiat. Meas.*, 45, 406 (2010).

[189] V. Jarý, E. Mihóková, M. Nikl, P. Boháček, A. Lauria, A. Vedda. *Opt. Mater.*, 33, 149 (2010).

[190] M. Nikl, P. Bohacek, B. Trunda, V. Jary, P. Fabeni, V. Studnicka, R. Kucerkova, A. Beitlerova. *Opt. Mater.*, 34, 433 (2011).

[191] E. Mihóková, N. Chiodini, M. Fasoli, A. Lauria, F. Moretti, M. Nikl, V. Jarý, A. Vedda. *Phys. Rev. B*, 82, 165115 (2010).

[192] G. Fabricius, E. L. Peltzer, Y. Blanca, C. O. Rodriguez, A. P. Ayala, P. De la Presa, A. Lopez Garcia. *Phys. Rev. B*, 55, 164 (1997).
[193] R. Vali. *Solid. State Commun.*, 147, 1 (2008).
[194] L. Feng, Z. Liu, Q. Liu. *Physica. B-Cond. Mat.*, 407, 2009 (2012).
[195] B. P. Mandal, N. Garg, S. M. Sharma, A. K. Tyagi. *J. Solid. State Chem.*, 179, 1990 (2006).
[196] Y. Liao, D. Jiang, Y. Ji, J. Shi. *Key Eng. Mater.*, 280-283, 643 (2005).
[197] Y. Liao, D. Jiang, Y. Xu, J. Shi. *Key Eng. Mater.*, 336-338, 640 (2007).
[198] E. V. Van Loef, Y. Wang, J. Glodo, C. Brecher, A. Lempicki, K. S. Shah. *MRS Proceedings*, vol. 1038; Materials Research Society: Warrendale, US, 2008; pp 1038-O06-02 (2007).
[199] A. Borisevich, M. Korzhik, P. Lecoq. *Nucl. Instrum. Methods Phys. Res. A*, 497, 206 (2003).
[200] Y. Ji, D. Jiang, Y. Liao, J. Shi. *Key Eng. Mater.*, 280-283, 577 (2005).
[201] Y. Ji, D. Jiang, T. Fen, J. Shi. *Mat. Res. Bull.*, 40, 553 (2005).
[202] A. Chaudhry, A. Canning, R. Boutchko, M. J. Weber, N. Grønbech-Jensen, S. E. Derenzo. *J. Appl. Phys.*, 109, 083708 (2011).
[203] L. Havlak, P. Bohacek, M. Nikl, R. Kucerkova, V. Studnicka. *Opt. Mater.*, 32, 1372 (2010).
[204] A. Lauria, N. Chiodini, M. Fasoli, E. Mihóková, F. Moretti, A. Nale, M. Nikl, A. Vedda. *J. Mater. Chem.*, 21, 8975 (2011).
[205] D. J. Singh. *Phys. Rev. B*, 82, 155145 (2010).
[206] Y. T. Viday, B. V. Grinyov, L. B. Zagarij, N. D. Zverev, V. V. Chernikov, V. A. Tarasov, A. M. Kudin. *IEEE NSS/MIC Conference Record.*, 762-763 (1995).
[207] S. Kh. Batygov, L. S. Bolyasnikova, V. A. Demidenko, E. M. Garibin, M. E. Doroshenko, K. V. Dukel'skiĭ, A. A. Luginina, I. A. Mironov, V. V. Osiko, P. P. Fedorov. *Dokl. Phys.*, 53, 485 (2008).
[208] A. A. Demidenko, E. A. Garibin, S. D. Gain, Yu. I. Gusev, P. P. Fedorov, I. A. Mironov, S. B. Michrin, V. V. Osiko, P. A. Rodnyi, D. M. Seliverstov, A. N. Smirnov. *Opt. Mater.*, 32, 1291 (2010).
[209] D. M. Seliverstov, A. A. Demidenko, E. A. Garibin, S. D. Gain, Yu. I. Gusev, P. P. Fedorov, S. V. Kosyanenko, I. A. Mironov, V. V. Osiko, P. A. Rodnyi, A. N. Smirnov, V. M. Suvorov. doi:10.1016/j.nima.2011.11.080 (2012).
[210] I. Holl, E. Lorenz, G. Mageras. *IEEE Trans. Nucl.*, 35, 105 (1988).
[211] P. Aubry, A. Bensalah, P. Gredin, G. Patriarche, D. Vivien, M. Mortier. *Opt. Mater.*, 31, 750 (2009).
[212] J. Fu, M. Kobayashi, S. Sugimoto, J. M. Parker. *J. Am. Ceram. Soc.*, 92, 2119 (2009).
[213] P. Samuel, H. Ishizawa, Y. Ezura, K. Ueda, S. M. Babu. *Opt. Mater.*, 33, 735 (2011).
[214] A. S. Shcheulin, A. I. Ryskin, A. E. Angervaks, P. P. Fedorov, V. V. Osiko, A. A. Demidenko, E. A. Garibin, A. N. Smirnov, K. V. Dukel'skii, I. A. Mironov. *Opt. Spectrosc.*, 110, 604 (2011).
[215] J. Glodo, H. Lingertat, C. Brecher, K. S. Shah, A. Lempicki. *IEEE NSS/MIC Conference Record.*, 112 (2005).
[216] A. Edgar, M. Bartle, S. G. Raymond, G. V. M. Williams, C. Varoy. *IEEE Trans. Nucl. Sci.*, *55*, 1160 (2008).

[217] V. V. Nagarkar, S. R. Miller, R. Sia, V. Gaysinskiy. *Nucl. Instrum. Methods Phys. Res. A*, 633, S286 (2011).
[218] V. V. Nagarkar, S. R. Miller, V. Gelfandbein, U. Shirwadkar, V. Gaysinskiy. *Nucl. Instrum. Methods Phys. Res. A*, 652, 271 (2011).
[219] E. V. D. van Loef, P. Dorenbos, C. W. E. Van Eijk, K. W. Kraemer, H. U. Guedel. *Appl. Phys. Lett.*, 79, 1573 (2001).
[220] C. M. Combes, P. Dorenbos, C. W. E. Van Eijk, K. W. Kraemer, H. U. Guedel. *J. Lumin.*, 82, 299 (1999).
[221] P. Yang, F. P. Doty, M. A. Rodriguez, M. R. Sanchez, X. Zhou, K. S. Shah. The synthesis and structures of elpasolite halide scintillators. In *Nuclear Radiation Detection Materials-2009.* D. L. Perry, A. Burger, L. Franks, K. Yasuda, M. Fiederie, M.; Ed.; Materials Research Society Symposium Proceedings; Materials Research Society: Warrendale, US, 2010; Vol. 1164, pp 185-192.
[222] M. Moszynski, M. Kapusta, M. Mayhugh, D. Wolski, S. O. Flyckt. *IEEE Trans. Nucl. Sci.*, 44, 1052 (1997).
[223] O. Barnouin, A. Procoli, H. Chung, G. H. Miley. *Nucl. Instrum. Methods Phys. Res. A.*, 299, 382 (1990).
[224] A. Procoli, O. Barnouin, G. H. Miley. *Opt. Eng.*, 31, 112 (1992).
[225] H. Kim, A. J. Moorhead. *J. Am. Ceram. Soc.*, 73, 496 (1990).
[226] R. Hofstadter. U. S. Patent No. 3,373,279 (12 March 1968).
[227] C. M. Wilson, E. V. Van Loef, J. Glodo, N. J. Cherepy, G. Hull, S. A. Payne, W. –S. Choong, W. Moses, K. S. Shah. *Proc. SPIE.*, 7079, 7079171 (2008).
[228] N. J. Cherepy, B. W. Sturm, O. B. Drury, T. A. Hurst, S. A. Sheets, L. E. Ahl, C. K. Saw, M. A. Pearson, S. A. Payne, A. Burger, L. A. Boatner, J. O. Ramey, E. V. Van Loef, J. Glodo, R. Hawrami, W. M. Higgins, K. S. Shah, W. W. Moses. *Proc. SPIE.*, 7449, 74490F (2009).
[229] S. R. Podowitz, R. M. Gaumé, W. T. Hong, A. Laouar, R. S. Feigelson. *IEEE Trans. Nucl. Sci.*, 57, 3827 (2010).
[230] E. A. McKigney, R. E. Del Sesto, L. G. Jacobsohn, P. A. Santi, R. E. Muenchausen, K. C. Ott, T. M. McCleskey, B. L. Bennett, J. F. Smith, D. W. Cooke. *Nucl. Instrum. Methods Phys. Res. A*, 579, 15 (2007).
[231] M. Nikl, K. Nitsch, I. Dafinei, P. Lecoq, G. P. Pazzi, P. Fabeni, M. Gurioli. Photoluminescence and scintillation properties of Pb^{2+} based quantum dots in CsCI host crystal. In *Scintillator and Phosphor materials*. P. Lecoq, W. M. Yen, R. C. Ruchti, M. J. Weber, C. Woody, R. –Y. Zhu,; Ed.; Materials Research Society Symposium Proceedings; Materials Research Society: Pittsburgh, US, 1994; Vol. 348, pp 155-162.
[232] A. Voloshinovskii, G. Stryganyuk, G. Zimmerer, P. Rodnyi, O. Antonyak, S. Myagkota, P. Savchyn. *Phys. Status Solidi. A*, 202, R101 (2005).
[233] K. Shibuya, M. Koshimizu, H. Murakami, Y. Muroya, Y. Katsumura, K. Asai. *Jpn. J. Appl. Phys.*, 43, L1333 (2004).
[234] S. E. Létant, T. –F. Wang. *Nano Lett.*, 6, 2877 (2006).
[235] T. B. de Queiroz, C. R. Ferrari, D. Ulbrich, R. Doyle, A. S. S. de Camargo. *Opt. Mater.*, 32, 1480 (2010).
[236] Z. Fu, B. Yang, L. Li, W. Dong, Ch. Jia, W. Wu. *J. Phys. Condens. Matter.*, 15, 2867 (2003).
[237] H. Qiu, Ch. Cao, D. Wang, H. Zhu. *Mater Chem Phy.*, 96, 296 (2006).

[238] A. Santana-Alonso, A. C. Yanes, J. Mendez-Ramos, J. del-Castillo, V. D. Rodrıguez. *Phys. Status Solidi. A*, 206, 2249 (2009).
[239] D. Niznansky, A. Begnamini, J. Vejpravova-Poltierova, R. Kucerkova, M. Nikl. *IOP Conf. Ser.: Mater. Sci. Eng.*, 18, 102020 (2011).
[240] M. Nikl, D. Niznansky, J. Ruzicka, C. Cannas, T. Yanagida. Silicate glass-based nanocomposite scintillators. In: *Advances in Nanocomposites Technology*, Hashim, A; Ed.; InTech: Rijeka, CI, 2011; pp 201-226.
[241] C. Dujardin, D. Amans, A. Belsky, F. Chaput, G. Ledoux, A. Pillonnet. *IEEE Trans. Nucl. Sci.*, 57, 1348 (2010).
[242] J. A. Johnson, S. Schweizer, B. Henke, G. Chen, J. Woodford, P. J. Newman, D. R. MacFarlane. *J. Appl. Phys.*, 100, 034701 (2006).
[243] H. Zheng, D. Gao, Z. Fu, E. Wang, Y. Lei, Y. Tuan, M. Cui. *J. Lumin.*, 131, 423-428 (2011).
[244] E. T. Goldburt, B. Kulkarni, R. N. Bhargava, J. Taylor, M. Libera. *J. Lumin.*, 72–74, 190 (1997).
[245] G. Stryganyuk, D. M. Trots, A. Voloshinovskii, T. Shalapska, V. Zakordonskiy, V. Vistovskyy, M. Pidzyrailo, G. Zimmerer. *J. Lumin.*, 128, 355 (2008).
[246] E. D. Bourret-Courchesne, S. E. Derenzo, M. J. Weber. *Nucl. Instrum. Methods Phys. Res. A*, 601, 358 (2009).
[247] V. A. Demidenko, E. I. Gorokhova, I. V. Khodyuk, O. A. Khristich, S. B. Mikhrin, P. A. Rodnyi. *Radiat. Meas.*, 42, 549 (2007).
[248] S. Nomura, K. Toyama, K. Kaneta. *Jpn J Appl Phys.*, 21, L624 (1982).
[249] H. Matsumoto, H. Tamura, K. Wakino. *Jpn. J. Appl. Phys.*, 30, 2347 (1991).
[250] Y. Kintaka, S. Kuretake, N. Tanaka, K. Kageyama, H. Takagi. *J. Am. Ceram. Soc.*, 93, 1114 (2010).
[251] T. Yanagida, Y. Fujimoto, Y. Yokota, A. Yoshikawa, S. Kuretake, Y. Kintaka, N. Tanakac, K. Kageyama, V. Chani. *Opt. Mater.*, 34, 414 (2011).
[252] M. Ito, K. Shimamura, D. A. Pawlak, T. Fukuda. *J. Cryst. Growth.*, 235, 277 (2002).
[253] L. M. Feng, L. Q. Jiang, M. Zhu, H. B. Liu, X. Zhou, C. H. Li. *J. Phys. Chem. Solids.*, 69, 967(2008).
[254] V. R. Kumar, P. R. S. Wariar, J. Koshy. *Cryst. Res. Technol.*, 45, 619 (2010).
[255] J. –P. Moy. *Nucl. Instr. Meth. Phys.Res. A*, 442, 26 (2000).
[256] G. K. von Schulthess. *Clinical Molecular Anatomic Imaging: PET, PET/CT, and SPECT/CT;* Lippincott WilliamsandWilkins: Philadelphia, US, 2003.
[257] J. –P. Moy. *Thin Solid. Films*, 337, 213 (1999).
[258] K. –P. Hermann, S. Obenauer, M. Funke, E. H. Grabbe. *Eur. Radiol.*, 12, 2188 (2002).
[259] U. Schiebel, N. Conrads, N. Jung, M. Weibrecht, H. Wieczorek, T. Zaengel, M. J. Powell, I. D. French, C. Glasse. *Proc. SPIE*, 2163, 129 (1994).
[260] C. W. E. van Eijk. *Radiat. Prot. Dosim.*, 129, 13 (2008).
[261] M. Conti. *Phys. Medica.*, 25, 1 (2009).
[262] M. W. Moses. *Nucl. Instrum. Methods Phys. Res. A*, 580, 919 (2007).
[263] E. R. Siciliano, J. H. Ely, R. T. Kouzes, B. D. Milbrath, J. E. Schweppe, D. C. Stromswold. *Nucl. Instrum. Methods Phys. Res. A*, 550, 647 (2005).
[264] Z. W. Bell, L. A. Boatner. *IEEE Trans. Nucl. Sci.*, 57, 3800 (2010).
[265] K. J. Riley, L. Ovechkina, S. Palamakumbura, Z. Bell, S. Miller, V. V. Nagarkar. *IEEE NSS/MIC Conference Record.*, 1777 (2010).

[266] R. Giacconi, H. Gursky, F. Paolini, B. Rossi, *Phys Rev Lett 9*, 439 (1962).
[267] H. V. Bradt, J. H. Swank, R. E. Rothschild *Adv Space Res 10*, 297 (1990).
[268] L. Piro, L. Scarsi, R.C. Butler, *Proc. SPIE 2517*, 169 (1995).,
[269] M. Kokubun, et al,. *Publ Astron Soc Jpn 59*, 53 (2006).
[270] K. Makishima, et al, *Publ Astron Soc Jpn 60*, 585 (2008)
[271] P. Olivier, *Nucl Instrum Methods Phys Res A 527*, 4 (2004).
[272] L. Eriksson, C.L. Melcher, M. Zhuravleva, M. Eriksson, H. Rothfuss, M. Conti, *Nucl Instrum Methods Phys Res A 648*, S288 (2011).
[273] S. Kishimoto, K. Shibuya, F. Nishikido, M. Koshimizu, R. Haruki, Y. Yoda, *Appl Phys Lett 93,* 261901 (2008).
[274] C. W. E. van Eijk, *Radiat Meas 38*, 337 (2004).
[275] C.W.E. van Eijk*, IEEE Trans. Nucl. Science 59*, 2242 (2012).

In: Recent Advances in Ceramic Materials Research
Editors: Joan J. Roa Rovira and Merce Segarra Rubi
ISBN: 978-1-62417-729-3

Chapter 7

MICROSTRUCTURE EVOLUTION AND MECHANICAL PROPERTIES OF SILICON NITRIDE BASED CERAMICS SINTERED BY SPARK PLASMA SINTERING (SPS)

Jow-Lay Huang[1,2,3]*and Pramoda K. Nayak[1]

[1]Department of Materials Science and Engineering,
National Cheng Kung University, Taiwan
[2]Center for Micro/Nano Science and Technology,
National Cheng Kung University, Taiwan
[3]Research Center for Energy Technology and Strategy,
National Cheng Kung University, Taiwan

ABSTRACT

Silicon Nitride is one of the major structural ceramics that possess excellent mechanical properties such as high flexural strength, good facture resistance and high hardness. In general, microstructural changes of Si_3N_4 in the matrix greatly influence the mechanical properties. In order to control the final microstructure and properties, the powder synthesis and sintering process play important roles in fabricating Si_3N_4-based ceramics. Recently developed Spark Plasma Sintering (SPS) is regarded as an energy saving technology due to short process time and fewer processing steps. The powder preparation and densification of this nano-powder using SPS technique have been reviewed. In the present study, a commercially available nanosizedβ-Si_3N_4 doped with sintering additives (6 wt% Al_2O_3 and 8 wt% Y_2O_3) has been used as initial power and consolidated by an SPS technique at different heating rates (50, 100 and 200 °C/min) by varying temperature from (1550 °C to 1700 °C).At slower heating rates (50 and 100 °C/min), the nanosized grains are maintained after sintering at 1600° C for 5 min, while anisotropic grain growth is accelerated above 1500 °C by applying a rapid heating cycle (200 °C/min).

In addition to the dynamic Ostwald ripening that occur during the sintering process, the presence of Morie′ fringes and dislocations is attributed to grain rotation and misfit strain between the subgrains and elongated, large grains. Grain coalescence is found to

* Corresponding Author's e-mail address: jlh888@mail.ncku.edu.tw (Jow-Lay Huang).

occur in the rapid heating process. The microstructures of the developed Si_3N_4 based ceramics were achieved, and the grain length, grain width, and aspect ratio for these sintering bulks were found to increase gradually with elevating sintering temperature. The fracture toughness for these β-Si_3N_4 based nanoceramics also has been evaluated for specimens sintered at different sintering temperatures. Furthermore, Si_3N_4 based nanocomposites incorporating electroconductive nanoparticles are regarded as promising materials for cutting tools and making electrical discharge machining possible. In this regards, conductive TiCnanopowder has been incorporated into nanosizedβ-Si_3N_4-based powder and consolidated by a SPS technique with a rapid heating process. The influence of this conductive phase on the microstructure development of the Si_3N_4 matrix has been demonstrated. This chapter would be interesting for scientists and engineers concerned with the novel processing of ceramics using SPS and use of these ceramics for engineering applications.

Keywords: Spark Plasma Sintering, Consolidation, β-Si_3N_4 based nanoceramics, Fracture Toughness

1. INTRODUCTION

Silicon nitride ceramics is a class of structural materials for high temperature applications. It has been showing tremendous research interest since last four decades in response to the challenge for developing a suitable ceramic to be used in gas turbine engine [1] to satisfy the needs for higher operating temperature, greater efficiencies in automotive [2], space [3] and electric power generating applications [4-6]. This material was first developed in 1950 for use as thermocouples tubes, crucible for molten metals and also rocket nozzles [7]. The material was formed by nitriding silicon powder compacts in the temperature range 1100-1450 °C and was regarded as reaction bonded silicon nitride (RBSN) [8]. One advantage in this process was that a very little or no shrinkage occurred during the nitriding process and therefore, these ceramics could be machined to final size and shape using conventional tooling. Furthermore, this material showed lots of interest for potential use in gas turbines. But, the major limitation to the use of RBSN for engine applications was due to its low mechanical strength (~ 200-250 MPa) as a result of 20-30 % microporosity. In order to get an intrinsically high strength and high hardness silicon nitride materials, this synthesis process was seemed to be inadequate. Thus, alternative approaches were developed during 1060's to get higher density silicon nitride resulted from hot-pressing (HP) of silicon nitride powder with various sintering additives [9] such as magnesium oxide (MgO) or nitride at 1850 °C under 23 MPa. Using MgO as sintering additive, full density material was produced and strength was substantially improved over that of RBSN. Magnesia was also used as the catalyst in the first commercial hot-pressed silicon nitride (HPSN) [10]. Ce_2O_3 and other rare earth oxides have been tested as densification additives [11-13]. Similarly, other quaternary system such as ZrO_2 has also been suggested as sintering aid [14, 15]. From the above attempts, it can be concluded that, a dense material is produced only in the presence of another oxide, which forms a liquid together with impurities from Si_3N_4 and enables fast densification.

Since 1070's, the search for improved materials has led to a better understanding of the role of sintering additives in the densification and microstructure development of silicon nitride based ceramics [16]. The amount of sintering aids used affects the density of the ceramics and the chemical composition of the grain boundary phase. The chemistry of the liquid phase, formed between sintering aids and oxide layer (SiO_2) on the surface of the Si_3N_4 particles are the key factor in determining the microstructure and mechanical properties of Si_3N_4 ceramics [17]. In 1984, Woetting and Ziegler [18] showed that the presence of Al_2O_3 and Y_2O_3 has a strong effect on density and grain shape of Si_3N_4 ceramics. Recently, the effect of grain width and aspect ratio on the mechanical properties of Si_3N_4 ceramics has been effectively studied by Krstic*et al.* [19] for different ratios of (Y_2O_3/ Al_2O_3).

1.1. Crystal Structure of Silicon Nitride

Silicon nitride exists in two major crystallographic configuration called as α and β phase. The structure of both phases is hexagonal [20, 21] with the c dimension of α approximately twice that of β. The lattice parameters for α-Si_3N_4 are a = 7.75 Å and c = 5.168 Å with an atomic layer sequence of ABCD, where as the corresponding parameters for β Si_3N_4 are a = 7.608 Å, c = 2.911 Å with an atomic layer sequence of ABAB. Both structures are built from SiN_4 tetrahedra and can be transformed into each other by a 180° rotation around an axis perpendicular to c. The atomic layer sequencing of α and β Si_3N_4 is given in Figure 1. [22]. β transformation in silicon nitride requires a lattice reconstruction. This type of process occurs usually only when the transforming material is in contact with a solvent.

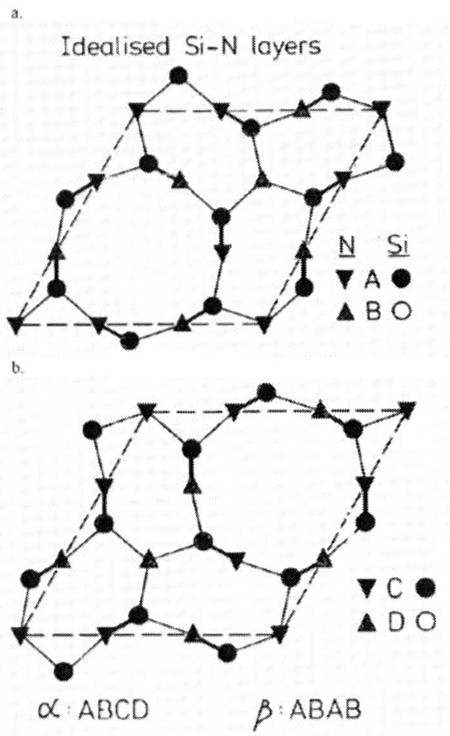

Figure 1. (a) AB layers of Si and N atoms in β-Si_3N_4 (b) CD layers of Si and N atoms in α-Si_3N_4 taken from [22].

The greater solubility of the more unstable form drives it into solution after which it precipitates as the less soluble, more stable form. The predominantly α silicon nitride powder used to produce dense silicon nitride ceramics was observed to transform to the β modification during the sintering process at temperatures in excess of 1400 °C, when the original α phase is in contact with a metal-silicon-oxynitride liquid phase.

1.2. Liquid Phase Sintering in Silicon Nitride

Generally, every powder particle of Silicon nitride is surrounded by a surface layer of silica. When oxide additives are added to silicon nitride, it react with silica and some of the nitride itself at a particular sintering temperature, and form an oxynitride liquid, which promotes densification by solution-precipitation. The α-Si_3N_4 dissolves in liquid and is precipitated as β-Si_3N_4, which grows in the longitudinal direction as hexagonal rod like crystals. The liquid cools as an intergranular phase, usually a glass according to the following equation.

$$\alpha - Si_3N_4 + SiO_2 + M_xO_y \rightarrow \beta - Si_3N_4 + M - Si - O - N \quad (1)$$

The systematic of pressureless sintering kinetics [23] applies to the liquid phase sintering model by kingery [24], which consists of three stages as summarized by the log-shrinkage/log-time plot shown in Figure 2. The stages are as follows:

1) Particle rearrangement within the initial liquid, where the rate and the extent of shrinkage depend on the volume and viscosity of the liquid.
2) Solution-diffusion-reprecipitation, where shrinkage can be expressed as

$$\Delta V / V_0 \propto t^{\frac{1}{n}} \quad (2)$$

where$t$ is the time and n depends upon the rate of diffusion of additive into the liquid. The value of n = 3 for MgO and 5 for Y_2O_3.

3) Final elimination of closed porosity during which the liquid acts to form more rounded grain morphology. Final density is greater than 95% of the theoretical value.

1.3. SiAlON System

SiAlON is formed when oxygen replaces nitrogen and Al replaces Si in the silicon nitride system. This is called as the solid solution based on the silicon nitride structure [25].

It exhibits in two types such as α and β-SiAlON. Just like α to β- Si_3N_4 transformation, α to β- SiAlON transformation is also possible by controlling heat treatment procedure and amount of sintering aids.

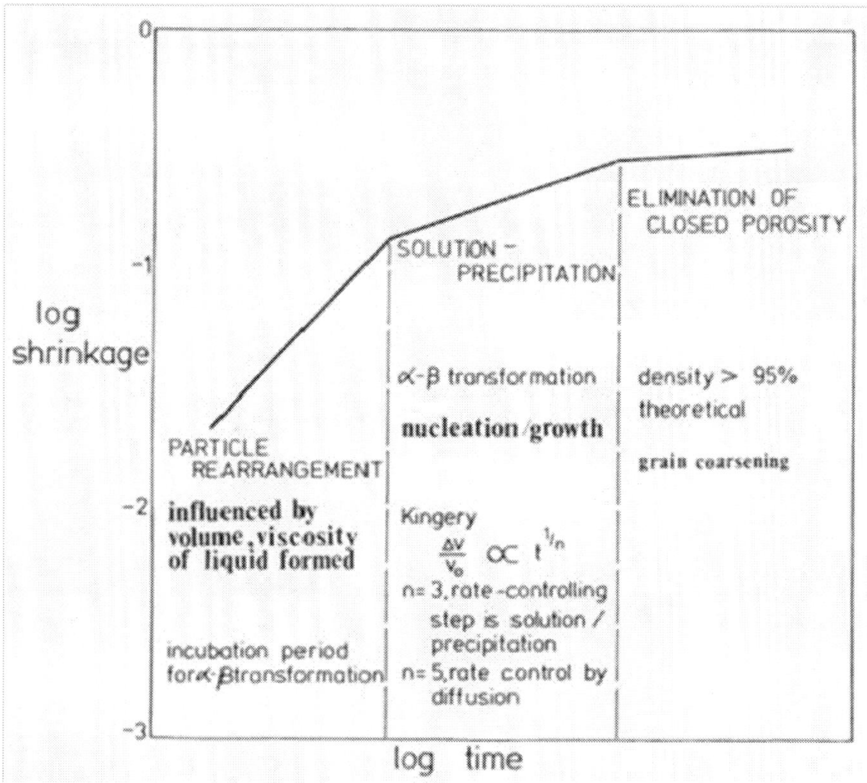

Figure 2. Schematic plot of three stages of liquid phase sintering of silicon nitride [22].

1.4. Spark Plasma Sintering (SPS)

Spark plasma sintering (SPS) is a novel sintering technique that uses the idea of pressure driven powder consolidation under pulsed direct electric current passing through a sample compressed in a graphite matrix. It is also known as the field assisted sintering technique or pulse electric current sintering.

This newly developed sintering technique is regarded as an energy-saving technology due to the short process time and fewer processing steps. This technique was first described by Raichenko in 1987 [26] and the key characteristics of this SPS is given as follows: *(i)* the generation of local electric discharge plasma and its effect on the material; *(ii)* the combined effect of external fields, such as a force field and electric field, on the densification and phase formation in a particulate system; *(iii)* the influence of electric current in the near surface layers of conductors and semiconductors (the so-called ''skin-effect''); and *(iv)* the rapid and nonuniform heating/cooling throughout the sample, causing large temperature gradients. The schematic of SPS process is shown in Figure 3 [27].

The SPS is commonly related to a uniaxial pressing mode. Here, heat isolators or heating elements are not required and the graphite matrix plays their roles when direct current is passed through it.

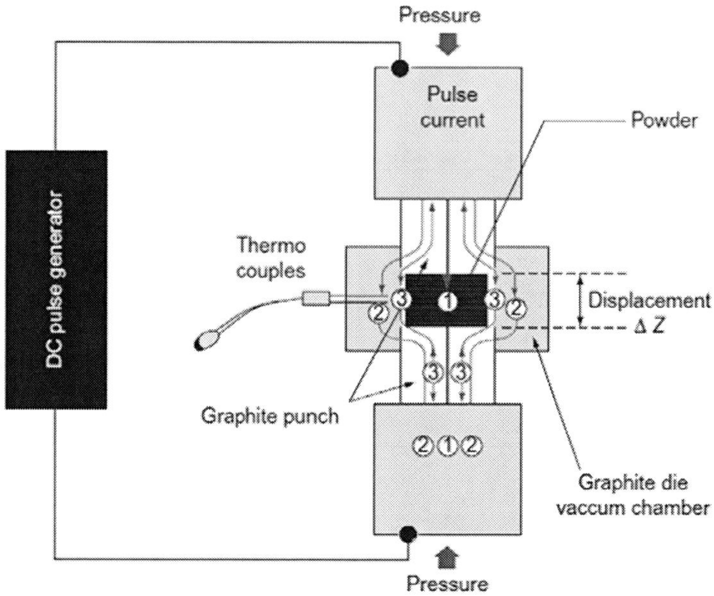

Figure 3. Schematics of Spark Plasma Synthesis (SPS) processtaken from [27].

Figure 4. Vacuum chamber of SPS apparatus with graphite matrix between punches (a) and nonuniformly heated matrix and pistons (b) [27].

A high-density graphite die and pistons are used for sample consolidation. The application of graphite significantly limits the range of work pressures by 100 MPa. SPS apparatus is usually equipped with a vacuum chamber (Figure 4(a)), which allows operation in vacuum and inert gas media, as well as a computer-aided design system to control pressure and temperature inside the die. The SPS is fed by pulsed direct current (in various apparatuses, the impulse periodicity varies from 1ms to 1s and the current amplitude from 0.1

kA to 10 kA), which is used from the very beginning till the end of heating process. Consolidation proceeds through two stages:

- First stage includes initiation of sintering by a current impulse through the sample and subsequent heating under direct or alternative impulse current. A typical impulse of current in SPS machines initiates a magnetic field. The interaction of the field and the current is accompanied with radial compression of the powder. Powders of conductors are heated up due to electroresistive heating. In nonconductive powders, heat flows from the graphite matrix and pistons by thermal conductivity. In this case, the matrix and pistons are heated up due to their own electric resistance.
- Second stage, the pressure can be changed in accordance with the given law, depending on the sintered substance. Sometimes the initial pressure is quite low (10–15 MPa) and gradually increases with increase in temperature. The full cycle of sintering takes less than 10 min. Heat transfer from matrix to sample is very effective as the matrix itself serves as a heating element. Nevertheless, a temperature gradient between the top and the bottom pistons (Figure 4(b)) appears when the heating rate is high. Usually, the SPS uses very high heating rates (to 600 °C/min and higher) and short isothermal holds (up to several minutes) to obtain samples with density close to the theoretical one at relatively low temperatures, normally several hundred degrees lower than temperatures during hot pressing.

1.4.1. Advantages of SPS over Other Techniques

The most impressive advantage of SPS is its applicability to sinter materials of various types of chemical bonding and electric conductivity. Novel materials have been prepared from powders of ceramic dielectrics, conductors, semiconductors, amorphous alloys, and, sometimes, polymers. The traditional driving force involved in commonly used consolidation techniques such as solid state sintering and hot press sintering are: surface tension, external pressure, chemical potential due to the gradient of surface curvature, concentration gradient in multicomponent system etc. In SPS technique, the additional driving forces include electromechanical stress, high local temperature gradients creating thermal stresses intensifying thermal diffusion, and dislocation creep. These additional driving forces are responsible for much faster transport mechanism, that accelerates rapid sintering, which is observed in SPS.

1.5. SPS for Silicon Nitride Based Ceramics

Silicon Nitride based ceramics possess excellent mechanical properties in combination with good corrosion and thermal shock resistance [28, 29]. In these applications, microstructural changes of Si_3N_4 in the matrix greatly influence the mechanical properties. For instance, elongated grains in the resulting ceramics cause higher toughness among the Si_3N_4-based materials [30-32], whereas fine-grained Si_3N_4 grains are expected to exhibit noticeable superplasticity and improved wear resistance [33-35]. In order to control the final microstructure and properties, the powder synthesis and sintering process play important roles in fabricating Si_3N_4-based ceramics. Notably, the use of Si_3N_4nanoprecusor powder is required to consolidate Si_3N_4nanoceramics with superior wear resistance [36]. As described earlier, SPS is an efficient sintering technique i.e. commonly used for structural ceramics like Si_3N_4, SiC, Al_2O_3, and ZrO_2[37-41]. The nanosized Si_3N_4-based materials prepared by SPS, such as Si_3N_4[42, 43] and SiAlON [44, 45] have been reported in literature. In these earlier studies, the rapid grain coarsening of Si_3N_4 was suppressed by very fast heating in a short time. In another case, Shen*et al.*[46] also reported that SPS was a reliable process to rapidly consolidate α or β phase SiAlON-based ceramics, and that a tough interlocking microstructure could be obtained by a dynamic Ostwald ripening mechanism, resulting from a rapid heating rate during sintering. Thus, the sintering behavior of Si_3N_4-based materials is very complex and unclear, because the factors involving phase transformation, composition of glassy phase, and sintering parameters during sintering would greatly influence the resulting microstructures and mechanical properties, especially in the SPS system.

In the present study, a commercially available nanosizedβ-Si_3N_4 doped with uniformly dispersed sintering additives has been used and consolidated by an SPS technique. Based on the classical liquid phase sintering theory [47], the sample is densified by using the wetting liquid phase in three overlapping stages referred to as *(1)* rearrangement, *(2)* solution–reprecipitation, and *(3)* grain coalescence. The advantages of the system are as follows:

- During the period when liquid forms and spreads by the stronger capillary force induced by nanosized particles, rearrangement events dominate the early response to reduce residual pores accompanied by the penetration and fragmentation of polycrystalline particles by the coexisting glassy phase. The rate of densification can thus be accelerated in this stage.
- Dissolution of finer grains, mass transfer through the liquid, and reprecipitation on larger particles, which is called Ostwald ripening, are the main features associated with the intermediate stage. This is because smaller particles have a higher solubility than larger ones [48]. Thus, the solubility difference causes matter to be transported down the resulting concentration gradient, from the smaller to the larger particles. Similar results can be observed with α or β-phase microsized Si_3N_4 in Lee *et al.* [49, 50].
- By applying the rapid heating technique, the specimen can be consolidated in a few minutes and prevent the decrease in sintered density associated with a prolonged time during the grain coalescence stage.

2. Experimental Details

2.1. Powder Preparation

Commercially available Si_3N_4nanopowder (SM131, Fraunhofer-Institutfür Keramische Technologien and Sinterwerkstoffe, Dresden, Germany) doped with sintering additives of 6 wt% Y_2O_3 and 8 wt% Al_2O_3 was taken as the starting material. It contains 90 wt% β-phase and 10 wt% α-phase with a manufacturer-determined average particle size of 70 nm by the Rietveld method. The as-received Si_3N_4 powder was further dispersed by planetary milling at a rotation speed of 300 rpm for 6 h using a 375 mL nylon bottle with Si_3N_4 balls.

The powder mixture was dried in a rotary evaporator, cold iso-statically pressed into round ingots at a pressure of 200 MPa, crushed and then passed through a #200 sieve for granulation.

2.2. Sintering by Spark Plasma Sintering

Approximately 3 gm of the treated powder was loaded into a graphite mold with a length of 50 mm and inner and outer diameters of 20 and 50 mm, respectively. A graphite sheet was inserted into the small gap between the punches and mold to improve the temperature uniformity effectively. The graphite mold was also covered with carbon heat insulation to avoid heat dissipation from the external surface of the die. After the chamber was evacuated to a pressure of 10 Pa, the sample was heated to the desired sintering parameters under a uniaxial pressure of 30 MPa by SPS (Dr. Sinter 1050, Sumitomo Coal Mining, Kawasaki, Japan).

All the SPS experiments were carried out at different heating rates, sintering temperatures and dwelling times, and the details of sintering conditions are described in Table 1. A 12 ms-on and 2 ms-off pulse sequence was used. The heating process was controlled using a monochromatic optical pyrometer that was focused on the surface of the graphite mold. The linear shrinkage behavior of the powder during sintering was recorded manually in every 15, 10, and 5 s for heating rate of 50, 100, and 200°C/min, respectively. During dwelling time, the overshooting temperatures were also recorded in every 15 s for each specimen. In these processing conditions, the final sample size was in the form of tablets of 19 mm in diameter and ~ 3 mm in thickness.

2.3. Microstructural Characterization

Phase compositions of the samples were identified using an X-ray powder diffractometer (XRD; Model D-MAX/II B, Rigaku, Tokyo, Japan). Samples were scanned from 25° to 75° with a scanning rate of 1°/min. Cell dimensions were determined from XRD peak data using UNITCELL [51] with a Si standard. Samples for the Vickers indentation technique and microstructural observation by scanning electron microscopy (SEM, FEI Quanta 400F, Eindhoven, Netherlands) were cut, ground, and polished by the standard diamond polishing technique, down to a diamond size of 1 μm. Typical microstructures were revealed by plasma etching (OMNI-RIE, Tainan, Taiwan) with CF_4 and O_2 sources at 120 W for 30 s. The ratio of CF_4 to O_2 was 95:5 and the total gas pressure was maintained at 85 mTorr. To avoid surface discharging under the electron beam, a conductive Au film was coated on the etched surface of Si_3N_4.

Furthermore, cross-sectional samples with 5 mm in length, 5 mm in width and under 200 nm in thickness were prepared with a focus ion beam (FIB, SMI3050, Seiko, Tokyo, Japan). Structure, orientation, and composition analysis were performed in a region vertical to the upper surface of the bulk.

Table 1. Specimen Designation and Sintering Conditions of SPS-Sintered Nanosized β-Si_3N_4-Based Powder

Designation	Heating Rate (°C/min)	Sintering temperature (°C)	Dwelling time min)
S1450	50	1450	5
S1600	50	1600	5
M1450	100	1450	5
M1600	100	1600	5
F1300	200	1300	3
F1400	200	1400	3
F1500	200	1500	3
F1600	200	1600	3

This study was carried out on a transmission electron microscope (FEG-TEM, Tecnai G2 F20, Philips, Eindhoven, the Netherlands) operated at 200 kV. To characterize the grain morphology of Si_3N_4-based composites quantitatively, the length and area of each developed grain were measured for over 2000 grains with an image analyzer using software (Sigmascan Pro 5.0, SPSS Inc.) and the average grain size was calculated following the statistical derivation by Woetting et al. [52]. The grain diameter and aspect ratio of each grain were determined from the area/length and length/diameter, respectively. The ''true'' aspect ratio was obtained from the frequency distribution of the aspect ratio and selecting the value as a_{95}.

2.4. Analyzing Mechanical Characteristics

The apparent densities of sintering specimens were measured by the Archimedes method. Hardness was measured with a Vickers hardness tester (AKASHI AVK-A, Akashi Seisakusho Ltd., Japan) and by applying a micro-hardness indent at 196 N for 15 s on polished surfaces.

Fracture toughness was measured by the Vickers surface indentation technique [53]. Each data point for fracture toughness was measured for at least five times.

3. Spark Plasma Sintering of a Nano-Sizedb-Si_3N_4-Based Powder

3.1. Exact Temperature Estimation in SPS

It is very important to estimate the exact temperature of the specimen during SPS process because there is probability of overshooting of temperature due to high heating rate (≥100 °C/min). The densification temperature for SPS varied considerably, depending on not only the dimensions and arrangement of the heating elements, and heating rate (related to the power supply) but also the dimension and properties of sintering compact [54–57].

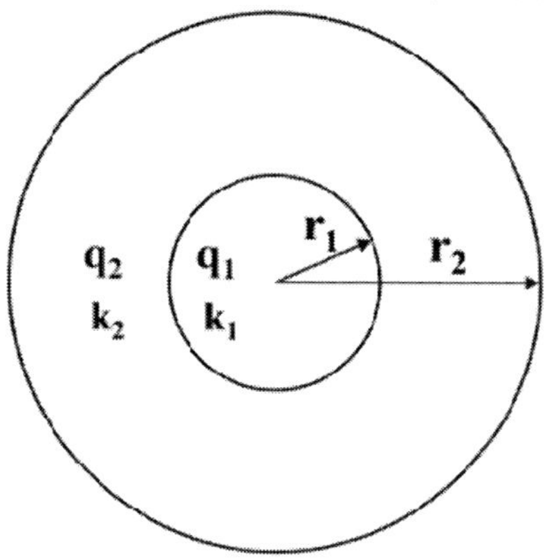

Figure 5. A composite system of the sample (inner circle) and graphite die (outer region).

In the present study, the densification temperatures were assumed to be evaluated by heating rate, because the other factors affecting the temperature distribution were almost the same. Wang and Fu [58] used analytical solutions of the one-dimensional Fourier equation to obtain the radial temperature distribution of the inner cylinder (shown in Figure 5), i.e. given by

$$T(r) = \frac{q_1}{4k_1}(r^2 - r_1^2) + \frac{q_2}{4k_2}(r_2^2 - r_1^2) + \frac{r_1^2(q_1 - q_2)}{2k_2}\ln(\frac{r_2}{r_1}) + T_2 \tag{3}$$

whereq_1 and q_2 refer to the heating rates of the sample and graphite die, respectively, and they are related to the power input. k_1 and k_2 are the heat conductivity of the sample and graphite, respectively, and r is semi diameter. r_1 and r_2 refer to the distance from the center of the circle to the border of sample-die and die-vacuum, respectively. The calculated temperature difference in the die and sample is thus

$$T(r_1) - T(r_2) = \Delta T = \frac{q_1}{4k_2(r_2^2 - r_1^2)} \propto \frac{q_1}{k_2} \tag{4}$$

Salamon*et al.*[59] indicated that the real temperature of the sample is 120 °C higher than the one measured on the die surface at 1600–1700 °C with a heating rate of 100 °C/min in the heat-up process. Other studies also note that the temperature difference is in the range of 70–140 °C, despite the different heating conditions [37, 46]. As a consequence, the temperature difference of 120 °C was chosen for the specimen M1600 in the present work. According to Eq. (4), the estimated ΔT deduced from power input, multiplied by the recorded current and voltage, is presented in Figure 6. (The thermal conductivity of graphite: k_2=< 65–0.017 T [56]. The value of *T* was introduced by the surface temperature of the graphite die.

Figure 7 shows the overshooting temperature collected from the recorded sintering schedule. It is known that applying a high heating rate means achieving greater overshooting at the final sintering temperature. The value of the overshooting temperature is 14, 24, and 54 °C for samples S1600, M1600, and F1600, respectively. After the holding time of over 90 s, the detected temperature falls to 1600 °C for all specimens. Wang *et al.*[58] found that the temperature difference between the sample and die surface is 60 °C when the sintering temperature was held at a constant value. Thus, the power input was similar for the samples with different heating rates, and the actual temperatures for sintering compacts with different heating rates were assumed to be identical. Henceforth, the temperature refers to die temperature unless otherwise noted.

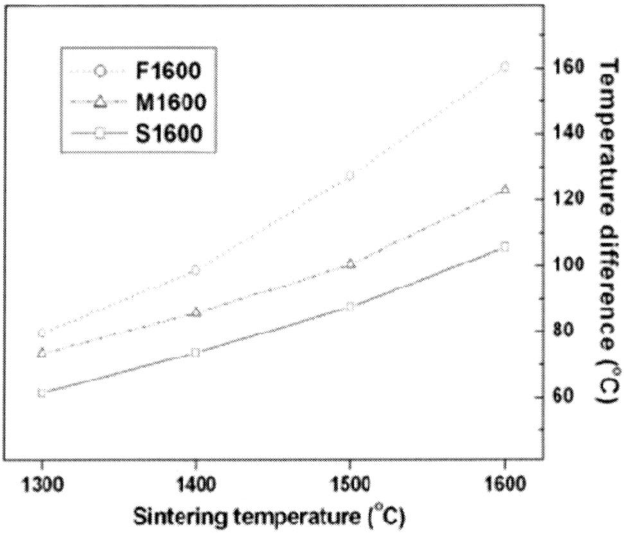

Figure 6. Estimated temperature difference between sample and die during the heat-up schedule. The heating rate was controlled at 50°C/min (□), 100°C/min (Δ), and 200°C/min (○) up to 1600°C under a uniaxial pressure of 30 MPa in vacuum.

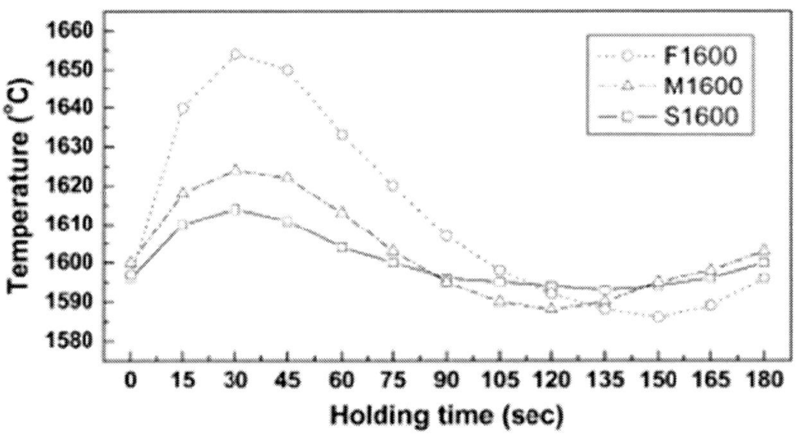

Figure 7. Overshooting temperature for the sintering compacts at their dwelling stage.

3.2. Densification and General Characteristics

Figure 8 reveals the in situ shrinkage curves of sintering compacts at different temperatures ranging from 1200 to 1600 °C with a heating rate of 50, 100, and 200 °C/min. Before 1200 °C, the displacement curves at different heating rates are similar and these data are not shown in this Figure. It is observed that the displacement curve shifts to a higher temperature region with an increased heating rate, which reflects that a larger amount of residual pores exist for a rapidly heated specimen at the same sintering temperature. The sintering time for specimen F1600 was only 2 min from 1300 to 1600 °C during the heat-up process, whereas it was 4 and 8 min for M1600 and S1600, respectively. Following the fast densification, the displacement of graphite punches stops shrinking at around 1480 and 1560 °C for specimens S1600 and M1600, and then exhibits a thermal expansion with the increasing sintering temperature, whereas specimen F1600 reaches its terminal of shrinkage within 10 s in its holding time. In the final stage of densification, no further shrinkage can be detected, and consequently the sample is believed to be near fully dense. The physical properties for specimens with different sintering conditions are summarized in Table 2. The value of 3.25 g/cm^3 for the apparent density of as-sintered materials was assumed to be near full densification (no obvious pores were observed on the polished surfaces of these sintering bulks). With a slower heating rate (50 and 100°C/min), a fully dense sintering bulk can be achieved at 1450 °C for 5 min. However, with a faster heating rate, nearly full densification can be obtained at 1400°C for 3 min. As it can be expected from Figure 6, the higher heating input accompanied by a higher overshooting temperature results in the higher real temperature of specimen F1400. Although a faster heating rate is beneficial for yielding a fully densified sample, the real temperature around the sintering bulk might be similar for the F1400, M1450, and S1450 specimens.

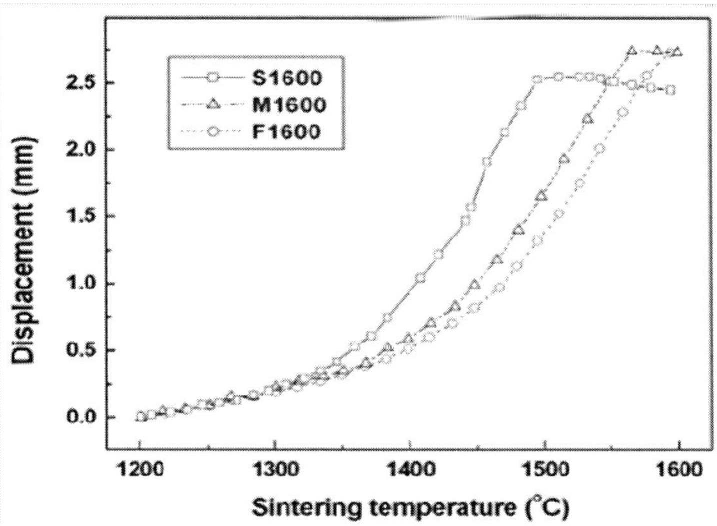

Figure 8. In situ displacement curves for the sintering compacts during the rise of temperature in SPS. The heating rate was controlled at 50°C/min (□), 100°C/min (Δ), and 200°C/min (○) up to 1600°C under a uniaxial pressure of 30 MPa in vacuum.

Table 2. Apparent Density, the Unit Cell Dimensions of Formed Phase, and Fracture Toughness after Compacting Under Various Sintering Conditions

β-phase unit cell (Å)				
Designation	Apparent density (gm/cm^3)	a_0	c_0	Fracture toughness (MPa.m$^{1/2}$)
S1450	3.25	7.620	2.918	-
S1600	3.25	7.620	2.919	3.98 ± 0.51
M1450	3.25	7.617	2.919	-
M1600	3.25	7.620	2.918	3.86 ± 0.12
F1300	2.66	7.624	2.921	-
F1400	3.25	7.620	2.921	1.93 ± 0.26
F1500	3.25	7.621	2.920	4.60 ± 0.26
F1600	3.25	7.621	2.920	5.44 ± 0.43

Phase compositions detected by XRD of as-sintered materials at different sintering conditions show the presence of a β-Si_3N_4-based pattern. The unit cell dimensions for the samples are shown in Table 2. The values for a_0 and c_0 for a β-phase unit cell are ~ 7.62 and ~ 2.92 Å, respectively, which are somewhat deviated from those for pure β-Si_3N_4. This result suggests that a tiny amount of Si–N may be replaced by Al–O in the particle dissolution and coarsening stages of liquid-phase sintering [60]. The fracture toughness obtained for a series of fully densified samples sintered at 1400–1600 °C with different heating rates are also summarized in Table 2, and are discussed in relation to the microstructure below.

3.3. Microstructure and Fracture Behavior

The microstructural changes in as-sintered specimens for various sintering conditions are shown in Figures 9–13. Figures 9(a) and (b) show the typical SEM micrographs of specimens S1600 and M1600 sintered at 1600 °C for 5 min with a heating rate of 50 and 100°C/min, respectively, which indicates that the uniform and ultrafine grains with a low aspect ratio can be observed in the cases of slower heating rate. Although both micrographs exhibit similar grain shape and grain size, some larger crystallites can be seen in specimen M1600. From the bright-field/TEM(BF/TEM) image of specimen M1600 (Figure 10), some smaller, rough grains under 70 nm marked by arrows and larger, edged particles with ~200 nm in width can be observed. This suggests that a dissolution–reprecipitation process might occur during sintering. Similar characteristics are also generally consistent with observations of equiaxial grains sintered with a slower heating rate in SPS [46] and hot-pressed nanosizedβ-Si_3N_4 samples at 1600 and 1700 °C [61]. The features of the rapidly heated specimens (200 °C/min) at different sintering temperatures for 3 min are shown in the SEM images of the plasma-etched polished surface with the same magnification (Figs. 11(a)–(c)). For these near fully densifiedβ-Si_3N_4-based materials, microstructural characterization as a function of sintering temperature reveals that there is an increase in size and number of large reinforcing grains as the die temperature increases from 1400 °C to 1600 °C. Figure 12(a) reveals the morphology of specimen F1400 observed by BF/TEM. Most crystallites show a general rounded morphology with a grain size in the nanosized scale without significant lattice defect contrast.

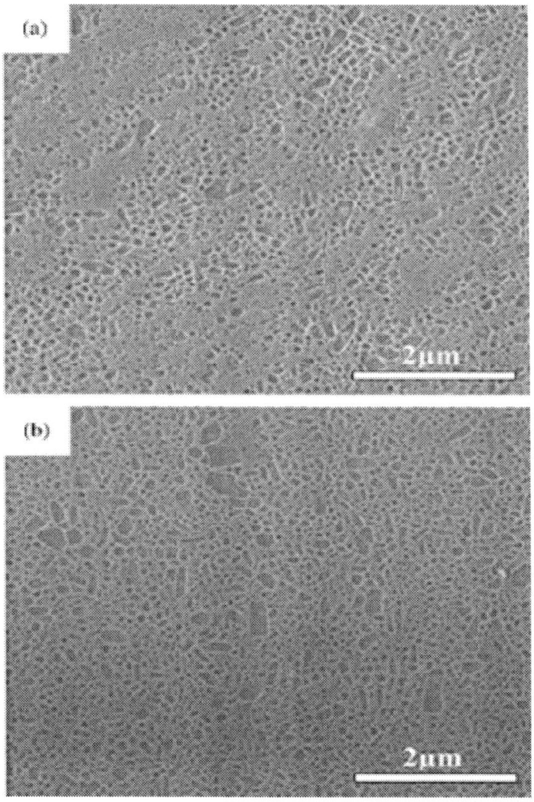

Figure 9. Typical SEM micrographs showing the plasma-etched surface of as-sintered Si_3N_4-based materials; samples were sintered at 1600°C for 5 min with a heating rate of (a) 50°C/min and (b) 100°C/min in vacuum.

Figure 10. BF/TEM image of specimen M1600, showing smaller, rough grains accompanied by larger, edged grains of B200 nm in width; the sample was sintered at 1600°C for 5 min with a heating rate of 100°C/min in vacuum.

Because of the overlapping of nanosized Si_3N_4, the noncrystalline interphase is hard to be identified in most of the observed regions. Figure 12(b) shows the typical diffraction pattern recorded from a larger area of the sample F1400. The measured values of d-spacing reveal a small deviation from that of β-Si_3N_4, suggesting that Si–N may be substituted by Al–O and form a β-SiAlON phase in the sample [60]. This is in line with XRD data and confirms a substitution in the Si_3N_4 phase. As described earlier in Section 3.2, a fully densified specimen F1400 implies that the liquid phase consists of homogeneously distributed sintering aids, which might be formed, leading to an enhanced fragmentation of nanosized grains and consolidation.

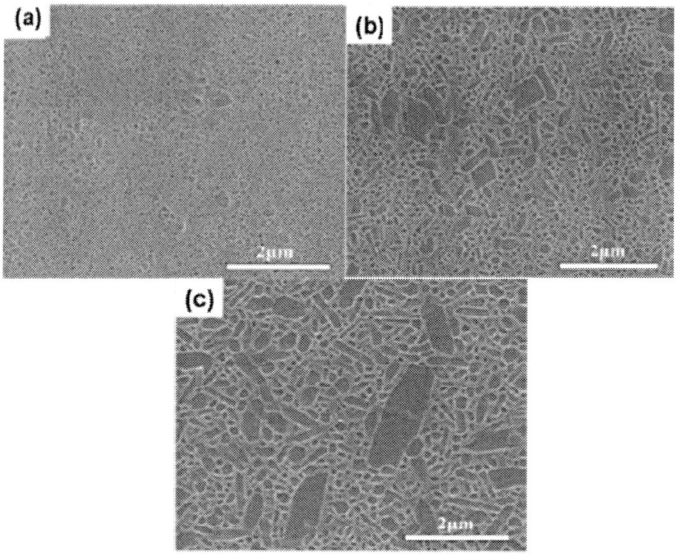

Figure 11. SEM micrographs showing the plasma-etched surface of as sintered Si_3N_4-based materials; samples were sintered at (a) 1400°C; (b) 1500°C; and (c) 1600°C for 3 min with a heating rate of 200°C/min in vacuum.

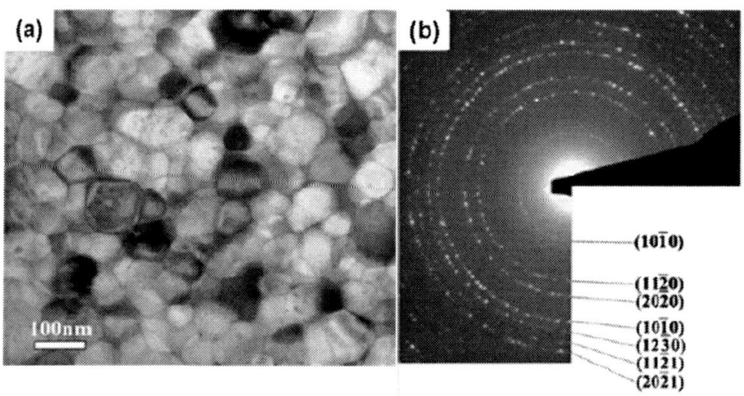

Figure 12. TEM micrographs of specimen F1400; the sample was sintered at 1400°C for 3 min with a heating rate of 200°C/min in vacuum: (a) bright-field image, showing most nanosized crystallites with rough morphology, and (b) typical diffracting pattern recorded from a larger area of the sample.

Figure 13. BF/TEM image of specimen F1500 showing large, elongated grains are embedded in a finer grained matrix; the sample was sintered at 1500°C for 3 min with a heating rate of 200°C/min in vacuum. Note some smaller, rough grains are observed and positioned in glass pockets, as indicated by arrows.

It is apparent that grains of this specimen are of comparable size to the as-received nanosizedβ-Si_3N_4-based powder, indicating that no appreciable grain growth occurs at the low sintering temperature of 1400 °C. The nanosized crystallites with rough morphology also suggest that a highly viscous glass phase exists at lower sintering temperature, inhibiting mass transfer of Si_3N_4 through the liquid medium and then grain growth. At the higher sintering temperature of 1500 °C, the corresponding BF/TEM image of specimen F1500 reveals that rapid grain growth occur, as can be seen in Figure 13. Many large β-Si_3N_4-based grains (showing a certain degree of edging or large, elongated grains) are embedded in a fine-grained matrix. Some smaller, rough grains are observed and positioned in glass pockets (marked by arrows).

With further sintering, they grow steadily into larger grains, and impingement on neighbor grains becomes more frequent (Figure 11(c)). The characteristic morphology suggests that there is fast mass transport from the smaller particles to larger grains in a few minutes by applying a rapid heating process. The value of toughness for each specimen is shown in Table 2. The highest toughness value of 5.4 MPa.m$^{1/2}$ is achieved for specimen F1600, whereas the medium toughness value of ~ 3.9 MPa.m$^{1/2}$ is obtained for specimens S1600 and M1600 with a slower heating rate at the similar sintering temperature of 1600 °C. The lowest measurable toughness is 1.93 MPa.m$^{1/2}$ for specimen F1400, which has ultra-fine crystallites, as indicated in Figure 11(a). In two independent experiments, Kawashima *et al.*[62] and Mitomo [63] showed that the fracture toughness of silicon nitride increases with the diameter of the larger elongated grains. Therefore, the resulting toughness has a strong dependence on the obtained microstructure of the consolidated specimens. The value of toughness is always < 4MPa.m$^{1/2}$ for samples with equiaxial grains and > 4MPa.m$^{1/2}$ for those with elongated grains [33, 36, 46].

3.4. Sintering Behavior

In general, SPS of as-received powder appears to follow the classic mechanism of liquid phase sintering, and the sintering behavior can be classified into three stages such as slow densification, fast densification, and solid-state sintering, according to the slope of these curves. In the stage of slow densification, the rearrangement of particles takes place in the sintering compact. During the second regime of fast densification, the shrinkage begins at about the die temperature of 1300, 1350, and 1360 °C for specimens S1600, M1600, and F1600, respectively, as the wetting liquid (or noncrystalline) phase of Y–Si–Al–O(–N) begins to form near 1350 °C [64]. This stage is responsible for the majority of the total shrinkage, and is believed to be the elimination of the pores inside the sintering compact. With regards to the time issue during the heat-up process, the densification of the rapidly heated specimen appears to be faster than that of the slowly heated sample. The elapsed time from 1300 to 1600 °C for the sintering compact with a heating rate of 200 °C/min is only 2 min, which is much shorter than 4 and 8 min for samples with heating rates of 100 and 50 °C/min, respectively. Therefore, the densification is thought to be enhanced by a rapid heating schedule. Moreover, from the estimation of the temperature difference (Figure 6) and shrinkage curves (Figure 8), a higher temperature was required for the rapidly heated sample to reach full densification. It can thus be inferred that the generated heat from the graphite component does not have enough time to transfer from the external surface to the center of the sintering compact, due to the low thermal conductivity of Si_3N_4-based materials [65]. Anisotropic grain growth of β-Si_3N_4 occurs above 1750 °C with its heating rate lower than 20 °C/min for a long dwelling time [66]. Grain growth is thermally activated and recognized as a process of static ripening. However, as exemplified in Figure 9 through Figure 13, equiaxial crystallites were obtained in samples S1600 and M1600 with a slower heating rate, while anisotropic grain growth for the β-Si_3N_4-based grains was encouraged for specimen F1600 and F1500 in a short time. Notably, the grain size for specimen F1500 with a lower temperature and shorter dwelling time was even greater than that of M1600 (comparing the BF/TEM images of Figs. 10 and 13). Even when the temperature difference between the sample and die accompanied by the effect of overshooting temperature are taken into consideration, the estimated actual temperature for F1500 does not exceed that for specimen M1600. As a consequence of the above observations, the rapid heating process seems to provide a unique mass-transport route for Si_3N_4-based materials during sintering.

Shen*et al.*[46, 67] correlated the rapid features of the grain growth that occur in the SPS process to a dynamic ripening mechanism by applying a high heating rate. The chemical composition of the liquid phase deviated from the state of thermodynamic equilibrium within minutes during a sintering cycle. The dissolution of small grains was promoted in a short time at the desired sintering temperature, and a momentarily oversaturated liquid was then formed. After a momentary extra chemical driving force was created between the finer and larger grains, the grains were coarsened by reducing the high surface interfacial energy between the liquid and solid phases. We thus consider that the kinetic aspects during rapid heating sintering, referred to as dynamic Ostwald ripening, may be one of the main factors that have a significant influence on the resulting microstructures of as-sintered materials. As the sintering compact stops shrinking, a marked difference in microstructure is observed for samples with different heating rates. The results for samples with different heating rates are essentially similar to those obtained by Shen and colleagues [46, 67, 68] using microsized Si_3N_4, AlN, and Y_2O_3 as raw materials. According to their reports, a high heating rate yields elongated grains and a slow heating rate gives rise to equiaxial grains. In addition, anisotropic grain growth occurs after full densification is achieved by the rapid heat-up process. However, the present work reveals that elongated grains for specimen F1500 are formed before a die temperature of 1600 °C for nearly full densification.

This is most likely because the nanoparticles have a higher solubility than submicrometersized particles during a liquid-phase sintering process [48]. Hence, by utilizing a nanopowder precursor and controlling the heating schedule, the growth of elongated grains was enhanced in a rapidly heated process, whereas Si_3N_4-based nanoceramics were obtained in a slowly heated process. Shen and colleagues sintered lithium-doped duplex α–βSiAlON at a heating rate of 100 °C/min [69]. The oxide-based liquid formed at the lowest eutectic in the Li_2O–SiO_2–Al_2O_3 system, and fully dense samples were obtained at 1450–1500 °C. The resulting α- or β-SiAlON grains had equiaxial morphology in samples consolidated at a die temperature of 1600 °C. The Si-AlON grains were submicrometer in size with short dwelling time (< 3 min) at 1600 °C; holding times of 5 min or more yielded microsized grains. As a result, our microstructure of specimen M1600, sintered at 1600 °C for 5 min, remained submicrometer in size, probably due to the lack of transformation. We also consider that the delayed densification behavior in a rapidly heated specimen (200 °C/min), as shown in Figure 8, greatly influences the abnormal grain growth behavior (indicated by arrows in Figure 11). The shrinkage curve (Figure 8) implies that there are uneliminated pores inside the sintering body beyond a die temperature of 1600 °C during the heat-up process, and the grains may be constrained in a localized region.

This phenomenon also suggests that a much higher degree of oversaturation may occur, and the driving force for grain growth along the basal and prismatic planes is accelerated. Hence, the abnormal grains can be seen in the rapidly heated specimens. Moreover, the coexisting α-Si_3N_4 in the starting material with a higher dissolving rate in the transient liquid phase might also increase the oversaturation. As a consequence, oversaturation at a localized region may occur and cause anisotropic grain growth, even abnormal grains, in the resulting microstructure.

3.5. Anisotropic Grain Growth by Coalescence

Extensive efforts have been made to identify the possible mechanisms for anisotropic grain growth in this study. This has been carried out mainly by observations from BF/TEM images at different tilting angles and selected area electron diffraction (SAD) patterns. Lattice defects can be observed frequently in the larger or elongated grains in specimen F1500 and F1600. One of the typical BF/TEM images of specimen F1500 sintered at 1500 °C for 3 min with a heating rate of 200 °C/min is shown in Figure 14. Some defect image characteristics, such as Moiré fringe or dislocation loops in the large, elongated grain indicate that at least four ultrafine grains, labeled to W, X, Y, and Z, are involved. In order to confirm why the lattice defect appears, it is important to identify the crystal orientation of these finer grains and the large, elongated grain. Figure 15(a) shows an enlarged image of particle W, and a typical Moiré fringe lies to the left of it, implying a small rotational misfit between the finer subgrain W and the neighboring larger grain.

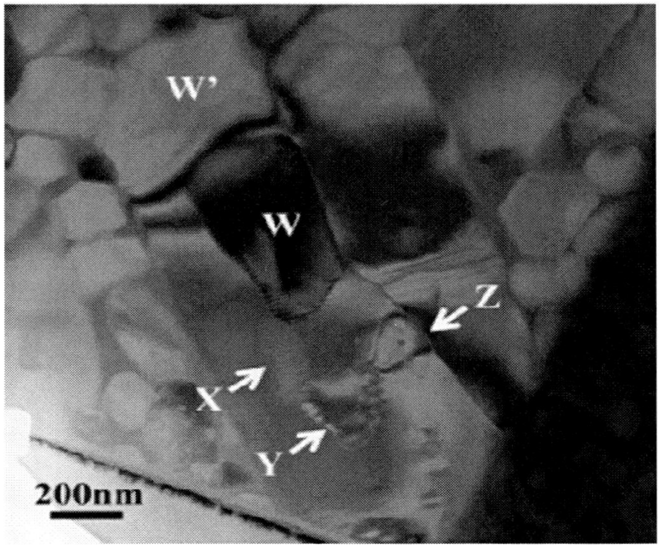

Figure 14. BF/TEM image of a large, elongated grain (specimen F1500), showing at least four ultrafine subgrains (labeled to W, X, Y, and Z) were involved; the sample was sintered at 1500°C for 3 min with a heating rate of 200°C/min in vacuum.

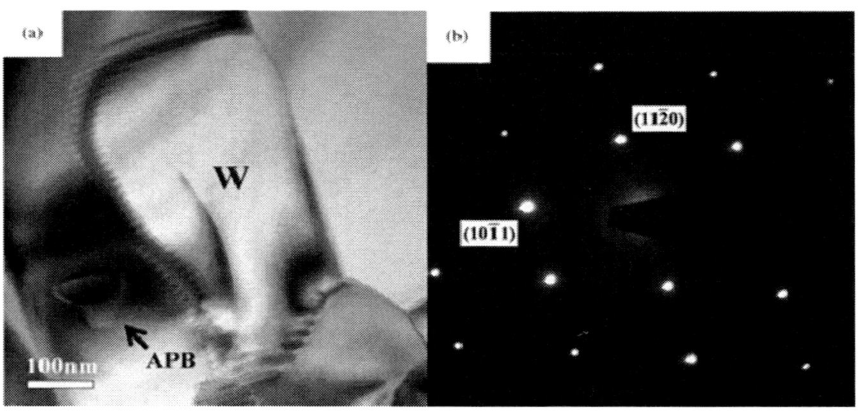

Figure 15. TEM images of specimen F1500 showing coalescence of two subgrains; the sample was sintered at 1500°C for 3 min with a heating rate of 200°C/min in vacuum: (a) bright-field images, (b) diffraction pattern at region W. Note the Moiré fringe around particle W and antiphase boundary (indicated by arrow) were involved in the large, elongated grain.

In order to distinguish the misorientation across the interface, the specific crystallographic orientation for regions W and W′ (shown in Figure 14) was identified by the misalignment of diffraction pattern. Figure 15(b) shows the [1 $\bar{1}$ 0 $\bar{1}$] SAD pattern recorded from region W. The closest reflections to the direct beam are the four [10 $\bar{1}$ 0] planes, and the next closest reflections are the two [11 $\bar{2}$ 0] reflections. By aligning the exact zone of [1 $\bar{1}$ 0 $\bar{1}$] as precisely as possible, the misorientation between the two subgrains (W and W′) comes out to be ~ 3.14 Å, which is a direct evidence for grain coalescence.Another lattice defect, the

π-boundary or antiphase boundary (APB) (indicated by alternate black/white fringes), can be seen under an appropriate diffraction condition, as indicated by an arrow in the left side of the elongated grain. An APB occurred when the atoms sheared through a displacement and formed an --- ABBA --- arrangement, which offers possible evidence of grain coalescence during sintering [70].

Figure 16(a) shows the BF/TEM micrograph in the middle of the large, elongated grain. The Figure reveals that a periodically arranged lattice fringe (subgrain X) is involved in the large, elongated grain. The lattice fringe is regarded as a Moiré fringe. Figure 16(b) reveals the [1 $\bar{1}$ 0 $\bar{1}$] SAD pattern recorded from subgrain X, and it is in the same crystallographic orientation as the large, elongated grain. In the meanwhile, the splitting diffracting spots can also be observed. Rediffraction within the same crystal or by passing into a second crystal is the cause of the double diffraction, and is directly responsible for the Morié effect in the electron images [71]. Therefore, the results strongly suggest that an ultrafine Si_3N_4-based crystallite is involved in the large, elongated grain. In Figure 16(a), the dislocation loops that encircle two subgrains (marked as Y and Z) inside the elongated grain can be seen. Lee and Hilmas[72] proposed explanations for the appearance of these dislocations, including a volume change between two grains, a difference of thermal expansion coefficient between two grains, and stacking defects. Typical stacking defects around the two subgrains can be observed in this Figure . In addition, the crystallographic orientation is the same for the two subgrains and this suggests that the large, elongated grains were possibly formed by the coalescence of ultrafine grained Si_3N_4. Based on the assumption and explanation of forming elongated Si_3N_4-based grains presented by Shen*et al.* [46], we have demonstrated that different sintering behavior can be observed with nanosizedβ-Si_3N_4-based powder by controlling the heating schedule. The TEM technique was first applied to observe the grain coarsening behavior and revealed that the elongated grains are formed either by a dissolution–reprecipitation process or by grain coalescence in a SPS process. The typical grain growth phenomenon of an elongated, large grain can also be seen in specimens F1500 and F1600.

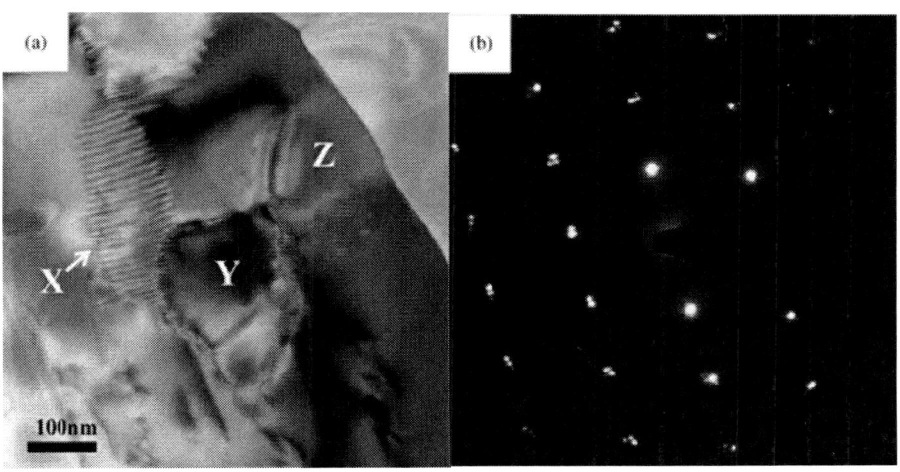

Figure 16. TEM images of specimen F1500 showing coalescence of three subgrains in the large, elongated grain; the sample was sintered at 1500°C for 3 min in a vacuum with a heating rate of 200°C/min: (a) bright-field image, (b) SAD pattern of region X in (a). Note the spot splitting observed in the SAD pattern obtained from subgrainX.

4. Influence of Conductive Nano-TiC on Si_3N_4Fabricated via SPS

Si_3N_4-based nanocomposites incorporating electro-conductive nanoparticles are regarded as promising materials for cutting tools and making electrical discharged machining possible. It has been reported that conductive TiN/Si_3N_4 [73, 74], TiN/ SiAlON [75, 76] and TiC_xN_{1-x}/SiC/Si_3N_4nanocomposites [77] were successfully fabricated in a short time by SPS. However, the powders for sintering were prepared through chemical routes, and are not suitable from an economic point of view. In one TiN/Si_3N_4 composite system produced via a conventional powder processing route, the results indicated that the SPS process enhances the grain coarsening of TiN compared with hot pressing, leading to the lower electrical conductivity for the SPS samples [78]. In order to obtain Si_3N_4-based nanocomposites containing nanosized grains of TiN through the powder processing route, the SPS technique seems to be avoided due to the pulsed current effect during sintering. NanosizedTiC was chosen as a conductive phase instead of TiN owing to two important characteristics: *(1)*TiC (or its solid solution) possesses similar mechanical and conductive properties as TiN[79, 80]; *(2)* The sintering temperature for nanosizedTiC (1700° C) [81] is much higher than that of nanosizedTiN (1400°C) [82]. Hence, using the hard-to-sinter TiC as a conductive phase is expected to yield semiconductive or conductive Si_3N_4-based nanocomposites without harming the mechanical properties.

The starting materials used in this study is the commercially available Si_3N_4nanopowder doped with sintering additives (6 wt% Y_2O_3 + 8 wt% Al_2O_3) and having average particle size of 70 nm as specified in the earlier section. Nanosized$TiC_xO_yN_z$ with a size of ~ 60 nm (Hefei Kiln Nanometer Technology Development, Hefei, China) was used as the conductive phase. Actually, the as-received TiCnanopowder has been confirmed as $TiC_xO_yN_z$ from the following experiments. It is reported that titanium carbide exhibits a minimum of 3% carbon vacancies and a maximum of 50% in its fcc carbon sublattice[80]. The content of C was determined by a carbon sulfur determinator (CS-244, Leco, LECO Corp., St. Joseph, MI). A nitrogen/oxygen analyzer (TC-436, Leco) was used to determine oxygen and nitrogen contents of the nano-$TiC_xO_yN_z$ powder. As a result, an amount of about 72.0 at.% C, 17.0 at.% O, and 11.0 at.% N was found in this nanopowder. It was probably because the existence of oxygen and nitrogen in the atmosphere lead to a titanium oxycarbonitride solid solution. Various compositions of nano-$TiC_{0.72}O_{0.17}N_{0.11}$ added to a mixture of Si_3N_4nanopowders and ethanol were ultrasonically dispersed for 15 min. The powders were mixed by planetary milling at a rotation speed of 300 rpm for 6 h using a 375 mL nylon bottle with Si_3N_4 balls. The powder mixture was dried in a rotary evaporator, cold-iso-statically pressed into round ingots at a pressure of 200 MPa, crushed, and then passed through a 200-mesh sieve for granulation. As-received Si_3N_4nanopowder in the same processing procedure without adding nano-$TiC_{0.72}O_{0.17}N_{0.11}$ was used for comparison. The composites with 0, 5, 10, and 20 wt% conductive phases in the matrix are designated as SN, 5T/SN, 10T/SN, and 20T/SN, respectively. Approximately 3 g mixture powder was loaded into a graphite Mold and the sample was heated to the desired sintering parameters under a uniaxial pressure of 30 MPa by SPS (Dr. Sinter 1050, Sumitomo Coal Mining, Kawasaki, Japan). Samples were heated to 600 °C for 3 min and then ramped to 1600 °C at a fixed rate of 200 °C/min. The temperature was held at 1600 °C for 3 min before the power was turned off and the cooling rate was 200

°C/min in the temperature interval of 1600–1000 °C. When Si_3N_4-based materials are sintered at their densification temperature, the temperature difference between the samples and die surface is in the range of 70–180 °C, depending on the dimension and arrangement of the heating element as well as sintering condition [37, 59, 83]. Henceforth, the temperature refers to the die temperature unless noted.

4.1. General Characteristics of the Si_3N_4 Based Composites

The apparent density of all Si_3N_4 based composite is shown in Table 3. These composites can be regarded as nearly full-dense because no obvious pores have been observed on polished surfaces from the SEM images (not shown here). The typical XRD patterns of sintered materials are shown in Figure 17. The composites consist of the β-Si_3N_4 matrix as a major phase and coexist with a titanium-based solid solution. The unit cell parameters for the two phases are shown in Table 3. The values for a_0 and c_0 for β-Si_3N_4-based material are ~ 7.620 and ~ 2.918 Å, respectively, which are somewhat deviate from the values for pure β-Si_3N_4 (+0.02 Å). This result suggests that a tiny amount of Si–N may be replaced by Al–O in the dissolution–reprecipitation process of liquid-phase sintering [60]. By considering the XRD result of the titanium-based phase, it can be observed that the lattice constant of as-received nano-$TiC_{0.72}O_{0.17}N_{0.11}$ powder is 4.3149 ± 0.0004 (Å). After sintering at 1600 °C with a heating rate of 200 °C/min (the sample temperature might be over 1750 °C), the lattice constant for composites 5T/SN, 10T/SN, and 20T/SN was 4.2775 ± 0.0010, 4.2761 ± 0.0008, and 4.2750 ± 0.0001 (Å), respectively (as indicated in Table 3). Based on the investigation of a TiC–TiN–TiOternery system by Newmenn*et al.*[84], the composition of as received nano-$TiC_{0.72}O_{0.17}N_{0.11}$ located near the TiC corner, would change toward the TiN–TiO boundary.

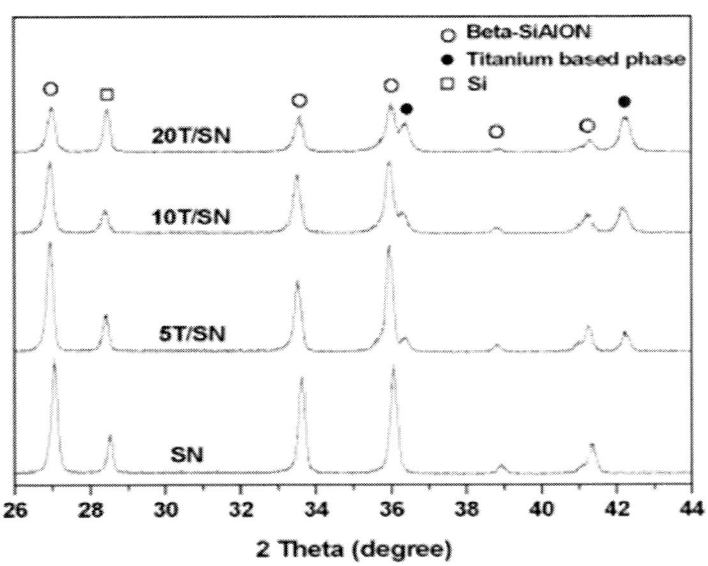

Figure 17. XRD patterns of the Si_3N_4-based composites with various contents of nano-$TiC_{0.72}O_{0.17}N_{0.11}$ addition.

Table 3. The Apparent Density and Unit Cell Parameters of the Formed Phases after Sintering

		β-phase unit cell (Å)		Titanium-based phase unit cell (Å)
Designation	Apparent density (gm/cm^3)	a_0	c_0	a_0
SN	3.25	7.6220 ± 0.0022	2.9182 ± 0.0010	-
5T/SN	3.30	7.6188 ± 0.0012	2.9171 ± 0.0005	4.2775 ± 0.0010
10T/SN	3.34	7.6184 ± 0.0013	2.9178 ± 0.0006	4.2761 ± 0.0008
20T/SN	3.39	7.6190 ± 0.0020	2.9184 ± 0.0009	4.2750 ± 0.0001

Thus, a smaller lattice constant of titanium oxycarbonitride for these sintered composites than that for as-received nano- $TiC_{0.72}O_{0.17}N_{0.11}$ can be found. Subsequently, from the investigation of the phase diagram of the glassy phase in the Si_3N_4-based system, oxide liquid would form at a eutectic temperature of 1370°C [64], and then oxynitride liquid would form at a higher temperature of 1700°C [85]. (In the present work, the actual temperature might be over 1750°C when the die temperature was 1600°C). There is a possibility that the atoms of O and N in the noncrystal interphase would react with titanium-based particles at the sintering temperature, i.e. cubic titanium oxide, titanium nitride, and titanium carbide have similar NaCl structures, and a mutual exchange of the atoms of O, N, and C at the interstitial sites in the crystal structure might occur [86]. As mentioned above, the side reactions in the present study can be written in the following equations:

$$TiC_{0:72}O_{0:17}N_{0:11}(s) + (Si, Al, Y, O)\ (l) \rightarrow Ti(C,O,N)\ (s) + C_x(s) + (Si, Al, Y, O)'(l) \quad (5)$$

(1370°C< T < 1700°C)

$$Ti(C,O,N)\ (s) + (Si, Al, Y, O, N)\ (l) \rightarrow Ti(C,O,N)' + C'_x(s) + (Si, Al, Y, O, N)'(l) \quad (6)$$

(> 1700°C)

Therefore, nano-$TiC_{0.72}O_{0.17}N_{0.11}$ as the starting powder has been transformed into a solid solution of titanium oxycarbonitride in a few minutes, probably due to the high specific surface area of the nanopowders. The lattice constant of titanium oxycarbonitride for these composites gradually decreases with the increase in nano-$TiC_{0.72}O_{0.17}N_{0.11}$ addition (as indicated in Table 3). This probably results from the different reactions between the titanium-based phase and noncrystal interphase. However, the exact mechanism for the formation of a solid solution titanium oxycarbonitride in this investigation is not completely clear, and needs to be investigated further. In order to reveal the effect of titanium oxycarbonitride on the resulting microstructures of Si_3N_4-based grains, we have discussed the morphology and distribution of titanium oxycarbonitride in the matrix first, and then considered the microstructural development of β-SiAlON. Figure 18 shows the element mapping for the Si, C, Ti, and back scattered electron (BSE) image of composite 5T/SN fabricated from Si_3N_4 composite powders and nanosized $TiC_{0.72}O_{0.17}N_{0.11}$ mixture and sintered by SPS. The other mapping results for elements, such as Al, Y, O, and N, have not been shown here because their distributions were similar and homogeneously spread over the whole detected area. The concentration bars for each element are shown in the right-hand side of Figure 18. The concentration for the given element in a localized volume is higher when the color is close to red or pink. In contrast, if the color is close to blue or black, the concentration is rare or nonexistent. Combining these mapping images of element distribution and the XRD results, the brighter phase can be identified as titanium oxycarbonitride, the darker region as β-SiAlON, and the gray region with a continuous network as a noncrystal interphase, as indicated in the BSE image. Some of the titanium oxycarbonitride have a rough and large morphology, near or above 500 nm in grain size, while others show agglomerates of the smaller titanium oxycarbonitride grains in the Si_3N_4-based matrix. Another important characteristic is that Si and C are present in the whole detection area, whereas Ti has an isolated-island distribution. The results imply that Si and Ti rarely diffuse in the system, whereas C is easier to diffuse through the whole sintering compact.

Element mapping and BSE image were also used to analyze the element distribution of composite 20T/SN (Figure 19). As mentioned in Figure 18, the brighter region can be identified as the titanium oxycarbonitride, and the darker phase is the β-SiAlON matrix. The BSE image indicates that most of the titaniumbased crystallites with a finer grain size, close to the starting powders, were embedded in the β-SiAlON and noncrystal interphase. The element mapping for titanium also reveals that there was no significant detectable Ti in the larger, dark region and the element of titanium seems to form a nearly connected network that is unlike the isolated-island distribution of titanium for composite 5T/SN. The change in electrical resistivity as a function of the amount of nano-$TiC_{0.72}C_{0.17}N_{0.11}$ in the starting materials is shown in Figure 20.

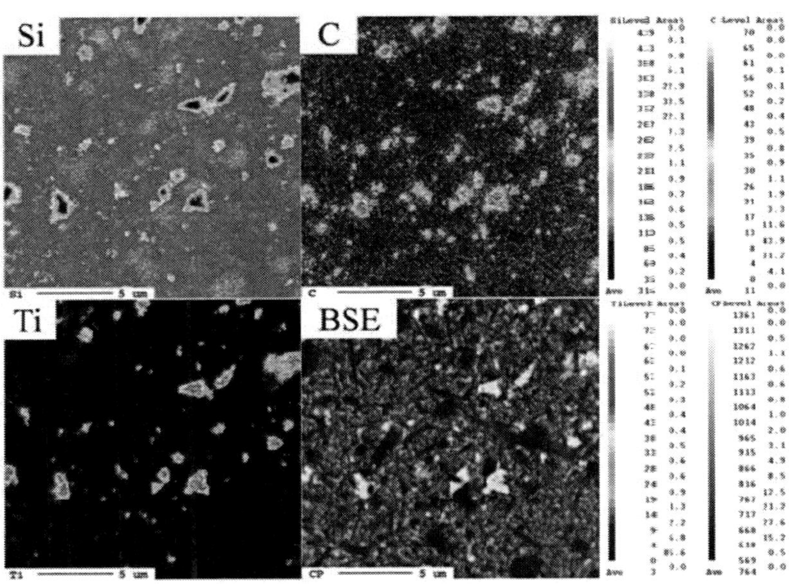

Figure 18. Element mapping (Si, C, and Ti) and back scattered electron micrographs of 5T/SN composite; the sample was sintered at 1600°C for 3 min with a heating rate of 200°C/min in a vacuum. The brighter part shows the titanium oxycarbonitride and the darker part is the β-SiAlON matrix.

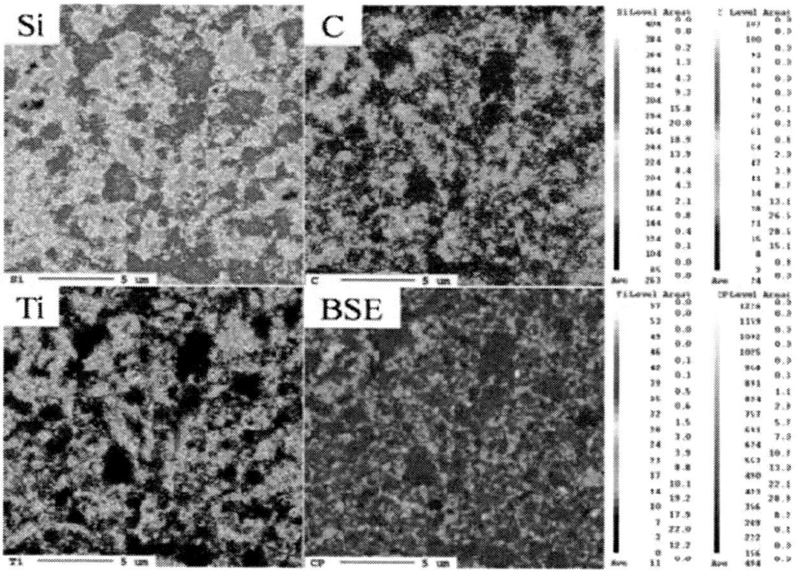

Figure 19. Element mapping (Si, C, and Ti) and BSE micrographs of 20T/SN composite; the sample was sintered at 1600°C for 3 min with a heating rate of 200°C/min in a vacuum. The brighter part shows the titanium-based phase and the darker part is the Si_3N_4-based matrix.

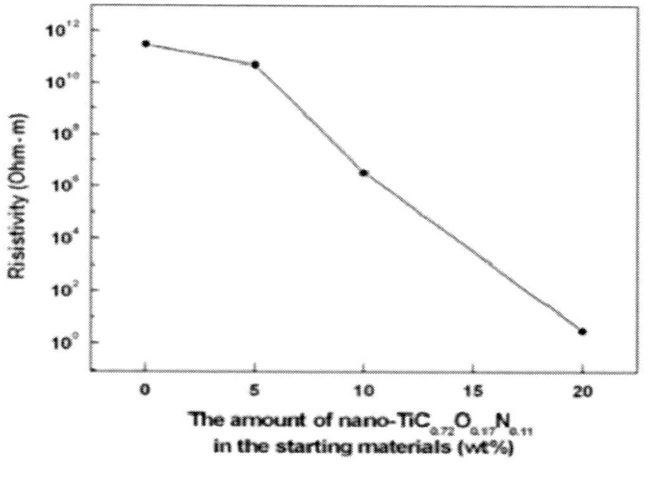

Figure 20. Change of electrical resistivity of Si_3N_4-based composites with various contents of nano-$TiC_{0.72}O_{0.17}N_{0.11}$ addition; the sample was sintered at 1600°C for 3 min with a heating rate of 200°C/min in a vacuum.

The monolithic β-SiAlON and composite 5T/SN have an insulating property. The electrical resistivity is decreasing with the increasing amount of nano-$TiC_{0.72}O_{0.17}N_{0.11}$ addition. Composite 20T/SN has an electrical resistivity of 3.13 (Ω.m) that is 10 orders lower than that for composite 5T/SN (5.09×10^{10}(Ω. m)). In two independent studies on the electric properties of the conductive particles mixing with insulating matrix, Deepa*et al.* [87] and Abdurakhmanov*et al.* [88] suggested that the electrical resistivity was significantly reduced by introducing a higher concentration of a conductive phase. The probability of electron hopping increases with the decreasing interparticle distance, delocalizing the electron wave within the conductive grains and promoting the tunneling effect between the conductive particles [87, 89, 90].Hence, in the present work, the resulting composites containing the conductive titanium-based phase show trends toward conducting nature. By evaluating the results for titanium and carbon mapping by EPMA (Figures 18 and 19), it can be seen that the morphology and distribution of titanium oxycarbonitride with the conductive property translate from large, isolated grains (5T/SN) to finer, near-connected crystallites (20T/SN), contributing to the lower electrical resistivity of resulting composites with the increasing content of titanium oxycarbonitride. This result provides important information about the microstructural development under sintering conditions, because a pulsed direct current might go through the compact during sintering. The current strength is directly dependent on the electrical resistivity of the sintering compact, and would affect the temperature distribution of the graphite mold and sample [91, 92].

4.2. Microstructural Development of Si_3N_4-Based Grains

Some representative microstructures for the composites sintered at 1600 °C for 3 min with a heating rate of 200 °C/min are shown in Figure 21. The typical microstructure for monolithic β-SiAlON is shown in Figure 21(a), in which elongated grains can be observed by applying a rapid heating schedule [93]. At a low content of titanium oxycarbonitride (composite 5T/SN), as indicated in Figure 21(b), most of the large, elongated β-SiAlON grains can be seen in the resulting microstructure as well.

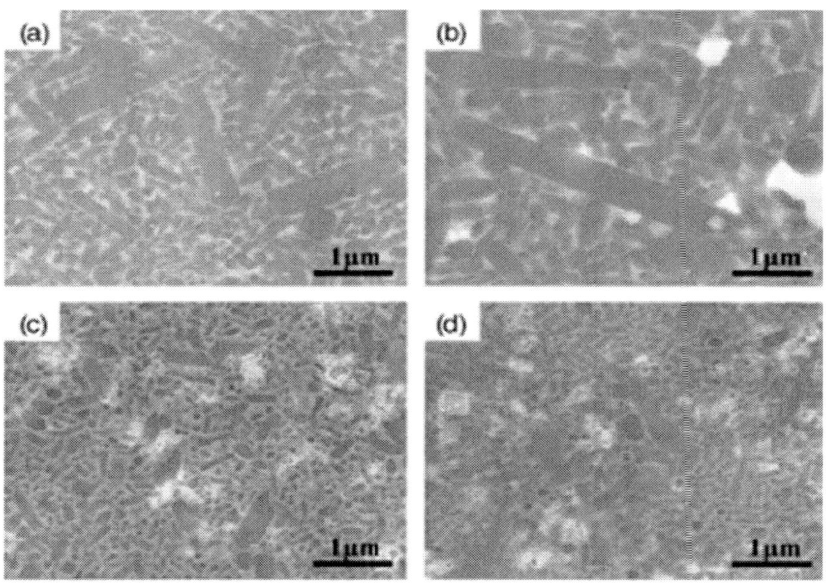

Figure 21. Representative SEM micrographs of Si_3N_4-based composites with (a) 0, (b) 5, (c) 10, and (d) 20 wt% nano-$TiC_{0.72}O_{0.17}N_{0.11}$ addition; the samples were sintered at 1600°C for 3 min with a heating rate of 200°C/min, applying a pressure of 30MPa in a vacuum by spark plasma sintering.

On the other hand, for the composites with a higher amount of titanium oxycarbonitride, the representative SEM microstructures can be observed in Figures 21(c) and (d) for 10T/SN and 20T/SN. These SEM results show that the areas for titanium oxycarbonitride (brighter region) and β-Si_3N_4-based matrix (darker region) are aggregates of many nanosized grains (certain etched Si_3N_4-based crystallites were embedded in unetched titanium oxycarbonitride agglomerates), which is significantly smaller than the β-SiAlON grains in specimens SN and 5T/SN. The frequency distributions of grain length, grain diameter, and aspect ratio of β-SiAlON grains for specimens SN and 5T/SN (shown in Figure 22), were evaluated by image analysis. Because the frequency distribution of grain length and diameter for composites 10T/SN and 20T/SN were almost all gathered in a small range, the grain characteristics are not shown here.

Hence, the results reveal that grain length and grain diameter at maximum frequency for composite 5T/SN were slightly greater than those of specimen SN, while the two specimens (10T/SN and 20T/SN) had a similar tendency for the frequency of aspect ratio. Owing to the relatively small number of large grains for the two specimens, their frequency distribution could not be clearly differentiated from the matrix grains. The distribution of area percentage of β-SiAlON grains as a function of grain diameter for specimens SN and 5T/SN was plotted, based on the results of image analysis (as shown in Figure 23). It is apparent that the composite 5T/SN exhibits a distinct bimodal distribution and exaggerated grains with diameters >1000 nm can be observed in the Figure . However, the monolithic specimen SN only shows a broad normal distribution.

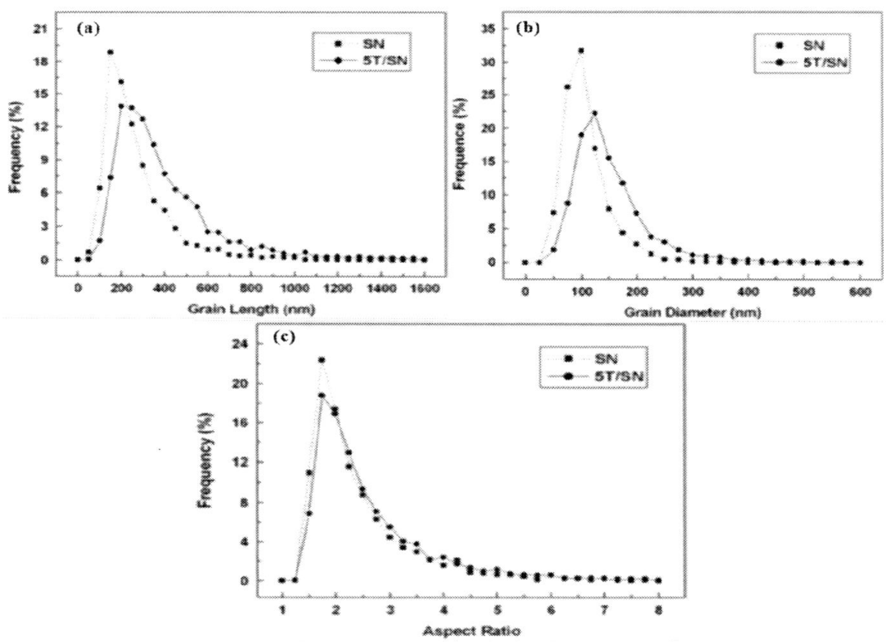

Figure 22. Frequency distribution of (a) grain length, (b) grain diameter, and (c) aspect ratio of monolithic β-SiAlON (SN) and composite with 5 wt% nano-$TiC_{0.72}O_{0.17}N_{0.11}$ addition (5T/SN). Samples were sintered at 1600°C for 3 min with a heating rate of 200°C/min in a vacuum by spark plasma sintering.

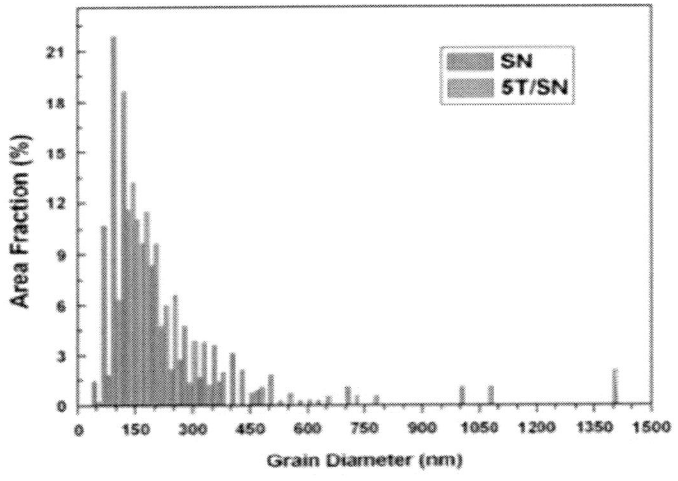

Figure 23. Distribution of surface area as a function of grain diameter in specimen SN and 5T/SN; samples were sintered at 1600°C for 3 min with a heating rate of 200°C/min in a vacuum.

The calculated results of image analysis for specimens SN, 5 T/SN, 10 T/SN, and 20 T/SN, sintered at 1600 °C for 3 min in a vacuum by SPS, are summarized in Table 4. The large grains were chosen from the 50 largest areas for each sample, and the average large grain length and diameter were also calculated. The results indicate that the average grain length and the average large grain length for the composite 5 T/SN are greater than those of monolithic sample SN.

A similar trend was also observed in the average grain diameter, average large grain diameter, and aspect ratio (a_{95}). However, for 10 T/SN and 20 T/SN composites, the grain characteristics are significantly smaller than those of specimens SN and 5 T/SN, and they are close to nanostructured features, which implies that no significant grain growth occurred after the SPS process. As a consequence, elongated grains of β-SiAlON could grow to larger grain size and higher aspect ratio, when a small amount of conductive titanium oxycarbonitride was contained in the Si_3N_4-based matrix. On the other hand, the nanostructured characteristics for most β-SiAlON grains with a lower aspect ratio seem to be maintained in the cases of the 10 T/SN and 20 T/SN composites. The composites with higher amounts of titanium oxycarbonitride (10 T/SN and 20 T/SN), show aggregates of many nanosized grains with a grain width of 70 – 200 nm, as demonstrated in Figs. 21(c) and (d).

Table 4. Characteristic Measurements of Microstructure Developed in Specimens SN and 5T/SN; the Samples were Sintered at 1600°C for 3 min with a Heating Rate of 200°C/min in a Vacuum

Samples	SN	5T/SN	10T/SN	20T/SN
Average grain diameter (nm)	108.3	153.9	73.6	83.7
Average grain length (nm)	271.3	409.6	149.7	156.7
Aspect ratio (a_{95})	4.60	5.07	3.18	2.97
Average large grain diameter (nm)	251.0	488.0	175.6	173.1
Average large grain length (nm)	996.5	1796.2	437.3	443.4

This kind of special microstructure, consisting of aggregated nanosizedβ-SiAlON-based grains surrounded by a strengthening or conductive phase, is consolidated without sintering additives via SPS by the several reports [73, 77, 92]. Hence, the observation suggests that a higher amount of titanium oxycarbonitride significantly suppresses the grain growth of β-SiAlON-based materials during a rapid heated sintering cycle, even with the presence of sintering additive in the sintering compact and the occurrence of a possible Joule heating effect at a localized temperature inside the sintering compact. In our previous study, we demonstrated that nanosized crystallites would grow into large, edge grains by applying a fast heating process [93]. We now find that average grain size and the resulting aspect ratio for a β-Si_3N_4-based composite (5 T/SN) are greater than those for monolithic β-Si_3N_4-based ceramic. Owing to the similar powder composition of Si_3N_4-based powders and sintering conditions for the specimens, the microstructural difference possibly results from the sintering temperature inside the compact during the SPS process, as indicated by Lai and Tien[66] i.e. the sintering temperature greatly influences the anisotropic growth rate of Si_3N_4-based materials. The special sintering behavior for the Si_3N_4-based composite containing nanosized titanium oxycarbonitride is illustrated in Figure 24 and described in the following sentences. During the SPS process, a voltage might be generated between the upper and lower punches due to the changes in electromagnetic field during pulsed current sintering process [94]. The existence of titanium oxycarbonitride embedded in the matrix would likely accumulate charges at the boundary and then introduce percolation between the adjacent nanosized grains of titanium oxycarbonitride. Subsequently, it is reasonable that a pulsed current may hop across the conductive titanium-based particles and then pass through the sintering compact.

A similar phenomenon was observed by Tung *et al.* [95] and Tang *et al.* [96] in the ultrathin gate oxide of a semiconductive device. Various kinds of microstructural damage and changes occur within the transistor structures and are likely to induce a thermal effect at an area, depending on the localized current density and resistivity of the leakage path. These heat generation mechanisms, such as dielectricbreakdown-induced epitaxy or dielectric-breakdown-induced migration, can lead to meltdown, recrystallization, or regrowth under appropriate circumstances [95, 96].

In general, the voltage drop over the sintering compact is expected to be very small. The calculated current for insulating composite 5T/SN was only six times larger than that of the specimen SN. However, in the dielectric regime of a system of nanometric conductive particles embedded in a continuous insulating matrix, the conductive particles do not touch each other and it is more suitable to apply interparticle tunneling mechanism [97]. Recently, Ambrosetti*et al.*[97] examined the tunneling conductance between two conductive particles (indicated as *i* and *j*) in an insulating matrix, and the equation is given below:

$$g_{ij} = g_0 \exp(-\frac{2\delta_{ij}}{\xi}) \tag{7}$$

where g_0 is a constant, ξ is the characteristic tunneling length, and δ_{ij} is the minimum distance between the adjacent particle surfaces. Although ξ is an uncertain value in the present study, $\delta_{ij}$for composite 5T/SN is much shorter than specimen SN, i.e. the $\delta_{ij}$for the composite is < 5 μm at least (as estimated from Figure 18), and for specimen SN is > 3 mm (the distance between the upper and lower punches). As a consequence, for composite 5 T/SN, a tunneling process might occur during the SPS process with the increasing tunneling conductance at a localized region. A pulse current might penetrate through the insulating matrix, and then give rise to a Joule heating effect around the current paths, leading to the generation of a thermal gradient inside the sintering compact. Without a significant pinning effect by the titanium oxycarbonitride, exaggerated grain growth for β-SiAlON occurs, and greater average grain size for the composite 5 T/SN compared with specimen SN was thus achieved in the present study.

In addition, a dielectric breakdown mechanism might occur at a localized region. The amount and composition of sintering aids could determine the densification process and grain growth in the Si_3N_4-based matrix, because the features of the microstructure are controlled by the viscosity of the liquid phase [28, 98]. In our study, there is a substitution of O and N into titanium based grains for 5 T/SN composite in which the amount of liquid phase as well as the O and N content of the liquid phase is reduced. Although it is hard to evaluate the chemical composition of the coexisting liquid phase, there is another possibility. The possibility is that the viscosity of liquid phase with a lower N/O ratio is reduced [99] leading to the grain size for composite 5 T/SN to be larger than that of specimen SN.

4.3. Grain Growth Behavior of Exaggerated Grains in TiC/Si_3N_4 Composite

A typical BSE/EPMA micrograph of 5 T/SN composite is shown in Figure 25. The exaggerated grains (grain diameter ~ 1 μm) can be seen as indicated by arrows. TEM observation was conducted to indentify the growth behavior. The bright-field image for one of the exaggerated grains with a grain width of ~ 1.6 μm is shown in Figure 26(a). The typical [0001] SAD diffraction pattern for the grain is shown in Figure 26(b) and the closest reflection to the direct beam is the six [10 $\bar{1}$ 0] planes.

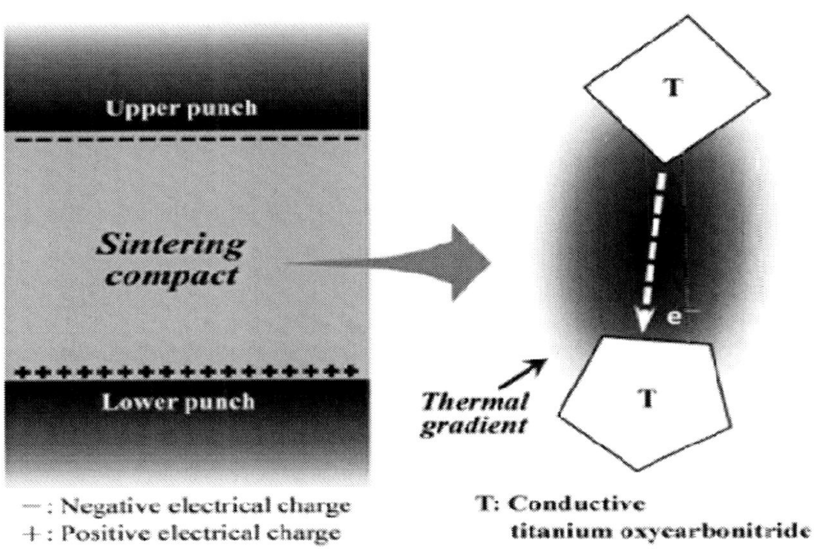

Figure 24. Schematic plot of a possible Joule heating mechanism around the current path during a SPS process.

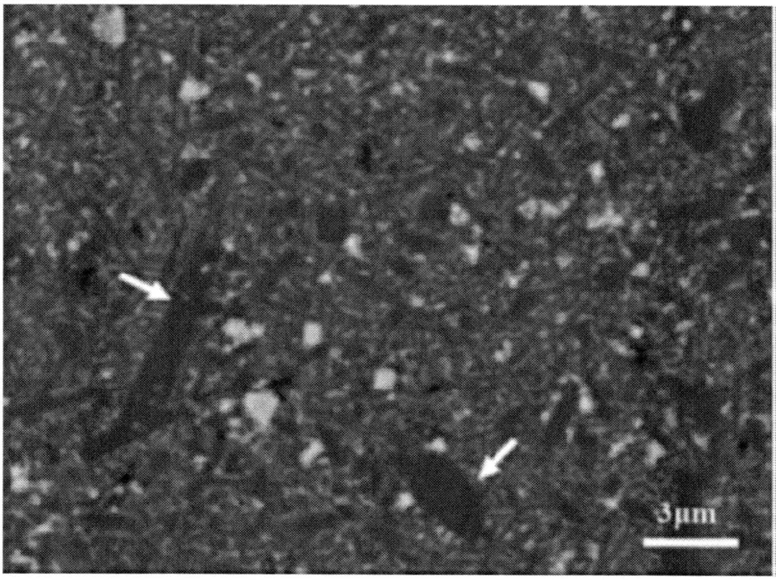

Figure 25. Typical BSE/EPMA image for composite 5T/SN showing exaggerated grains in the matrix; the sample was sintered at 1600°C for 3 min with a heating rate of 200°C/min in a vacuum.

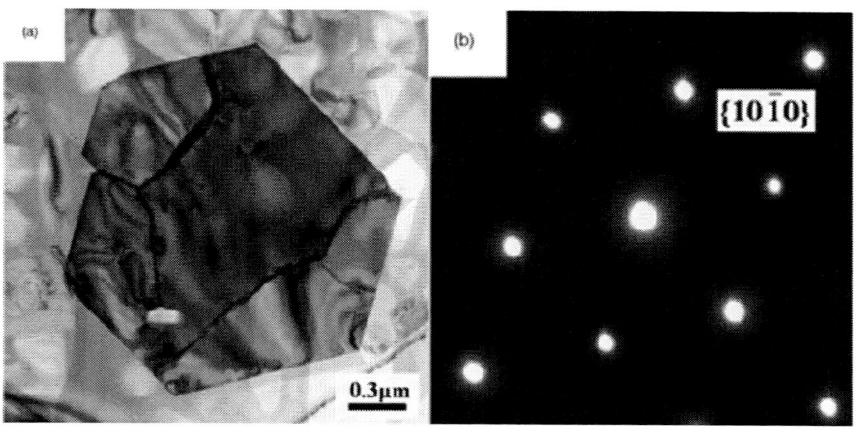

Figure 26. TEM images of exaggerated grain, showing coalescence of five β-SiAlONsubgrains; the sample was sintered at 1600°C for 3 min with a heating rate of 200°C/min in a vacuum: (a) BF/TEM image, (b) [0001] SAD diffraction pattern.

The *d*-space along the [10 $\bar{1}$ 0] is evaluated as 6.6 Å. This result implies that an exaggerated *β*-SiAlON grain was generated during the sintering process. At appropriate diffracted conditions, several subgrains, divided by a Morié fringe, in the edge and ultra-large grains are observed (Figure 26(a)). The phenomena also suggest grain coalescence occurs in the rapid heated schedule, as indicated in our previous work [93]. Figure 27(a) shows an enlarged image in one subgrain of exaggerated *β*-SiAlON. A darker core can be seen in the center of the subgrain. The high-resolution image near the core (as indicated by an arrow) is shown in Figure 27(b) and it reveals an epitaxial structure.

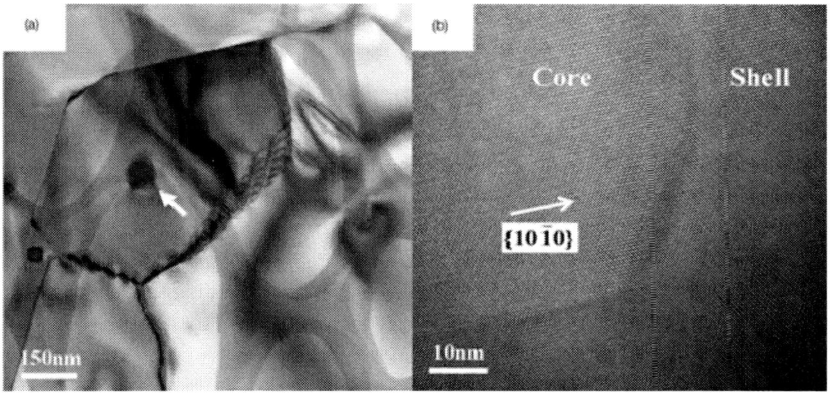

Figure 27. TEM images of the subgrain in the exaggerated grain, showing a core/shell structure in the subgrain and implying that a reprecipitation process occurs during sintering; the sample was sintered at 1600°C for 3 min with a heating rate of 200°C/min in a vacuum: (a) BF/TEM image, (b) high-resolution lattice image at the boundary with a core/shell structure.

The analyzed chemical composition of the core and shell analyzed by TEM/EDS with a nano beam of ~ 20 nm reveals that the Al/Si ratio for the core and shell is 6.7 and 8.8. Obviously, the Al/Si ratio is lower for the former. As a consequence, the different chemical composition between core and shell implies that dissolution-reprecipitation from nanopowder precursor to a grown subgrain occurs in a short time in the grain coarsening stage. In the traditional sintering technique, the average grain size of β-Si_3N_4-based grains always decreases with the increasing amount of conductive (or strengthening) phase [100]. In contrast in this study, the grain growth of β-SiAlON is further accelerated by a Joule heating effect, possibly resulting from percolation through the nanosized titanium oxycarbonitride particles [101]. As a consequence, an unusual phenomenon, that the average grain size for the composite is larger than the one for the monolithic material, can be seen in the composite of 5 T/SN with a lower amount of conductive titanium oxycarbonitride However, the nanostructured β-SiAlON and titanium oxycarbonitride are maintained after a high-temperature sintering process in the 10 T/SN and 20 T/SN composites.

CONCLUSION

The microstructural development of a commercially available nanosizedβ-Si_3N_4-based material has been investigated by applying different heating rates. The results show that a nanosizedβ-Si_3N_4-based ceramic could be obtained with slower heating rates (≤ 100 °C/min) and the grains are essentially equiaxed with an average grain size of 70–200 nm. In the cases of a faster heating rate (200 °C/min), bimodal distribution of grain size is observed. Anisotropic grain growth was enhanced at a relatively low temperature of 1500 °C, owing to dynamic Ostwald ripening accompanied by grain coalescence. Moreover, abnormal grain growth occurred due to oversaturation at some localized regions. Therefore, applying a nanosized powder precursor, which could enhance the solubility of the starting powder, accompanied with a rapid heating rate by SPS is an efficient method to yield highly toughened β-Si_3N_4-based ceramics in a short time. The resulting fracture toughness for each specimen sintered with different heating rates is indeed determined by the grain size of Si_3N_4. The toughness of samples sintered at 1600 °C for nanosized, equiaxed crystallites (slower heating rate) and elongated, large Si_3N_4 grains (high heating rate) is ~ 3.9 and 5.4 $MPa.m^{1/2}$, respectively. Furthermore, by utilizing nanosized powder precursors and SPS technique, the microstructural development of a series of Si_3N_4-based composites containing nanosized titanium oxycarbonitride has been demonstrated. The electrical resistivity for each specimen is greatly determined by the connectivity of titanium oxycarbonitride. It can be concluded that the nano-$TiC_{0.72}O_{0.17}N_{0.11}$ particles incorporated in the insulating matrix may induce a Joule heating effect and a pinning effect during the SPS process. The former results in microstructural coarsening of β-SiAlON, whereas the latter inhibits the grain growth of Si_3N_4-based grains. For composite 5 T/SN, titanium oxycarbonitride revealed an isolated-island distribution with an insulating property, and β-Si_3N_4-basedgrain showed larger average grain size and aspect ratio compared with monolithic β-Si_3N_4-based ceramic, probably owing to a temporary temperature rise caused by the dielectric breakdown mechanism at a localized region. Moreover, dissolution–reprecipitation and coalescence occurred during sintering and were observed by TEM, contributing to exaggerated grains appearing in the present composite. When the amount of nano-$TiC_{0.72}O_{0.17}N_{0.11}$ added is up to 10 and 20 wt%, the grain growth behavior of Si_3N_4 is suppressed significantly, despite the existence of sintering aids. These results also suggest a simple method to fabricate the Si_3N_4-based nanocomposites.

ACKNOWLEDGMENTS

Authors are thankful to National Science Council of Taiwan for its financial support under the contract No: NSC 99-2923-E-006-002-MY3 to carry out the present work.

REFERENCES

[1] F. L. Riley, *J. Am. Ceram. Soc.* 83, 245 (2000).

[2] For current work on automotive applications see *6th International Symposium on Ceramic Materials and Components for Engines*, ed. K. Niihara, S. Kanzaki, K. Komeya, S. Hirano and K. Morinaga. Japan Fine Ceramics Association, Halifax Onarimon Bldg. 6F, 3-24-10, Nishi-Shimbashi, Minatoku, Tokyo 105-0003, Japan, Printed by Technoplaza Co., Ltd., (1998).

[3] T. P. Herbell, and W. A. Sanders, Monolithic ceramics. In *Flight-Vehicle Materials, Structure, and Dynamics, Assessment and Future Directions, Vol. 3: Ceramics and Ceramic-Matrix Composites,* ed. S. R. Levine. The American Society of Mechanical Engineers, New York, (1992), pp. 19-41.

[4] V. M. Parathasarathy, J. R. Price, W. D. Brentnall, G. A. Graves, and S. Goodrich, Material characterization of candidate silicon based ceramics for stationary gas turbine applications. ASME Paper 95-GT-249, *ASME International Gas Turbine and Aeronautical Congress and Exposition*, Houston, TX. June, 1995.

[5] V. Parthasarathy, M. V. Roode, J. Price and S. Gates, Review of Solar's ceramic stationary gas turbine development program. In *6th International Symposium on*

Ceramic Materials and Components for Engines, ed. K. Niihara, S. Kanzaki, K. Komeya, S. Hirano and K. Morinaga. Japan Fine Ceramics Association, Halifax Onarimon Bldg. 6F, 3-24-10, Nishi-Shimbashi, Minatoku, Tokyo 105-0003, Japan, Printed by Technoplaza Co., Ltd., (1998), pp. 259-264.

[6] T. Nakashima, T. Tatsumi, I. Takehara, Y. Ichidawa, and H. Kobayashi, Research and Development of CTGT 302. In *6th International Symposium on Ceramic Materials and Components for Engines,* ed. K. Niihara, S. Kanzaki, K. Komeya, S. Hirano and K. Morinaga. Japan Fine Ceramics Association, Halifax Onarimon Bldg. 6F, 3-24-10, Nishi-Shimbashi, Minatoku, Tokyo 105-0003, Japan, Printed by Technoplaza Co., Ltd., (1998), pp. 233-236.

[7] J. F. Collins, R. W. Gerby, Journals of Met. 7, 612 (1955).

[8] N. L. Parr, G. F. Martin and E. R. W. May, *Preparation, Microstructure, and Mechanical Properties of Silicon Nitride, in Special Ceramics 1*, ed. P. Popper, Heywood, London, (1960), pp. 102-135.

[9] G. G. Deeley, J. M. Herbert, N. C. Moore, *Powder Metall.,* 8, 145 (1961).

[10] R. F. Coe, R. J. Lumby, M. F. Pawson, *Some Properties and Applications of Hot-Pressed Silicon Nitride, in Special Ceramics 5*, ed. P. Popper, British Ceramics. Association, Stoke-on-Trent, U.K. (1972), pp. 361-376.

[11] K. H. Jack, *Phase Diagrams: Materials Science and Technology,* ed. A. M. Alper, New York: Academic (1978), pp. 241- 85.

[12] R. R. Wills, J. A. Cunningham, *J. Mater. Sci.* 12, 208 (1977).

[13] F. F. Lange, *Bull. Am. Ceram. Soc.* 59, 239 (1980).

[14] J. M. Finkel' shtein, *Poroshk. Metall.* 138, 96 (1976).

[15] R. W. Rice, W. J. McDonough, *J. Am. Ceram. Soc.* 58, 264 (1975).

[16] F.F. Lange, J. *Ceram. Soc. Japan* 114, 873 (2006).

[17] Z. Krstic, V. D. Krstic, *Mater. Sci. Forum* 413, 129 (2003).

[18] G. Woetting, G. Ziegler, *Ceram. Int.* 10(1),18 (1984).

[19] Z. Krstic, Z. Yu, V. D. Krstic, *J. Mater. Sci.* 42, 5431 (2007).

[20] D. Hardie and K. H. Jack, *Nature* 180, 332 (1957).

[21] S. Wild, P. Grieveson, K. H. Jack, *Spec. Ceram.* 5, 385 (1972).

[22] S. Hampshire, *Journal of Achievements in Materials and Manufacturing Engineering,* 24 (1), 43 (2007).

[23] S. Hampshire and K.H. Jack, *Proc. Brit. Ceram. Soc.,* 31, 37 (1981).

[24] W. D. Kingery, *J. Appl. Phys.* 30, 301(1959).

[25] D. P. Thompson, *Nature* 417, 237 (2002).

[26] A. I. Raichenko, *Basics of Electric Current-Assisted Sintering*, Moscow, Metallurgy (1987) pp.129.

[27] A.V. Ragulya, Fundamentals of Spark Plasma Sintering, *Encyclopedia of Materials: Science & Technology,* (2010) pp.1-5.

[28] J. J. Melendez-Martinez and A. Dominguez-Rodriguez, *Prog. Mater. Sci.,* 49 [1], 19 (2004).

[29] F. L. Riley, *J. Am. Ceram. Soc.,* 83 [2], 245 (2000).

[30] F. F. Lange, J. Am. Ceram. Soc., 56 [10], 518 (1973).

[31] P. F. Becher, E. Y. Sun, K. P. Plucknett, K. B. Alexander, C. H. Hsueh, H. T. Lin, S. B. Waters, C. G. Westmoreland, E. S. Kang, K. Hirao, and M. E. Brito, *J. Am. Ceram. Soc.,* 81 [11], 2821 (1998).

[32] H. H. Lu and J. L. Huang, *J. Am. Ceram. Soc.,* 85 [9], 2331 (2002).
[33] G. D. Zhan, M. Mitomo, T. Nishimura, R. J. Xie, T. Sakuma, and Y. Ikuhara, *J. Am. Ceram. Soc.,* 83 [4], 841 (2000).
[34] X. Xu, T. Nishimura, N. Hirosaki, R. J. Xie, Y. Yamamoto, and H. Tanaka, *Acta Mater.,* 54 [1], 255 (2006).
[35] M. Herrmann, I. Schulz, C. Schubert, W. Hermel, I. Zalite, and G. Ziegler, *CFI-Ceram. Forum Int.*, 75 [4], 38 (1998).
[36] J. H. Kim, B. V. M. Kumar, S. H. Hong, and H. D. Kim, *J. Am. Ceram. Soc.,* 93 [5], 1461 (2010).
[37] M. Suganuma, Y. Kitagawa, S. Wada, and N. Murayama, *J. Am. Ceram. Soc.,* 86 [3], 387 (2003).
[38] Y. Zhou, K. Hirao, M. Toriyama, and H. Tanaka, *J. Am. Ceram. Soc.,* 83 [3], 654 (2000).
[39] Z. J. Shen, M. Johnsson, Z. Zhao, and M. Nygren, *J. Am. Ceram. Soc.,* 85 [8], 1921 (2002).
[40] G. Bernard-Granger and C. Guizard, *ActaMater.,* 55 [10], 3493 (2007).
[41] R. Orrù, R. Licheri, A. M. Locci, A. Cincotti, and G. C. Cao, *Mater. Sci. Eng. R-Rep.,* 63 [4–6] 127 (2009).
[42] T. Nishimura, M. Mitomo, H. Hirotsuru, and M. Kawahara, J. *Mater. Sci. Lett.,* 14 [15], 1046 (1995).
[43] X. Xu, T. Nishimura, N. Hirosaki, R. J. Xie, Y. C. Zhu, Y. Yamamoto, and H. Tanaka, *J. Am. Ceram. Soc.,* 88 [4], 934 (2005).
[44] X. Xu, T. Nishimura, N. Hirosaki, R. J. Xie, Y. Yamamoto, and H. Tanaka, ''Fabrication of Alpha-Sialon Nano-Ceramics''; *in Science of Engineering Ceramics III, Key Engineering Materials,* Edited by T. Ohji, T. Sekino, and K. Niihara. Trans Tech Publications Ltd., Zurich-Uetikon, Switzerland, Vol. 317–318, (2006) pp. 629–32.
[45] X. Xu, T. Nishimura, N. Hirosaki, R. J. Xie, Y. Yamamoto, and H. Tanaka, *Nanotechnology* 16 [9], 1569 (2005).
[46] Z. J. Shen, Z. Zhao, H. Peng, and M. Nygren, *Nature* 417 [6886], 266 (2002).
[47] R. M. German, *Liquid Phase Sintering:* Plenum Press, New York, (1985).
[48] C. M. Wang, H. Emoto, and M. Mitomo, *J. Am. Ceram. Soc.,* 81 [5], 1125 (1998).
[49] C. J. Lee, D. J. Kim, and E. S. Kang, *J. Am. Ceram. Soc.,* 82 [3], 753 (1999).
[50] C. J. Lee, J. I. Chae, and D. J. Kim, J. *Eur. Ceram. Soc.,* 20 [14–15], 2667 (2000).
[51] T. J. B. Hollandm and S. A. T. Redfern, *Mineral. Mag.,* 61 [404], 65 (1997).
[52] G. Woetting, B. Kanka, and G. Ziegler, ''Microstructural Development, Microstructural Characeterization and Relation to Mechanical Properties of Dense Silicon Nitride''; *in Non-Oxide Technical and Engineering Ceramics,* Edited by S. Hampsire. Elsevier Applied Science, London, U.K., (1986), pp. 83–96
[53] A. G. Evans and E. A. Charles, *J. Am. Ceram. Soc.,* 59 [7–8], 371 (1976).
[54] N. Murayama and W. Shin, *J. Ceram. Soc. Jpn.,* 108 [9], 799 (2000).
[55] A. Zavaliangos, J. Zhang, M. Krammer, and J. R. Groza, *Mater. Sci. Eng. A-Struct. Mater. Prop. Microstruct. Process.,* 379 [1–2], 218 (2004).
[56] X. Wang, S. R. Casolco, G. Xu, and J. E. Garay, *Acta Mater.,* 55 [10], 3611 (2007).
[57] J. Rathel, M. Herrmann, and W. Beckert, *J. Eur. Ceram. Soc.,* 29 [8], 1419 (2009).
[58] Y. C. Wang and Z. Y. Fu, *Mater. Sci. Eng. B-Solid State Mater. Adv. Technol.,* 90 [1–2], 34 (2002).

[59] D. Salamon, Z. Shen, and P. Šajgalík, *J. Eur. Ceram. Soc.*, 27 [6], 2541 (2007).

[60] Z. J. Shen and M. Nygren, *J. Mater. Chem.*, 11 [1], 204 (2001).

[61] M. Herrmann, I. Schulz, and I. Zalite, *J. Eur. Ceram. Soc.*, 24 [12], 3327 (2004).

[62] T. Kawashima, H. Okamoto, H. Yamamoto, and A. Kitamura, *Nippon SeramikkusuKyokaiGakujutsuRonbunshi-J. Ceram. Soc. Jpn.*, 99 [4], 320 (1991).

[63] M. Mitomo, ''Toughening of Silicon Nitride Ceramics by Microstructural Control''; *in Proceedings of an International Symposium on the Science of Engineering Ceramics,* Edited by S. Kimura, and K. Niihara. Ceramic Society of Japan, Tokyo, Japan, (1991), pp. 101–107

[64] U. Kolitsch, H. J. Seifert, T. Ludwig, and F. Aldinger, *J. Mater. Res.*, 14 [2], 447 (1999).

[65] M. W. Barsoum, Fundamentals of Ceramics, *McGraw-Hill Series in Materials Science and Engineering,* International Edition, McGraw-Hill, New York, (2000), p. 505.

[66] K. R. Lai and T. Y. Tien, *J. Am. Ceram. Soc.*,76 [1], 91 (1993).

[67] H. Peng, Z. J. Shen, and M. Nygren, *J. Mater. Res.*, 19 [8], 2402 (2004).

[68] Z. J. Shen, H. Peng, J. Liu, and M. Nygren, *J. Eur. Ceram. Soc.*, 24 [12], 3447 (2004).

[69] H. Peng, Z. J. Shen, and M. Nygren, *J. Mater. Chem.*, 13 [9], 2285 (2003).

[70] H. H. Lu and J. L. Huang, *J. Am. Ceram. Soc.*, 84 [8], 1891 (2001).

[71] D. B. Williams and C. B. Carter, Transmission Electron Microscopy: *A Textbook for Materials Science*, Plenum Press, New York, (1996), pp. 278–80.

[72] W. E. Lee and G. E. Hilmas, *J. Am. Ceram. Soc.*, 72 [10], 1931 (1989).

[73] S. Kawano, J. Takahashi, and S. Shimada, *J. Am. Ceram. Soc.*,86 [4], 701 (2003).

[74] S. Kawano, J. Takahashi, and S. Shimada, J. *Mater. Chem.*, 12 [2], 361 (2002).

[75] E. Ayas, A. Kara, and F. Kara, *J. Ceram. Soc. Jpn.*, 116 [1355], 812 (2008).

[76] S. Shimada and K. Kato, *Mater. Sci. Eng. A-Struct. Mater. Prop. Microstruct. Process.*, 443 [1–2], 47 (2007).

[77] R. G. Duan, J. E. Garay, J. D. Kuntz, and A. K. Mukherjee, *J. Am. Ceram. Soc.*, 88 [1], 66 (2005).

[78] Z. Guo, G. Blugan, R. Kirchner, M. Reece, T. Graule, and J. Kuebler, *Ceram. Int.*, 33 [7], 1223 (2007).

[79] W. Lengauer, S. Binder, K. Aigner, P. Ettmayer, A. Guillou, J. Debuigne, and G. Groboth, *J. Alloy. Compd.*, 217 [1], 137 (1995).

[80] W. S. Williams, *Int. J. Refract. Met. Hard Mat.*, 17 [1–3], 21 (1999).

[81] M. S. El-Eskandarany, M. Omori, T. Kamiyama, T. J. Konno, K. Sumiyama, T. Hirai, and K. Suzuki, *Sci. Rep. Res. Inst., Tohoku Univ. Ser. A Phys.*, 43 [2], 181 (1997).

[82] L. J. Wang, W. Jiang, L. D. Chen, M. Yang, and H. M. Zhu, *J. Am. Ceram. Soc.*, 89 [7], 2364 (2006).

[83] Z. J. Shen, Z. Zhao, H. Peng, and M. Nygren, Nature, 417 [6886], 266 (2002).

[84] G. Neumann, P. Ettmayer, and R. Kieffer, *Monatsh. Chem.*, 103 [4], 1130 (1972).

[85] S. Hampshire, ''Oxynitride Glasses and Glass Ceramics''; *in Materials Research Society Symposia Proceedings. Silicon Nitride Ceramics: Scientific and Technological Advances,* Edited by I.-W. Chen, P. F. Becher, M. Mitomo, G. Petzow, and T.-S. Yen. Materials Research Society, Pittsburgh, PA, (1993), Vol. 287, pp. 93–104.

[86] T. Lièko, V. Figusch, and J. Púchyová, *J. Eur. Ceram. Soc.*, 5 [4], 257 (1989).

[87] K. S. Deepa, S. Nisha, P. Parameswaran, M. T. Sebastian, and J. James, *Appl. Phys. Lett.*, 94 [14], 142902, (2009).

[88] U. Abdurakhmanov, S. Sharipov, Y. Rakhimova, M. Karabaeva, and M. Baydjanov, J. *Am. Ceram. Soc.,* 89 [9], 2946 (2006).
[89] V. Ambegaokar, B. I. Halperin, and J. S. Langer, *Phys. Rev. B-Solid State,* 4 [8], 2612 (1971).
[90] P. Sheng, B. Abeles, and Y. Arie, *Phys. Rev. Lett.,* 31 [1], 44 (1973).
[91] U. Anselmi-Tamburini, S. Gennari, J. E. Garay, and Z. A. Munir, *Mater. Sci. Eng. A-Struct. Mater. Prop. Microstruct. Process.,* 394 [1–2], 139 (2005).
[92] J. L. Wan, M. J. Gasch, and A. K. Mukherjee, J. *Am. Ceram. Soc.,* 86 [3], 526 (2003).
[93] C.H. Lee, H. H. Lu, C. A. Wang, P. K. Nayak, J. L. Huang, *J. Am. Ceram. Soc.* 94[4], 1182 (2011).
[94] M. H. Fang, ''Mechanism Of Spark Plasma Sintering Of Non-Conductive Materials''; *Doctoral Dissertation of Tsinghua University*, (2005).
[95] C. H. Tung, K. L. Pey, L. J. Tang, M. K. Radhakrishnan, W. H. Lin, F. Palumbo, and S. Lombardo, *Appl. Phys. Lett.,* 83 [11], 2223 (2003).
[96] L. J. Tang, K. L. Pey, C. H. Tung, M. K. Radhakrishnan, and W. H. Lin, *IEEE Trans. Device Mater. Reliab.,* 4 [1], 38 (2004).
[97] G. Ambrosetti, C. Grimaldi, I. Balberg, T. Maeder, A. Danani, and P. Ryser, *Phys. Rev. B,* 81 [15], 155434, (2010).
[98] M. Belmonte, A.de Pablos, M. I. Osendi, and P. Miranzo, *Mater. Sci. Eng. A-Struct. Mater. Prop. Microstruct. Process.,*475 [1–2], 185 (2008).
[99] N. Saito, K. Kai, S. Furusho, K. Nakashima, K. Mori, and F. Shimizu, *J. Am. Ceram. Soc.,* 86 [4], 711 (2003).
[100] H. L. Chiu and J. L. Huang, *Ceram. Int.,* 20 [1], 49 (1994).
[101] C. H. Lee, H. H. Lu, C. A. Wang,P.K. Nayak, andJ. L. Huang, *J. Am. Ceram. Soc.,*94[3], 959(2011).

In: Recent Advances in Ceramic Materials Research
Editors: Joan J. Roa Rovira and Merce Segarra Rubi
ISBN: 978-1-62417-729-3

Chapter 8

FRICTION, HARDNESS AND ELASTIC MODULUS DETERMINED BY AFM-FS AND NANOINDENTATION TECHNIQUES FOR ADVANCED CERAMICS MATERIALS

***G. Oncins*[1,2,*], *J. J. Roa*[3], *E. Rayón*[4], *J. Díaz*[1], *M. Morales*[5], *M. Segarra*[5] *and F. Sanz*[2]**

[1]Scientific and Technological Centers of the University of Barcelona (CCiTUB), Barcelona, Spain

[2]Institut of Bioengineering of Catalonia, Barcelona, Spain

[3]Institut Pprime Département de Physique et Mécanique des Matériaux, CNRS-Université de Poitiers, Futuroscope Chasseneuil Cedex, France

[4]Universidad Politécnica de Valencia. Materials Science Institute, Valencia, Spain

[5]Departement de Ciència dels Materials i Eng. Metal·lúrgica, Facultat de Física i Química,Universitat de Barcelona, Barcelona, Spain

ABSTRACT

In the last years, nanoindentation by means of Atomic Force Microscopy-Force Spectroscopy (AFM-FS) or Nanoindenters has become a powerful tool to study the nanomechanics of all type of materials at micro-, nano- and also picometric scale, from soft metals, like copper, to brittle materials, as ceramics. The experimental basis of these techniques is the evaluation of the response of a material to an applied vertical load (in order to obtain the hardness, *H,* and the elastic modulus, *E*) or to a shear force (Lateral Force Microscopy, LFM) so as to obtain the friction coefficient, μ. In this work, the different methods to analyze friction, hardness and elastic modulus by means of AFM are explained for several examples as diamond single crystals, SiC single crystals, titanium dioxide coatings, mica and silicon oxide. Besides, examples of elastic deformation using nanoindentation are included. Moreover, several examples of results obtained by means

[*] Corresponding Author's e-mail address: goncins@ccit.ub.edu (Gerard Oncins).

of a Nanoindenter are also described in detail. A particular emphasis on ceramic coatings and advanced ceramic materials, such as YBaCuO superconductors, Yttria-stabilized zirconia, doped ceria electrolytes for fuel cells and yttria stabilized polycrystalline tetragonal zirconia, are reported and commented. The interest of these techniques is evidenced by the increasing quantity of nanomechanics-related papers published in the last decades, near a thousand of which appeared during the last five years. Unfortunately, a lot of practical information about nanomechanics of hard materials is still scarce in the literature (only one percent of the above mentioned publications are related to ceramic materials). This chapter aims to present the basic principles and methods applied to extract the different mechanical properties and also to review and comment real examples related to the cited techniques.

1. Introduction

1.1. Classical Mechanisms Using AFM-FS

1.1.1. Classical Mechanisms of Friction

The study of friction and its implications in our world formally began with the experiments of Leonardo da Vinci, who was the first scientist that studied friction phenomena in a quantitative way [1]. He concluded that friction force (F_f) does not depend on the area of contact and it is proportional to the mass of the slider (M_s). This assumption leads to the first definition of friction coefficient (μ) as the direct relationship between F_f and M_s:

$$\mu = \frac{F_f}{M_S} \tag{1}$$

Leonardo found a μ value of 0.25 for all tested materials and this was considered as a universal constant during several centuries. It was Guillaume Amontons (1663-1705) who rediscovered Leonardo's works about friction and took them a bit further, proposing that F_f is proportional to the vertical force (F_v) [2].

Finally, Charles Augustine de Coulomb (1736-1806) set the basis of the macroscopic friction theory as we know it today:

$$F_f = \mu F_v \tag{2}$$

He also established that F_f value is independent of velocity once motion starts, idea that was refined by Leonhard Euler (1707-1783), [3, 4, 5] who introduced the concept of *kinetic* and *static friction* for the first time; when the two sliding surfaces are in motion relative to each other, μ is called *kinetic friction coefficient* and the F_f value is the force that tries to counteract the pushing force that created the sliding movement. In the case of two surfaces that do not move relative to each other, μ is called *static friction coefficient* and it is usually higher than the *kinetic* one. Now, according to these new concepts, F_f values must be considered as a threshold force. As we can see, F_f defined by Coulomb does not depend on the contact area. In this sense, John Theophilius Desanguliers (1683-1744) previously proposed that adhesion force (F_a), which is the force needed to separate two bodies in contact, played a key role in the friction phenomena and that F_a is a function of the contact area.

In 1954, as an attempt to discern the dependence between F_f and the contact area, Bowden and Tabor proposed a solution for the area of contact dilemma[6]; not only did they establish that two bodies do not contact with their whole surface, but also that they do it by means of a finite number of small *asperities*. This new concept redefined the way of studying contacting surfaces and drove tribology towards the micrometric world. As a matter of fact, they theoretically demonstrated that:

$$F_f = F_v^{2/3} \tag{3}$$

As a result, Amontons' law was called into question again but not for long because in 1957 Archard found the way to unify both the old and new theories [7]; in their works, Bowden and Tabor had considered that the number of asperities, which are proportional to the real contact area value, were independent on the applied F_v value. Archard considered that this assumption was not right and modeled the relationship between the number of asperities and F_v. Finally, he concluded that F_f and F_v values were proportional.

1.1.2. Classical Mechanisms of Hardness and Elastic Modulus

Mechanical testing is extensively used for characterization and quality control of materials. Unfortunately, the results between measurements performed at different scales are not comparable and strongly depend on the experimental method [8]. AFM has become a powerful tool to detect structural changes at the nano- and picoscale and its popularity between materials science researchers keeps on rising [9].

Recently, AFM was used to study surface deformation during cycling loading [10, 11, 12, 13, 14, 15] and also to study and extract the intrinsic properties of ceramic materials [16]. Between the main drawbacks of AFM as a nanoindentation tool, we can consider the difficulty in measuring the real radius of the tip [3]. Nevertheless, nanoindentation is widely used to investigate the mechanical properties of a vast range of samples [17, 18]; for crystalline materials, it can be used to characterize the nucleation of dislocations during the initial stages of plastic deformation. Some dislocation events can be identified by characteristic discontinuities (pop-in events) in the loading-unloading cycles, the so-called force-displacement curves [19]. Atomistic simulations were performed in order to understand the mechanisms behind dislocation nucleation and indentation size effects (ISE) [20, 21]. However, there is a gap in length scale between the limited size of atomistic simulations, which usually address nanometer-scale indenters and only a few dislocations, and nanoindentation experiments, where micrometer-scale indenters are used and a remarkable number of dislocations are produced. AFM-based nanoindentation experiments can contribute to bridge the gap between atomistic simulations and experiments by addressing the nanoscale issue and providing experimental data about the smallest features in plasticity. In these regions, plasticity must be accommodated by a homogeneous nucleation of dislocations rather than activated by Frank-read sources [22]. Incipient plasticity was recently studied using AFM-based nanoindentation and imaging on a number of surfaces, including MgO (100) [23], Au (100) and (111) [24], KBr (100) [25], and Cu (100) [25], in both ambient and at ultra-high vacuum conditions.

1.2. Classical Mechanisms Using Nanoindentation

In traditional indentation tests (macro- and/or microindentation), a hard tip whose mechanical properties are well known (frequently made of a hard material as diamond) is pressed into a sample whose properties are unknown. The load placed on the indenter tip is increased as the tip penetrates further into the specimen to reach user-defined values. At this point, the load may be held constant for a period of time or directly removed. The area of the residual indentation in the sample is measured and the hardness, *H*, is defined as the maximum applied load, P_{max}, divided by the residual indentation area, A_r, as follows:

$$H = \frac{P_{\max}}{A_r} \tag{4}$$

For most techniques (macro- and microhardness tests), the projected area may be measured directly using optical microscopy. This technique is limited due to large and varied tip shapes, with indenter rigs which do not have very good spatial resolution (the location of the indented area is very hard to specify accurately). Comparison across experiments, typically done in different laboratories, is difficult and often meaningless. Nanoindentation represents an improvement respect to these macro- and microindentation tests by indenting on the nanoscale with a very precise tip shape, high spatial resolution and by providing real-time load-displacement (into the surface) data while the indentation test is in progress. Nevertheless, when a standard Vickers technique is used instead a nanoindenter machine, the acquired results show lower values than those obtained by nanoindentation. The difference between the nano- and micro indentation results is known as the Indentation Size Effect (*ISE*). The ISE predicts an increase of hardness when is decreased the indention load or indentation depth [17]. The *ISE* is usually caused by several mechanisms which can act alone or combined: the strain gradients effects, the structural non-uniformity of the deformed volume, the roughness of the surface, by differences in the friction between the indenter and the sample and by the different mechanisms that are activated in the elastic/plastic deformation. Note that impressions performed at very low loads could be localized in dislocation-free areas. Moreover, the calculus of the hardness in the instrumented indentation instruments is obtained under load in contrast to Vickers, which is calculated after release the load. Other difference between these techniques is that micro-hardness tester employs the true contact area and an error is made observing the residual imprint because of the effect of elastic relaxation, emphasizing the rather complicated nature of *ISE* effect. In this sense, another mechanism that could be affecting to the observed *ISE* in ceramic materials is the induced phase transitions. Some materials, as the followed described tetragonal zirconia Y-TZP, suffers tetragonal to monoclinic t→m, transformation under forced stresses, being the m-phase softer than t-phase and thus the *H* measured could be affected.

1.2.1. Classical of Indentation Tests

The methods used for local mechanical testing of materials have often been formally divided into macroindentation, microindentation, nanoindentation and even picoindentation depending on the characteristic indentation. However, these methods can be classified in a physically more justified manner because an increase in the applied load, *P*, leads to a change not only in the characteristic size of a locally deformed region but also in the ratio between the contributions of the elastic and plastic deformations and also in the atomic mechanisms of deformation (dislocation movement depending on the crystallographic orientation [28], twinning process [29], kink band process [30] and others).

1.2.2. Deformation Mechanisms Induced under Nanoindentation Tests

The deformation of materials takes place via two different processes:

1) elastic (reversible), and
2) plastic (irreversible)

Since elastic deformation is a reversible process governed by angstrom scale, it exhibits virtually no size dependence unless a large population of pre-existing defects is involved [31]. Differently, plastic deformation response, which takes place as a result of the generation, annihilation, and motion of defects such as dislocations, displays marked size effects when those defect dimensions are in the range of microns or below.

1.3. Classical Mechanisms Using Scratch Tests

The aim of scratch tests is to determine the mechanical, tribology and wear resistance of bulk materials, coatings and thin films. Practically, scratch procedure consists on moving the sample underneath the indenter while the normal force, F_N, is ramped. The result of this process is the generation of a tangential force, F_T, which scratches the specimen. Scratch tests are attracting an increasing attention in applications such as protective polymeric coatings, thin films, windows, hard coatings, etc.

As an example, the automotive industry would benefit by an improvement of the adhesion of paints and varnishes to the substrates while assuring an adequate viscoelasticity and tribology properties.

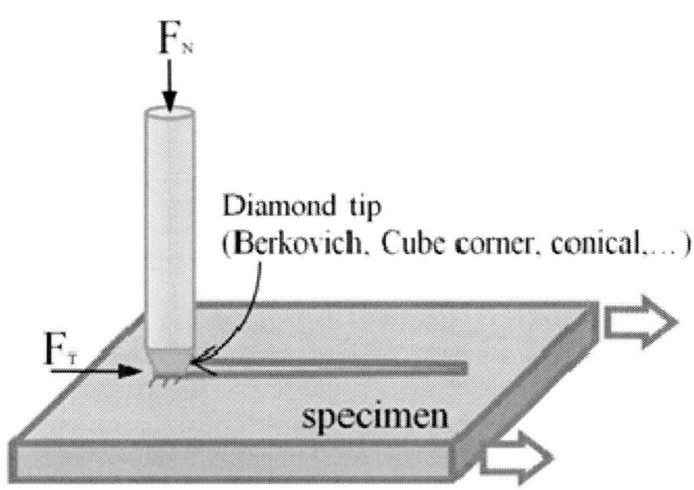

Figure 1. Schematic set-up representation of a scratch test.

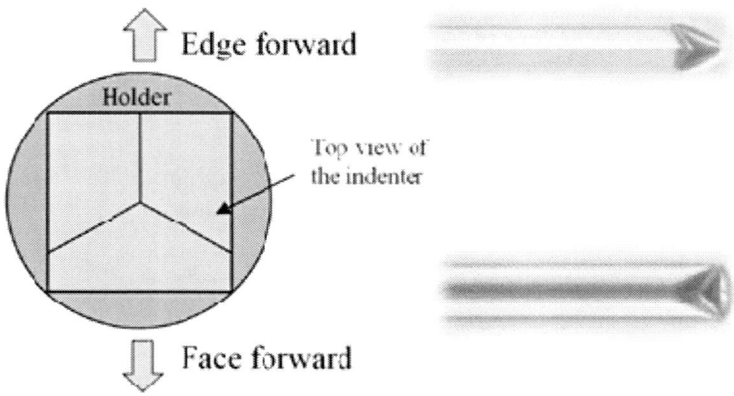

Figure 2. Schematic representation of a pyramidal tip. Face forward and edge forward directions raise plastic and elastic deformations respectively. In the first case, a lower attack angle promotes higher plastic deformations.

The track produced by scratch tests on a sample can be elastic or plastic and a variety of damage patterns can be observed. By means of a visual study of the wear track together with the penetration and friction curves recorded by the instrument it is possible to determine the scratch resistance (ability of the tested material to withstand abrasion without fracture), scratch hardness (track width divided by the diameter of the scratch tip), adhesion behavior and failure mechanism of bulk materials, thin films and coatings. In fact, it is also possible to identify changes in the friction coefficient, lamination and other type of failures supported by means of tracks images. In the case of adhesion measurements, the results depend crucially on the conditions used in the experimental test. In fact, nanoindentation instruments generally do not provide an adhesion test as such. Thus, it is recommendable to attempt to reproduce the in-service conditions to evaluate the obtained results. The parameters that have a great influence in the results obtained from a scratch test are:

- *The angle between the face of the indenter and the surface of the sample*: this angle is known as the 'attack angle' and is 25π for a Berkovich indenter and 45π for a Cube Corner geometry. When a pyramidal indenter is used, the direction of the scratch must be face forward for plastic deformation experiments and edge forward to work under elastic deformation (Figure 2).
- *The roundness of the indenter:* The tip of the conical and pyramidal indenters loses their ideal geometry along their service life due to mechanical wear. As a consequence, the radius of the tip can noticeably shift the critical load, which increases with the tip radius.
- *Scratch velocity orsliding speed:* Higher scratch velocities induce higher hardness behavior (*scratch contact pressure*) majority in a polymers and polymeric reinforced ceramics. Independently of the material essayed, several sliding speeds must be tested to discriminate between the different activated mechanism.
- *Applied normal load, F_N:* This parameter, together with the scratch distance, is one of the programmable parameters of the instrument. The user can define a constant F_N value to study the wear behavior) or make a ramp load along the sliding path to study

failure mechanisms. The optimal range of F_N it would be there that permit the user to visualize the elastic, plastic and failure deformation along the scratch path with enough clarity.

2. Contact Physics

2.1. Physics of friction

The "asperity" concept developed by Bowden and Tabor led their creators to propose that the physical process of friction consisted on the creation and melting of microscopic contacts as the two surfaces slid, ultimately considering that mechanical wear was the origin of friction. Nevertheless, this explanation did not endure the confrontation with reality: there was friction even between surfaces that did not show any appreciable wear. Israelachvili, a student in Tabor's lab, took the ideas of his advisor a step further [4]: in order to experimentally measure the real contact area value between two surfaces, he developed the surface force apparatus [4, 32, 33]. The idea consisted in bringing into close contact two pieces of mica bent in perpendicular directions. This setup would ensure that the apparent contact area value would equal the real contact area one, as the two contacting surfaces did not show any asperity. Israelachvili confirmed that friction truly depends on the real contact area value and also confirmed the close relationship between F_f and F_a values. In the 80s, Gary McClelland developed a model which successfully explained the wearless friction phenomena [34] that had been previously observed. He proposed that when two surfaces slide, the outer atoms vibrate due to the released energy. These vibrations, called phonons [35], travel along the surface in the form of a mechanical wave and act as a physical obstacle for the sliding interfaces. Then, wearless F_f value is the necessary force to overcome the mechanical wave of phonons, which is eventually dissipated as thermal energy. An interesting implication of this model is that the frequencies of the generated phonons play an important role; if the two surfaces resonate, F_f increases dramatically due to the consequent vibration amplitude enhancement. Furthermore, the phononic theory predicted a much more striking fact: if the phononic resonance between the two sliding bodies could be minimized in the absence of wear phenomena, F_f value would virtually drop to zero [36].

The development of AFM by Binning and Rohrer [37] let McClelland and his colleagues to test their theories. Briefly, AFM consists on a microfabricated probe ended with an extremely sharp tip that scans the sample, establishing an apparent contact area that can be in the nanometer range. McClelland tested his wearless friction theories just to find out that something was absolutely wrong; first of all, F_f was independent of F_v, which was in flagrant contradiction with all the previous theories. Besides, the shear stress (defined as the quotient between F_f and the contact area) was incomprehensibly high. Quartz Crystal Microbalance measurements (QCM) performed by Jacqueline Krim observed another striking effect: F_f between noble gases and metals increases a lot in the presence of liquid [38]. Far from being an intuitive idea, as liquid are commonly believed to reduce friction, it fits acceptably well with the previous thesis about high F_f values between commensurate surfaces; in fact, any liquid can be reasonably commensurate with a substrate due to the weak bonds between their molecules and its fluidity. Then, the liquid lets the two surfaces in contact to match more properly. A recent paper by Park *et al.* [39] shows how to control friction by tuning the potential landscape between the two sliding surfaces by applying an oscillatory F_v value, in an attempt to provide a way to control adhesion and friction in sliding micro- and nanoelectromechanical systems (MEMS and NEMS, respectively). This kind of experiments also explored the nature of electronic friction [40], an interaction due to the electrical surface charges in the two sliding bodies. In this field of knowledge is remarkable the contribution by Park *et al.* [8] who showed that friction can be increased between doped silicon surfaces by applying an electric field, giving rise to extremely high and tunable friction values that can be used to hold steady two surfaces in a controlled fashion with nanometric resolution.

2.2. Physics of Picoindentation

AFM-based nanoindentation is a powerful tool which lets the user to obtain quantitative, reliable and reproducible elastic modulus values. However, very little information about this theme is available in the literature [3]. Nowadays, it is widely accepted that a correct understanding and characterization of the elastic modulus values during the working lifetime of advance ceramic samples is critical to any desired structural applications. AFM-based nanoindentation and force spectroscopy have several advantages over standard methods (such as macro-, micro-, nanoindentation or bending) when it comes to the quantification of the elastic modulus values for ceramic samples. Firstly, the measurements are extremely local and can be performed on different areas so as to average the mechanical properties of the sample. Secondly, measured elastic modulus values are depth-sensing so it allows mechanical characterization of a material at different indentation depths. Finally, it is not necessary to observe the residual imprint with an AFM or scanning electron microscope (SEM). All of these advantages mean that nanoindentation performed by Nanoindenter (Agilent Technologies), CSM or Hysitron Triboscope allow us to extract hardness and elastic modulus for bulk and coatings materials with a limited thickness of 50 nm. However, nanoindentation with an AFM tip is a suitable and reliable technique for measuring elastic modulus value while applying F_v values ranging from a few nN down to the pN level. The main principle of nanoindentation technique using AFM-FS is widely explained in elsewhere [41, 42].

2.2.1. Contact Mechanisms Models

Due to the nanometric nature of the contacts between the AFM probe and the studied samples, the area of contact between both surfaces, A_c, cannot be directly measured by any imaging technique, as it would be possible in macro- [43] or microscopic experiments. Nevertheless, an accurate knowledge of the A_c value is extremely important if quantitative nanomechanical paramenters are to be measured from experimental data. Although the controversy about the real and the apparent A_c value is still a matter of debate, several contact mechanics models, mainly developed for macroscale modelling [44], can be successfully applied to AFM-based nanoindentation data as a function of the nature of the sample (i.*e.* metals, organic bilayers, ceramics, polymers, and others). A brief review of the most successful models is presented with both their differences and applicability range.

Hertz model: This model was developed in 1881 by Heinrich Hertz - German physicist- and presented in a pioneer paper titled "*On the contact of elastic solids*" [45]. Although this model provides with a calculated A_c value, Hertz model considers both the tip and the sample as two spheres of radius R_1 and R_2 and only assumes elastic deformation of the materials; thus, it does not model the interpenetration between surfaces during the indentation process nor any adhesion force acting between the surface and the AFM tip. The symmetry of the problem leads to a flat, circular contact area between the sample and the tip when they are in contact:

$$A_c = \pi\left(\frac{RF}{K}\right)^{2/3} \tag{5}$$

where R is the combined curvature radius of the contact and k corresponds to the combined elastic modulus of the different materials, given by the following equation:

$$K = \frac{4}{3}\left(\frac{1-\upsilon_i^2}{E_i} + \frac{1-\upsilon^2}{E}\right)^{-1} \tag{6}$$

being E and ν the elastic modulus and the Poisson's ratio (where the subindex i denotes the properties of the indenter). It is interesting to note that for an elastic single contact between a round tip and a flat sample, Ac value is a non-linear function of the applied load.

Johnson, Kendall and Roberts model (JKR model): Johnson, Kendall and Roberts [46, 47] developed a contact mechanics model that refined the somewhat crude Hertzian way of calculating A_c value. This model considers that γ is the work per unit of area necessary to separate the tip and the surface of study, which plays an important role in A_c value calculation. This parameter can be written as follows:

$$\gamma = \gamma_1 + \gamma_2 - \gamma_{12} \tag{7}$$

where γ_1 and γ_2 stand for the surface energies of the two contacting bodies and γ_{12} is the energy of the interface. As γ value takes into account all the different attractive forces arisen between the tip and the sample, this model is particularly suitable for tips with a large R value and for sticky surfaces, as these two factors will result in a noticeable increment of attractive Van der Waals forces. Nevertheless, this model considers that the attractive forces only take place when the two surfaces are in contact and no effect is noticed by the tip as it approaches the sample. Due to that, this model refines the considerations of the previous model but still fail to model surface interactions before the contact. JKR model can be extended to a huge variety of geometries [10] but in the general case of two bodies simulated as spheres, A_c value can be expressed by the following equation:

$$A_c = \pi\frac{R^{2/3}}{K^{2/3}}\left(F + 3\pi\gamma R + \sqrt{6\pi\gamma RF + (3\pi\gamma R)^2}\right)^{2/3} \tag{8}$$

Notice that in the case when γ value tends to 0 in equation 8, Hertz model is obtained. An interesting improvement regarding Hertz model is that JKR model predicts that a finite negative force is required to separate the surface. This value is often referred to as the critical load (L_c), and is given by the following expression:

$$L_c = -\frac{3}{2}\pi\gamma R \tag{9}$$

When L_c value is reached, a minimum finite contact area exists. We shall refer to this contact area as the critical area, A_c', which is given by:

$$A_C' = \pi\left(\frac{3\pi\gamma R^2}{2K}\right)^{2/3} \tag{10}$$

Afterwards, the JKR model predicts a particular value for the contact radius at zero applied load, called a_0, which is given by [48]:

$$a_o(JKR) = \left(\frac{6\pi\gamma R^2}{K}\right)^{1/3} \tag{11}$$

The most important parameter to take into consideration in the previous equation is γ, which icorresponds to the adhesive force between the sample of study and the indenter.

Derjaguin, Müller and Toporov model (DMT model): This model, proposed by Derjaguin, Müller and Toporov [49] takes into account the different adhesive behaviours between surfaces but in a different way than JKR model, as it considers the attractive interaction forces as an increase of the force value while maintaining the profile of the Hertzian contact area.

As a consequence, this model can also be considered as a modified Hertz model, so it works well for hard substrates with low penetration depth or low deformation rates. The main advantage over the other models is that γ is not only taken into account when two different bodies are in contact but the tip sample interaction is shaped as a kind of Lennard-Jones profile. Unlike the JKR model, the DMT theory assumes that the shape of the contact is not affected by interfacial forces. As a consequence, the overall Hertzian deformation profile is maintained, but the contact area is increased. Adhesion forces effectively act as an extra external force. According to this, A_c value is expressed as:

$$A_c = \pi\frac{R^{2/3}}{K^{2/3}}(F + 2\pi\gamma R)^{2/3} \tag{12}$$

Similar to the previous model, an expression for the critical vertical force (L_{co}) is proposed:

$$L_{co} = -2\pi\gamma R \tag{13}$$

Because this model is based on the Hertzian theory, the contact area is 0 at L_{CO}, so no jump-off-contact is predicted. Unlike the JKR model, stresses are finite at the contact zone edge and there is only a small tensile stress outside of the contact zone.

In DMT model, the contact radius at zero applied load, called a_o, is defined as:

$$a_o(DMT) = \left(\frac{2\pi\gamma R^2}{K}\right)^{1/3} \tag{14}$$

JKR-DMT transition: DMT and JKR theories vastly describe different contact properties. In fact, a rather acrimonious debate broke out in the literature regarding which theory is closer to reality. This was solved with the realization that both theories are correct, but for different situations. Compliant materials with strong, short-range adhesion forces and large radii are better described by the JKR model, whereas stiff materials with lower, longer-range adhesion forces and small radii follow the DMT model. Interestingly, cases that are in between will also occur. Thus, the JKR and DMT theories are actually limiting cases of the spectrum of contact mechanics behaviour. The transition between these two limits is discussed in more detail by Johnson et al [50], as well as by Maugis et al [51] and Müller et al [52].

2.3. Physics of Nanoindentation

During the last years, instrumented indentation testing (also known as depth-sensing indentation, continuous-recording indentation, ultra-low-load indentation or nanoindentation) has been widely employed and has proved to significantly expand the capabilities of traditional hardness testings.

The method was introduced in 1992 for measuring hardness, *H,* and elastic modulus, *E*, at nanometric scales [53, 54] and presents several advantages compared with the tradition hardness testers:

- It is a local probe that can evaluate the mechanical response of different materials in specific sample regions.
- It is depth-sensing, so it can characterize a material at different penetration depths.
- It can quantitatively measure the elastic modulus of the sample.
- It is well suited for the characterization of small volumes of material.
- It eliminates the need to visualize the imprint produced during the test for homogeneous materials, which makes the extraction of mechanical properties easier.

Due to all these advantages, this technique is a versatile candidate for characterizing the mechanical properties of a given component. As an example, *H* value, *E* value, toughness fracture, K_c, shear stress, τ and the stress-strain curve, σ–ε, can be directly obtained from the loading-unloading curve (or *P-h* curve) by using the suitable tip indenter.

2.3.1. Devices

The construction of depth-sensing indentation equipments is made possible by the inclusion of very sensitive displacement and load sensing systems. Load transducers must be capable of measuring forces in the micro-newton range and displacement sensors are very frequently capable of sub-nanometer resolution. It is worth to say that environmental isolation is crucial when it comes to operate the equipment; vibrations transmitted to the device, fluctuations in atmospheric temperature and pressure, and thermal fluctuations of the components during the course of an experiment can cause significant errors. The ability to conduct nanoindentation studies with nanometer resolution in depth and sub-nanoNewton force resolution is also possible using a standard AFM setup, which allows for nanomechanical studies to be conducted alongside topographic analyses. Load-displacement curves can be collected similarly for a variety of materials and mechanical properties can be directly calculated from these curves [55].

2.3.2. Contact Mechanisms

A. Elastic Contact

In the later part of the 19th century [56], Hertz formulated elastic contact equations for spherical surfaces. The model described in this section is essentially applied to study and better understanding of the elasto-plastic transition and it is partially based upon previous developments of others in the field, especially those reported by Hertbert *et al.* [57].

Hertz model can be expressed as follows:

$$P = \left(\frac{3}{4} E_{eff} \sqrt{R}\right) h_e^{3/2} \tag{15}$$

where P is the load, R is the radius of the indenter, h_e is the elastic penetration depth into the surface, and E_{eff} is the effective modulus given by the following equation:

$$\frac{1}{E_{eff}} = \frac{1-\nu^2}{E} + \frac{1-\nu_i^2}{E_i} \tag{16}$$

where ν is the Poisson's ratio for the test material. Subindex i denotes the values corresponding to the mechanical properties of the indenter. For a diamond tip indenter, the elastic constants E_i = 1141 GPa and ν_i = 0.07 are often used [12].

The model proposed by Sneddon [58] for a rigid spherical indenter exhibits that the elastic displacement of a plane surface above and below the contact region is equal and given by:

$$h_e = h_{\max} = \frac{a^2}{R} \tag{17}$$

where a is the contact radius between the indenter and the sample of study during indentation. This model verifies when the indentation is only in the elastic regime, the entire indentation depth is elastic and the total recovery is produced after the indentation process and determined by equation 17. The most well-known plot for nanoindentation is that of P vs h. Both the Oliver and Pharr [12] and Field and Swain [59] methods use the slopes of the initial portion of the unloading curves, dP/dh (also known as stiffness, S) to calculate h_e. The contact point or the contact radius of the circle, a (see equation 17), between the sample and the indenter is a function of P, S and R and it can be expressed as:

$$a = \sqrt[3]{\frac{3}{4}\frac{PR}{E_{eff}}} \tag{18}$$

In most cases of interest in nanoindentation testing, the indenter is in contact with what is essentially a semi-infinite half space, so R is the radius of the indenter alone. In a similar manner, E_{eff} value is the combined elastic modulus of the contacting bodies. Combining equation 15 and 17 yields:

$$\frac{P}{\pi a^2} = \frac{4}{3\pi} E_{eff}\left(\frac{a}{R}\right) \tag{19}$$

The left side of this equation represents Meyer's hardness, also known as the indentation stress or mean contact pressure, p_m [60]. The expression in parentheses on the right-hand side represents the indentation strain [14]. These are not the same as the stresses and strains measured in neither uniaxial compression test, nor those corresponding to other nanoindenter tip geometries [61]. The right part of equation 19 is constant so, for a completely elastic contact, the mean contact pressure, or indentation stress, is linearly proportional to the indentation strain. Let's consider the magnitude of the contact stresses, in particular, the shear stress in the specimen material as load is applied by a spherical indenter; For a fully elastic response, the principal shear stress for indentation with a perfect spherical indenter shows a maximum at around $0.47p_m$ at a depth of around $0.5a$ beneath the specimen surface directly beneath the indenter [62]. Moreover, when the plastic flow takes place at τ values around 0.5 times the yield stress (σ_{ys}), the Tresca and/or Von Misses shear stress criteria should be employed. Then, the plastic deformation in the specimen beneath a spherical indenter can be first expected to occur when $p_m \sim 1.1\sigma_{ys}$ [14]. For an initial elastic contact, the mean contact pressure increases with increasing load. In fact, this property is very difficult to define since it depends on how it is measured; when the penetration depth becomes larger than the tip radius, the pyramidal shape of the indenter becomes the dominant geometrical feature of the indentation, so the contact usually involves a fully plastic deformation within the specimen.

B. Elastic-to-Plastic Contact

Indentation tests of many materials result in both elastic and plastic deformation of the specimen material. In brittle samples, plastic deformation most commonly occur with pointed indenters such as the Vickers diamond pyramid. In ductile materials, plasticity may be readily induced with a blunt indenter, such as a sphere or cylindrical punch. Indentation tests are widely used to measure the hardness of different materials, but some specific ones as Berkovich, Vickers, and/or Knoop indenters may be used to investigate other mechanical properties of solids such as strength, fracture toughness or internal residual stresses. During the indentation tests, the indenter is eventually dominated by tip-rounding effects and the deformation has the property of geometrical similarity [63]. The analysis of force-displacement or load-unload curves obtained by instrumented indentation systems is often based on the works by Doener and Nix [64] and Oliver and Pharr [12]. H and E values are calculated as a function of penetration depth using the method proposed by Oliver and Pharr [12]. The contact depth, h_c, can be defined as the distance from the circle of contact to the maximum penetration depth, h_{max} as follows:

$$h_c = h_{\max} - \varepsilon \frac{P}{S} \tag{20}$$

where ε is a parameter approximately equal to 0.74 for a Berkovich indenter, and S is the contact stiffness. Based on the contact depth, the contact area $A(h_c)$ is calculated using the form factor of the indenter. Once the correct area function is known, H value is calculated by dividing P_{max} value by the projected contact area, as follows:

$$H = \frac{P_{\max}}{A(h_c)} = \frac{P_{\max}}{24.56 h_c^2} \tag{21}$$

H value, defined as instrumented indentation hardness, differs from that of a Vickers indentation. The previous equation employs the projected area of contact and is calculated at maximum applied load. In contrast, Vickers hardness is measured upon unloading. Measuring Vickers hardness using the residual imprint leads to error because of the elastic relaxation that occurs during or after the unloading process. This phenomenon is particularly important in ceramic materials, and may explain the difference observed in the literature between instrumented indentation and Vickers indentation hardness [65]. The effective elastic modulus, E_{eff}, is calculated as follows:

$$E_{eff} = \frac{1}{\beta} \frac{\sqrt{\pi}}{2} \frac{S}{\sqrt{A(h_c)}} \tag{22}$$

where β is the geometry shape factor and equal to 1.034 for a Berkovich indentation [12, 13, 66]. Equation 22 is found in elastic contact theory [13], and holds for any indenter that can be described as a body of revolution of a smooth function [12]. As this equation was derived for an axysimetric indenter, it is formally applied only to circular contacts, for which the indenter geometry parameter β equals to 1. However, it has been shown that this model works also well when the geometry is not axysimetric, provided that different values of β are employed [15]. For indenters with square cross sections, like the Vickers pyramid, β equals to 1.012; for triangular cross sections, like the Berkovich and the cube-corner indenters, β equals to 1.034

[15]. This factor plays a very important role when accurate property measurements are desired. The value of this parameter affects not only the elastic modulus calculated from the contact stiffness, but also the hardness.

The analysis used to determine the hardness and elastic modulus, is essentially an extension of the method proposed by Doener and Nix [67], which accounts for the fact that unloading curves are distinctly curved in a way that cannot be accounted for by that flat punch approximation. Doener and Nix [16] considered that the contact area remains constant as the indenter is withdrawn, resulting in a linear unloading curve. Oliver and Pharr [14] method begins by fitting the unloading portion of the load-unload curve to the power-law relation. This can be re-written as follows:

$$P = BA\left(h - h_f\right)^m \tag{23}$$

where B and m are power-law fitting constants [12] and h_f is the final displacement after a complete unloading process, also determined from the curve fit. S value is established by analytically differentiating the previous equation and evaluating the results at the maximum penetration depth, $h = h_{max}$:

$$S = \left.\frac{dP}{dh}\right|_{h=h_{\max}} \tag{24}$$

By differentiating equation 23, S value can be obtained thanks to the following equation:

$$S = Bm\left(h_{\max} - h_f\right)^{m-1} \tag{25}$$

It is wise prudent, when S value is calculated with equation 25, to fit only the upper portion of the unloading curve; S value determined from this fit should be checked by comparing the curve fit to the data but fitting the upper 25 up to 50 % of the data is usually sufficient. According to this, the contact depth is calculated, which for an elastic contact is less than the total depth of penetration, h_{max}. The basic assumption is that the contact periphery sinks in a manner that can be described by models of indentation of a flat elastic half-space by a rigid punch of simple geometry [13]. This assumption limits the applicability of the method because it does not account for the piled-up material present at the contact periphery, which occurs on some elastic-plastic materials.

The projected contact area is calculated by evaluating an empirically determined indenter area as a function of $A = f(h)$ [14] at the h_c; The expression can be re-written as:

$$A = f(h) = f(h_c) \tag{26}$$

The area function, also known as the shape function or tip function, must be carefully calibrated with standard fused silica, which is an amorphous material with a constant E value of 72 GPa [14]; Thus, the possible deviation from non-ideal indenter geometry is taken into account. This deviation can be quite deep near the tip apex of the Berkovich indenter, where some rounding inevitably occurs during the grinding process. The equation employed to perform the correct contact area calibration can be re-written as follows:

$$A(h) = \sum_{n=0}^{8} C_n (h_c)^{2-n} = C_o h^2 + C_1 h + C_2 h^{1/2} + C_3 h^{1/4} + \ldots + C_8 h^{1/128} \quad (27)$$

where C_0 to C_8 are constants determined by curve-fitting procedures. A perfect pyramid or cone is represented by the first term alone. The second term describes a paraboloid of revolution, which approximates to a sphere at small penetration depths and a perfect sphere of radius R is described by the first two terms with $C_o = -\pi$ and $C_1 = 2\pi R$ [14].

2.3.2. Limitations

Conventional nanoindentation methods for calculation of the elastic modulus value (based on the unloading curve) are limited to linear, isotropic materials. Problems associated with the '*pile-up*' or '*sink-in*' of the material on the edges of the indentation remain unsolved and are still under investigation. It is possible to measure the pile-up contact area using computerized analysis of AFM images of the indentations [68] but this process also depends on the linear isotropic elastic recovery for the indentation reconstruction.

Assuming, however, that pile-up processes are negligible, the elastic models show the amount of sink-in, h_s, which is given by the follow equation:

$$h_s = \varepsilon \frac{P_{\max}}{S} \quad (28)$$

where ε is a constant that depends on the geometry of the indenter. Typical values are 0.72 for a conical punch indenter, 0.75 for a parabolic of revolution, and 1 for a flat punch [14]. The depth of the contact between the indenter and the specimen, h_c, can be calculated by:

$$h_c = h_{\max} - h_s \quad (29)$$

2.4. Physics of Scratch: Scratch Testing Using Nanoindentation

As described early, a nanoindentation instrument is an instrumented indentation technique. This means that the hardness of a material is calculated without the observation of the indentation imprint. Thus, the equipment controls and records the penetration depths at the nanometer scale and the applied F_N (for that is an IT) with a very high accuracy, resolution and sensitivity.

Furthermore, the indenters designed for these equipments have a great accuracy in terms of geometry with typical radius values below 10 nm under a great variety of possible geometries. Because of these reasons, nanoindenter equipments are extremely useful to analyze thin films or thin coatings which usually have to increase their performance while decreasing their thickness. In this sense, the requirements to quantify the scratch resistance or adhesion under scratch conditions involve the finding of the critical load at which the material fails. The experimental sequence to perform a scratch test for any material is resumed as:

- The tip approaches the sample from a programmed distance above the surface.
- When the indenter is on surface, the original profile of the surface is scanned over the entire scratch path. This pre-profile curve is used to level all the subsequent scratch cycles and it is visualized as a profiling curve along scratch distance.
- The indenter is ramp loaded under a previously defined strain rate while a certain path (of several microns) is scratched at a preprogrammed velocity. The result obtained is a profile penetration curve along the scratch distance or load applied.
- When the scratch segment is completed, the indenter is unloaded, that is, it moves back to the initial position to perform a post-profile curve under load applied on sample.
- Subsequently, a residual scan of the scratch track is registered using a very low constant load, obtaining a residual profile curve along the scratch distance plot.
- Finally, a cross-profile is optionally scanned so as to assess the presence of pile-up material at the sides of the track.

In section 3.3, the most important data and curves acquired by a scratch experiment will be discussed in detail for three different materials: i) a bulk fused silica piece as an example of an amorphous material, ii) a Y-TZP ceramic material, and iii) a polymeric coating of a commercial optical storage DVD (digital Versatile Disc).

3. Examples

3.1. AFM-FS Examples

3.1.1. Friction in Diamond Single Crystals for MEMS and NEMS

In the last years, the awakening of MEMS and NEMS, has led to the development of new nanotribological tools in order to test the performance of these systems. Besides, the surface processes as friction is an extremely important design factor. MEMS and NEMS are usually made out of silicon due to the extensively studied microfabrication processes developed for this material but it turns out to be a bad candidate in terms of wear and adhesion, as mobile parts tend to get stuck and degrade very fast, partly due to the high hydrophilicity of silicon oxide. In this sense, diamond has become one of the best candidates and the first quantitative nanotribological studies of ultrananoscrystalline diamond (UNCD) were performed by Sumant *et al.* [69]. UNCD thin layers are formed by pitless 5 nm diameter diamond grains (mainly sp^3 bonds) separated by narrow boundaries where there is a certain amount of sp^2 C-C. Friction measurements were performed using an AFM on the flat underside of UNCD layers after etching away the silicon substrate, rendering surfaces with a RMS roughness of 0.9 nm. UNCD elastic modulus is similar to the value for single crystal diamond, as well as its friction coefficient value, μ, (0.03 in air). Friction experiments were performed scanning a WC-coated AFM tip in contact mode in a direction perpendicular to that of the main cantilever axis in order to observe the twisting of the cantilever, which is proportional to friction force. The work of adhesion between the tip and the silicon substrate was 115 mJ/m^2 and 59 mJ/m^2 for the UNCD-coated sample. Both friction and adhesion work were greatly reduced after treating the UNCD surface with hydrogen plasma, which resulted in a diminution of dangling bonds and in a reduction of the sp^2 hybridization in favor of the sp^3, as well as a reduction in C-O bonds [70].

The presence of sp^2 bonds had been previously proved to promote bonding between sliding surfaces by means of MD simulations [71]. After hydrogen plasma treatment, an adhesion work of 32-36 mJ/m^2 was measured, similar to the 10 mJ/m^2 observed for the same kind of AFM tips on *H*-terminated single crystal diamond (111) in UHV, where it was proposed that this energy was due to pure van der Waals interactions [72]. Diamond like carbon films [73] (DLC) have also been widely studied as surface coatings for their outstanding mechanical properties. Nevertheless, as they have different ratios of sp^2 and sp^3 bonding, tribological properties can be heterogeneous, being highly dependent on the degree of *–H* termination [74]. Experimental AFM results performed with diamond coated tips show that μ value increases with the presence of hydrogen and hardness and that wear performance of the coating worsens. Nevertheless, macroscopic experiments suggest that there can be an optimal *–H* content. Different methods have been devised to control the ratio of sp^3 bonding in DLC layers, such as the microwave plasma chemical vapour deposition (MPCVD [75]) and the pulsed laser deposition (PLD) [19]. With this last technique, DLC coatings with 34 and 53% of sp^3 bonding were achieved and AFM experiments with Si_3N_4 tips showed that, as elastic modulus increases with sp^3 content, μ value remains unchanged. Complementary MD simulations by Gao *et al.* [76] showed that not only the sp^3 bonding ratio is important, but also the 3D structure of the surface and the bulk chemistry of the sliding interfaces. It is worthy to say that there is an interplay between humidity and DLC nanotribological properties; When sliding a Si_3N_4 tip on a DLC surface, friction increases with humidity but adhesion remains more or less constant [77]. The authors proposed that water increases the shear strength of the contact but it was also suggested that the idea of a water meniscus being formed between the tip and the sample during sliding can be inaccurate. Besides, hydrogenated DLC films have proved to degrade in the presence of water [78] while single crystal diamond tribological performance is really poor in dry conditions [19].

A recent study by Konicek *et al.* dealt with the effect of humidity in the friction response of UNCD samples versus an UNCD slider [79]; It is well known that diamond shows a remarkably low friction in humidity environments [19], while showing a reverse behavior in dry conditions. The low friction response is attributed to two different processes: first of all, a graphitization of the sliding surfaces [80], i.e., rehybridization of sp^3 bonds to sp^2 (graphite), that is, the stable form of carbon at room conditions. This process can be promoted during friction because the energy barrier of the rehybridization is lowered by shear and heat released during sliding [19]; secondly, a passivation of dangling surface bonds due to the effect of water and H_2. By means of X-ray absorption fine structure and X-ray photoelectron emission microscopy, Konicek *et al.* concluded that passivation is the main responsible of the nanotribological behavior of a UNCD-UNCD interface because the chemical analysis of the sliding interface do not show signs of sp^2 bonding. These results were supported by simulation works showing that the dissociation of H_2 and H_2O molecules on a diamond surface are energetically favorable [81], fact that is experimentally confirmed by the increase of C-O and C=O bonds in the sliding surfaces after conducting the friction experiment.

3.1.2. Friction in Titanium Dioxide Coatings

Titanium dioxide has become an important material in a variety of fields that range from pigments to solar cells [82] but it is the field of biocompatibility [83] that gives to the nanotribological properties of this material a key performance role. Previous mechanical studies were performed in the past in the micro-/nanoscale [84, 85], as is the case of the work by Piwonski [86]. In that work, the authors measured the wear response and the μ value of TiO_2 layers deposited on silicon wafers as a function of the level of porosity by means of AFM. Basically, the scratch test showed that wear is maximum at a medium level of porosity and minimum when the porosity is low. In fact, no wear was detected for non-porous samples, which is in accordance with previous studies [20]. The authors proposed that the observed wear resistance versus porosity behavior is due to the thicker pore walls of low porosity samples, which increase their mechanical resistance. Besides, as the material becomes more plastic when it shows high porosity, μ value increases, mainly due to the higher area of contact between the AFM tip and the sample.

3.1.3. AFM Substrates: Mica and Silicon Oxide

Leite *et al.* [87] measured F_a value (mean pull-off force in force spectroscopy experiments) between an AFM tip and two types of substrates (muscovite mica and silicon wafers) and concluded that F_a strongly depends on both the surface roughness and type of material since the dispersion is higher for the rougher substrate. Indeed, it is well-known that mechanical surface properties in the nanoscale, such as friction and adhesion, are affected by the surface topography [88], with smooth samples showing lower F_a values [89]. Pöhlmann *et al.* [90] and Zum Garh *et al.* [91] studied in detail the tribological mechanisms of self-mated SiC sliding pairs under far different operating conditions with contact areas ranging from the nanoscale to the millimeter scale, different roughness and oxidation state of SiC surfaces and variable humidity conditions, both in vacuum and in air. The experimentally obtained μ values suggested a strong dependence on the chemical composition, the type of surface, and the amount of water surrounding the AFM-probe. Binggeli and Mate [92] carried out a study in order to quantify the F_a value of SiC as a function of the humidity for hydrophilic and hydrophobic components. This study revealed that strong capillary interactions take place around the tip for hydrophilic surfaces. However, an abnormal behavior occurs for hydrophobic surfaces of silicon oxide covered with a perfluoropolyether lubricant. In that case, water capillarity formation around the tip is suppressed.

3.1.4. Examples of Elastic Deformation Using Picoindentation in YBaCuO Materials

YBaCuO materials show high critical current density and high trapped magnetic field at cryogenic temperatures (around 77 K). However, practical applications of these types of materials are often limited by their poor mechanical performance, especially at working conditions. Hence, mechanical properties such as hardness, microhardness, elastic modulus, and others, are crucial for industrial applications of these kinds of materials. These parameters are really important to implement the use of this type of materials in the the fields of current transport, motors, flywheels, etc. $YBa_2Cu_3O_{7-x}$ presents a heterogeneous structure of Y-123 (superconducting phase) and Y-211 (non-superconducting phase). YBaCuO samples prepared by the Bridgman and TSMG technique render higher elastic modulus values for Y-211 than for Y-123 phases. This effect can be due to different reasons: Ionic bonds present in Y-211 phase are stronger than those present in Y-123, fact that is refelcted on the different melting point of these two different phases ($T_{Y\text{-}123}$ ~ 1010°C and $T_{Y\text{-}211}$ ~ 1200°C [93, 94]), and/or the high anisotropy confined onto the basal plane (001) of the Y-211 phase, which creates dislocations and residual stresses inside the particles due to a compressive state during the solidification process. As a consequence, microcracks are generated in the surrounding matrix [95, 96].

When YBaCuO samples textured by self-flux technique are doped with Ca and Sr, the elastic modulus value drastically decreases due to the ionic and crystallographic radius of the different doping agents [97]. Then, as Y^{3+} is substituted by a Ca^{2+} cation in the YBaCuO unit cell, an increment of the unit cell volume is produced. In this case, the elastic modulus value reduction can be attributed to different effects: the ionic and crystallographic radii are different, $r_{Ca}^{2+} > r_{Y}^{3+}$, and the different elements present a different coordination number [98]. Both phenomena promote the creation of vacancies, which in turn originate dislocations that are the origin of a high density of surface defects [21]. A similar phenomena takes place when the Ba^{2+} is substituted by Sr^{2+}; Nevertheless, as Ba^{2+} and Sr^{2+} share the same coordination number, the elastic modulus value for Sr^{2+} doped samples is a bit higher (around 6%) than in the case of Ca^{2+} doped ones. In conclusion, dopants modify YBaCuO crystallographic parameters and result in an elastic modulus reduction.

3.2. Examples Studied Using Nanoindentation Technique

3.2.1. YBaCuO Superconductors

YBaCuO presents a phase transition from tetragonal to orthorhombic that promotes a change in its electrical properties, that is, from insulating to superconductor, and that can be achieved by partially oxygenating the material [99]. On the other hand, the structure of the textured material presents an heterogeneous distribution of two different phases: $YBa_2Cu_3O_{7-x}$ (YBaCuO or Y-123), which is the matrix and Y_2BaCuO_5 (Y-211), corresponding with spherical or quasispherical inclusions. Moreover, the texture process induces an anisotropic structure, being the *ab*-planes (001) the ones with superconductor properties and the *c*-axis (110) as an insulator. Nowadays, effort is focused on the production of long length Superconductor Materials; In this sense, two different techniques are currently available: *silver welding technique* using Top Seeded Melt Growth samples and *coated conductors*.

In this section, the mechanical properties for YBaCuO bulk and coated conductors will be discussed in detail.

- *Hardness, H:* Normally, *H* value does not remain constant at different applied loads, which is a well-known physical phenomenon called Indentation Size Effect (ISE). Normally, *H* value presents a maximum at low loads as Roa *et al* demonstrated for YBaCuO samples textured by Bridgman technique [100]. However, when the applied load increases this property is widely modified by surface defects like roughness or macrocracks. When this phenomenon takes place, the contact area could be over- or underestimated, resulting in false *H* value. YBaCuO presents a heterogeneous microstructure in the Y-123 matrix and a minor phase around 30% of Y-211 [22]. Roa *et al* in 2007 [22], isolated the mechanical response for each phase by applying different loads. The minimum force applied to isolate the different phase's contribution was 10mN for each sample. With this applied load, the residual contact area is lower than the size of the different inclusions. Then, the mechanical property can be obtained. The *H* value for Y-211 are around twice of those obtained for Y-123. If the different residual imprints are performed near the grain boundary, the plastic deformation mechanism under the imprint could be affected by the Y-123 matrix, and *H* value would be modified, thus giving us an average of the studied property. When the applied load is around 100 mN, overall nanohardness for samples textured using the Bridgman and Top Seeded Melt Growth technique present a value of 8.9 and 7.9 GPa, respectively. The different results reported by Roa *et al.* [22] are in correct agreement with a previous work reported by Lim and Chaudhri [101], which studied H value of YBaCuO single crystals by Nanoindentation technique, obtaining values of 7.81 GPa. On the other hand, the hardness value reported for Bridgman samples [22] is in agreement with a previous work reported by Soifer *et al.* [102] and Cook et al. [103]. Verdyan *et al.* [104] reported nanohardness for orthorhombic YBaCuO thin films being around 8.5 GPa, when the applied load varies between 0.1 and 0.9 mN. However, the different H values reported in the literature exhibit a broad distribution that can be attributed to the different quality of the studied YBaCuO samples (i.e. grain structure, porosity, texturing process, etc) and also to the different measuring techniques (microindentation, nanoindentation, AFM-FS, etc.). Moreover, H values for coated conductors (like YBaCuO/LZO/Ni-5%W and YBaCuO/CeO_2/YSZ) in two different regions were observed by Roa *et al.* [105]: *i)* at low penetration depth values, the mechanical properties are not constant, presenting a high dependency with surface defects (i.e. dislocations, twins, porosity, grain boundary, etc) until 250 nm, and *ii)* at high penetration depths, the mechanical properties remain constant and similar to those of the bulk material, close to 13 GPa.

- *Elastic modulus, E: E* value is an intrinsic property of the studied materials and it only depends on their composition. Several measurements of *E* values for Y-123 resulted in values scattered within the range 40 - 200 GPa, variation that was due tocaused by residual porosity and bad contact between the grains [106]. Sakai *et al.* [107] reported a value of 370 GPa for 5 mm^3 specimens cut from a single-crystal of Y-123 phase prepared by Top Seeded Melt Growth Technique and using a single domain of Sm-123 as a seed. The different authors attributed the difference of *E* value to the 40 % excess of Y-211 phase present in the material. *E* values reported in ref [22] are in agreement with Johansen et al. [23] and Alford *et al* [108] when the applied loads are 30 and 100 mN. Moreover, *E* values for coated conductors (like YBaCuO/LZO/Ni-5%W and YBaCuO/CeO_2/YSZ) obtained using the Hertz equation remain constant and equal to 120 GPa [22].

3.2.2. Yttria-Stabilized Zirconia and Doped Ceria Electrolytes for Fuel Cells

Solid oxide fuel cells (SOFCs) are electrochemical devices for high efficiency energy production. Moreover, these devices are one of the alternative technologies for the production of clean and low-environmental-impact energy. SOFCs have a wide range of applications, sizes and configurations. Each assembled SOFC stack comprises several elements: electrolyte, anode, cathode, interconnections and sealing materials. The election of an appropriate candidate material for each component depends on its physical and chemical properties. During component fabrication, stack assembly and operation processes, SOFCs must resist relatively strong stresses such as those generated by thermal cycles, which are especially critical in the case of the electrolyte. Therefore, electrolyte materials must be tough enough. However, the study of the mechanical properties (hardness, elastic modulus and/or toughness fracture) of the most typical electrolytes used in SOFCs, that is $Gd_{0.2}Ce_{0.8}O_{2-x}$ (GDC) and $Y_{0.08}Zr_{0.92}O_{2-x}$ (YSZ), will be reviewed in this section. Roa *et al.* [109] studied the evolution of H value for YSZ at different penetration depths, showing a constant value of 15-17 GPa for depths equal or higher than 500 nm. This value is slightly lower than the one reported by Bekale *et al.* (around 16 GPa [110]), obtained by Vickers microhardness tests. E value was observed to be constant and equal to 200 GPa [23], which is in correct agreement with the different values reported in the literature [111, 112, 113, 114] for Zr stabilized with various contents of Yttrium oxide.

For GDC samples, measured H and E values were 10-12 and 140-180 GPa, respectively [23]. E values are about 15% lower than those reported by Sato *et al.* [115] using the Vickers microhardness test. This difference can be attributed to the different technique used to get the measurements. Morales *et al.* [116] studied the mechanical properties of different YSZ and GDC electrolytes with different stoichiometry, showing the different mechanical properties for each composition at different applied loads: from 5 to 500 mN, the results show that the mechanical properties decrease with lower dopant content, probably due to the increase of oxygen vacancies density, which could have other effects than just the reduction of the grain size. Comparing the E values obtained by nanoindentation, The different values reported in ref. [23] are about 20% higher than the values reported by Sato *et al.* [23] obtained with the small punch method. However, Wang *et al.* [117] obtained an E value around 250 GPa for 10GDC, which is similar to the value found in the ref. [23], by means of nanoindentation at an applied load of 5 mN. H values for GDC samples measured by using nanoindentation or other conventional techniques are not reported in the literature. In the case of 8YSZ, E values are in agreement with the nanoindentation and impulse excitation technique results reported by Bekale *et al.* [23].

3.2.3. Induce Phase Transition in Yttria Stabilized Polycrystalline Tetragonal Zirconia, Y-TZP Generated under a Hardness Test

The yttria stabilized polycrystalline tetragonal zirconia, Y-TZP is one of the best situated ceramic material for technological and biomedical applications due to their high fracture resistance (~ 1GPa), elevated toughness (4-5 MPa√m), hardness (HV=13GPa) and Young's Modulus (~240GPa). The zirconia presents a monoclinic structure at room temperature and the stabilization in a tetragonal phase, that is harder than monoclinic, is achieved by means of the addition of several elements, normally Yttria in 3 M, then named such as 3Y-TZP. Nevertheless, the tetragonal phase is then stable at room temperature except under two situations, (i) by induced pressures and (ii) under hydrothermal conditions. This transformation involves a volumetric expansion and that is the cause for which the Y-TZP is a material with higher toughness features than other ceramics; *(i)* the stresses generated by the crack propagation induces the t→m transformation, *(ii)* the generated expansion produces compressive forces on the tip of the cracks, *(iii)* the compressive force leads to stop the crack propagation. In the followed results, it will be show how the hardness test induces to a phase transformation. In this issue, the Raman microprobe spectroscopy is a suitable technique to detect the tetragonal (*t*) and monoclinic (*m*) phase in small volumes of zirconia ceramics. The *m*-phase is detected in the doublet 181 and 192cm^{-1} while the *t*-phase is visualized at 148 and 264cm^{-1}.

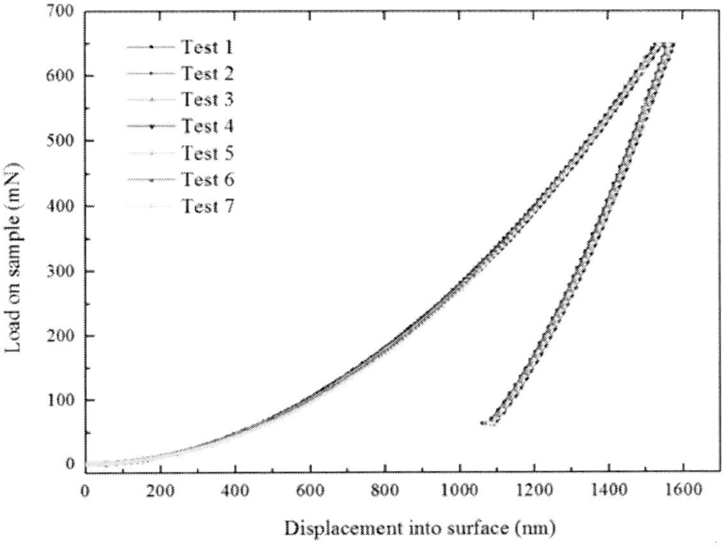

(a)

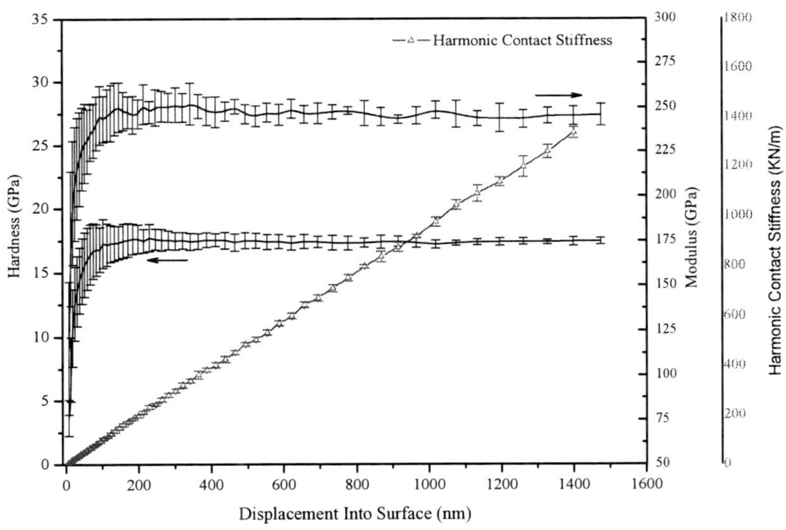

(b)

Figure 3. a) Loading-unloading recorded curves for a 3Y-TZP using the nanoindentation technique with Berkovich indenter, and b) Nanoindentation hardness and elastic modulus curves along penetration depth for a Y-TZP sample.

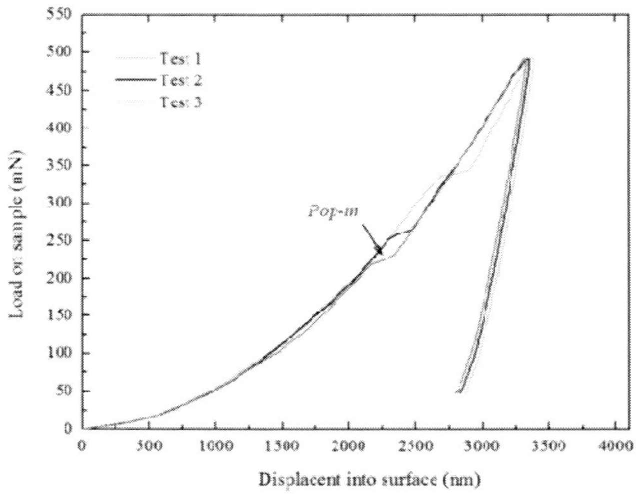

(a)

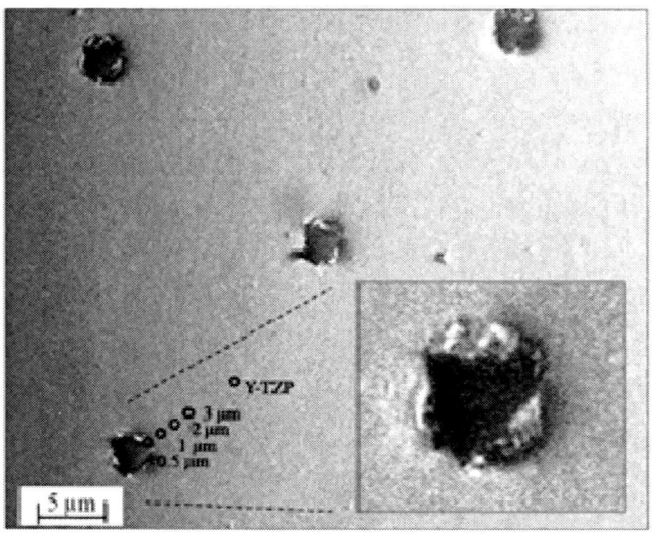

(b)

Figure 4. a) Nanoindentation P-h curves acquired with a Cube Corner indenter on 3Y-TZP, and b) Optical microscope image of the Cube Corner imprints on 3Y-TZP. The screened distances are indicating where the Raman spot was located.

The analyzed specimen was a sintered slip casting piece of 3Y-TZP. The hardness was measured by means of a micro-hardness Vickers tester and using a nanoindenter with two indenter geometries; (i) a previously calibrated diamond Berkovich tip and (ii) a Cube Corner tip. The continuous stiffness method (CSM) gives us the *H* and *E* profiles in depth. The indentations were conducted at max. load of 500 mN. The Figure 3a shows the typical *P-h* curves acquired by the nanoindenter with Berkovich tip for an elastic-plastic imprint. The recorded curves shown in Figure 3b revealed that the average *H* and *E* values were 17 GPa and 250 GPa, respectively. Nevertheless, when a Cube Corner indenter was used, the *P-h* curves showed several pop-in events (Figure 4a) and an optical inspection of the imprints revealed a pile-up emerged material around the imprint, Figure 4b. Figure 4b is an optical image of the Cube Corner imprints showing the localizations where the Raman analysis was performed and Figure 5 is the Raman spectra acquired on the Y-TZP surface for each indenter tested. In the case of the Berkovich and Vickers tests only the curve acquired on the middle of imprint is showed for clearer presentation. These curves show that when the Raman laser spot is moved on the Berkovich imprint, only the presence of the *t*-phase without content of *m*-phase is observed. Thus, a nanoindentation test using a Berkovich tip geometry, it enables us to determine the mechanical properties of a Y-TZP material without tc induce changes of phase.

Nevertheless, if a Cube Corner instead of a Berkovich indenter is used the 181 and 192 cm^{-1} double peak is revealed when the Raman laser spot is gradually moved towards the Cube Corner imprints. Furthermore, it is observable that the intensity of the *t*- phase peak decreases at the expense of an increased intensity of the *m*-phase band. The monoclinic phase was detected even 3 microns away from the side of the imprint, indicating that the $t \rightarrow m$- phase transformation occurs far away from the plastic deformation, confirming the catalytic character of this reaction. These results, indicates that the required applied load to achieve the $t \rightarrow m$ transformation is lower for a Cube Corner indenter geometry compared with the Berkovich. Similar results were found when a Vickers microhardness test was used on the Y-TZP, showing monoclinic phase far away the generated imprint even the tested load was reduced until 5g. These results evidenced that when the *H* and *E* values of a Y-TZP specimen are measured by indentation techniques an induced $m \rightarrow t$ transformation must be considered; (i) because the developed stresses generated beyond the indenter and (ii) because the softer character of the monoclinic generated phase.

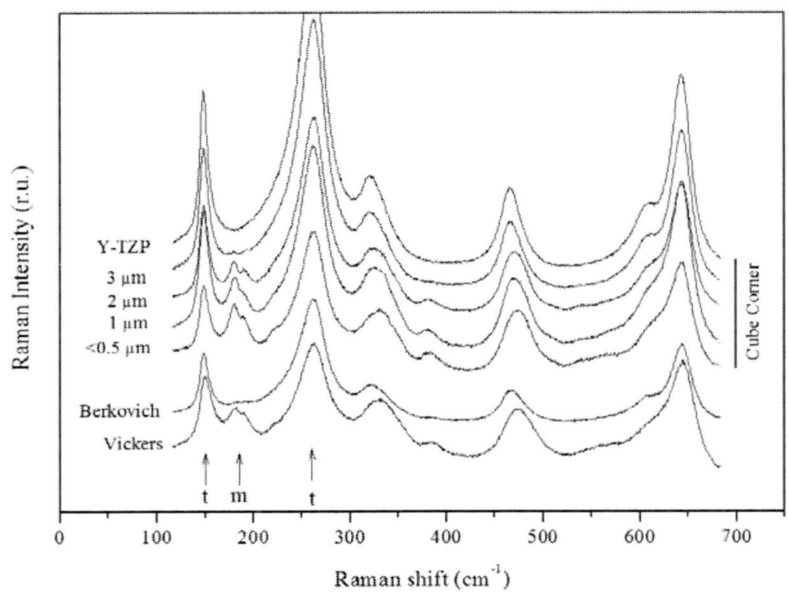

Figure 5. Raman curves obtained on a Berkovich and Vickers imprints and at several microns away from the side of the Cube Corner imprint as shown in Figure 4b.

3.3. Scratch Examples

Three different materials were analyzed using scratch tests by a nanoindenter; *(i)* a bulk piece of fused silica as an amorphous material, *(ii)* a tetragonal Y-TZP, and *(iii)* a polymer coating of a commercial optical storage disc. The experimental conditions were 3 tests per sample using a Berkovich indenter with a measured radius of 200 nm, and an edge-forward scratch direction. Load was ramped from 0 mN to 500 mN along a 2000 μm long scratch path. The cross-profile was acquired at 400 mN. In the following section, the most important data and curves are discussed.

3.3.1. Calculus of the Critical Load (P_c) by Means of Microscopy Wear-Track Visualization

Once a scratch test is finished, the first single operation is the visualization of the wear track. If the scratch load was enough to create a micrometer-deep trench, an optical microscope could be an interesting tool to acquire a first impression and to assess the deformation mechanism. Other type of topographic characterization instruments such as confocal microscopes, electron scanning microscopes, atomic force microscopes, profilometers or surface 3D profilometry reconstruction (Nanovision® from Agilent Tech. e.g.) are suggested to analyze the damage and plastic deformation mechanisms. Figure 6 shows an optical microscopy image of 3 wear tracks on the fused silica after the scratch experiment was finished.

The figure shows that track obtained at low loads follows a ploughing-wear deformation mechanism. Nevertheless, at a certain distance, about 300 μm, the material began to fail. From this image the mechanism of failure seems to be generated by radial cracks that extended around 100 μm from each side of the wear track.

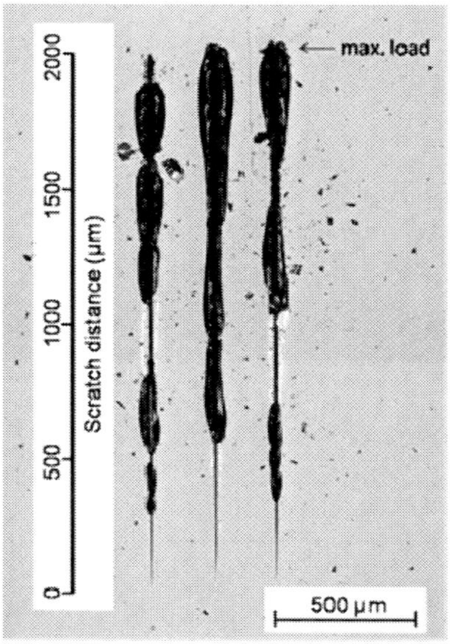

Figure 6. Optical images of 2000 μm long scratch paths on bulk fused silica using a Berkovich indenter under loads ramping from 0 mN to 500 mN.

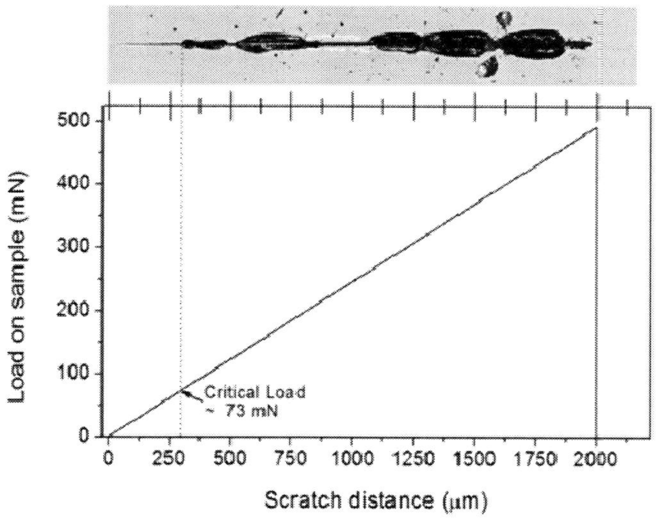

Figure 7. Load ramp applied along the scratch distance on a bulk fused silica sample.

When the visualized tracks are placed together on their load curves it is possible to determine the critical load, P_c. Figure 7 shows that the linearity of applied load ramp was kept along all the scratch test, even during the promotion of failure mechanisms. Here, the P_C is a measure of the scratch resistance and depends on the loading rate, scratching speed and tip geometry. Considering that the scratch resistance of a material is the ability to withstand abrasion without fracturing, the image of the acquired tracks reveals a scratch resistance of 73 mN.

3.3.2. Calculus of the P_c Value by Means of Analysis of the Friction Kinetics

As described below, the scratch track is generated applying a normal force, F_N, by the indenter on the specimen surface while the sample is moving. Subsequently a tangential force, F_T *is developed and* the coefficient of friction, μ, is calculated using equation 1. The F_T is acquired by specials spring sensors installed in the head assembly of the indenter that are working such as force transducers. The μ value can be calculated from constant or ramped load experiments. Thus, the kinetic behavior of μ it could help us to determine the P_c value. The change of the slope or the noise in the experimental curves evidences the failure mechanisms. Coming back to the fused silica, the Figure 8 shows the μ curve evolution. The Kinetic friction shows several mechanisms activated along the tested paths. Along the firsts 20 nm, μ value is very high because the static friction and adhesive forces must be overcome. The transition from elastic to plastic deformation is usually observed such a linear ramp of load. Note that the increase in friction with increasing scratch load along a sliding contact agrees with Nosonovsky and Bhushan's experiences [118, 119]; briefly, μ value increases at small contact sizes when contact size and applied load increase. Once the sliding and ploughing mechanisms dominate the scratching process, μ value is kept constant up to the end of the first 1000 μm of wear path. Along this scratch distance, tribological behavior is governed by the theory of multi-asperity contacts put forward by Zhang *et al.* [120, 121]; when two surfaces are in contact, a part of the applied load is distributed on the contact area of the asperities and on the contact area of debris particles.

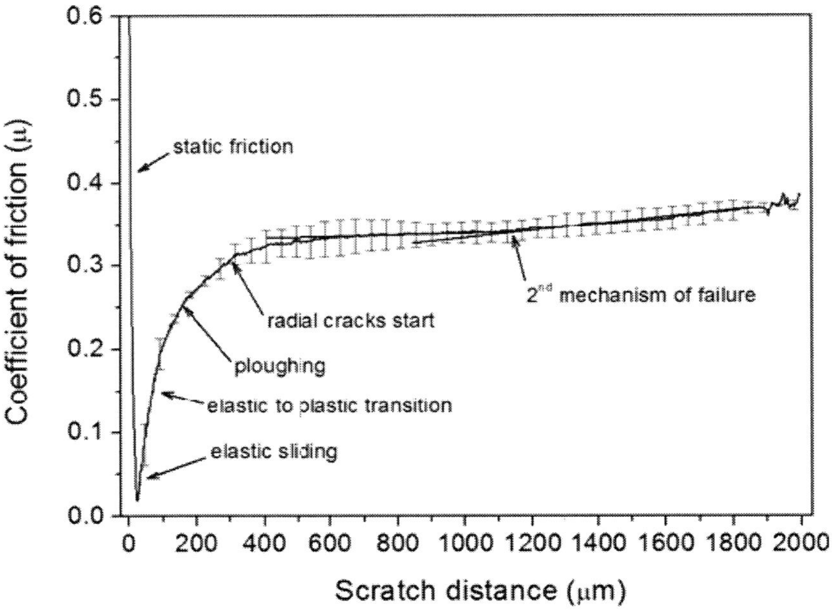

Figure 8. Averaged μ value obtained from the 3 scratch tests performed on fused silica.

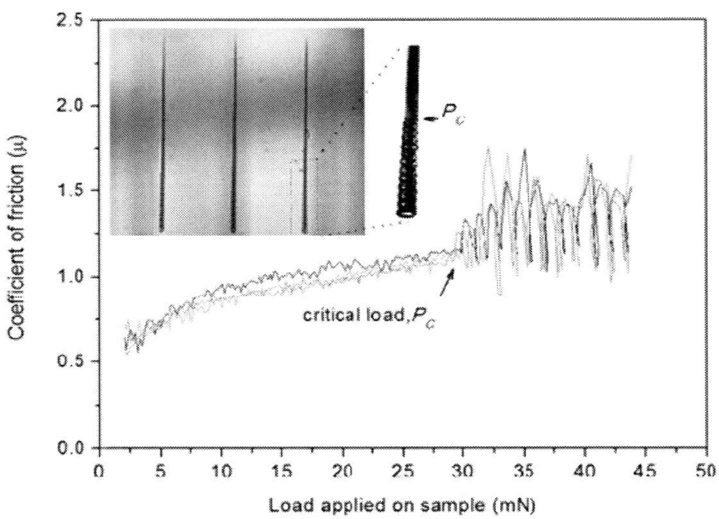

Figure 9. μ kinetic of the 3 scratch tests performed on a DVD protective polymeric coating. The inset figure is an optical microscope image of the resulting wear tracks.

Thus, the resultant μ value consists on the sum of three components: adhesion, asperity ploughing and debris ploughing. Finally, μ raises along the 1100 and 2000 μm range. This raise of the measured friction could be explained by a second failure mechanism that appears when 250 mN load is reached. As can be seen, the kinetic friction curves permit us to study other deformation mechanisms that take place along the scratch tests and that would be kept hidden with the single observation of the track morphology. Figure 9 shows a case where P_c can be clearly determined by the appearance of a significant noise level in the friction curve. In this case a polymeric coating is analyzed instead of a bulk material. It is worth to note the oscillatory behavior of μ value when P_C is overcome, failure which was probably due to adhesive ploughing formation mechanism. Other possible explanation is that the indenter reached the substrate after thoroughly penetrating the coating. This option could be studied by the analysis of the penetration curves as described in the next section.

3.3.3. Calculus of the P_c Value by Using Penetration Depth Curves

So far, we saw that visual inspection of the wear tracks and the subsequent analysis of the kinetic friction are powerful options to detect and quantify the P_c and the activated failure mechanisms. The next section pretends to be a basic guide to understand the different curves and graphs that can be obtained using a nanoindenter under a scratch test.

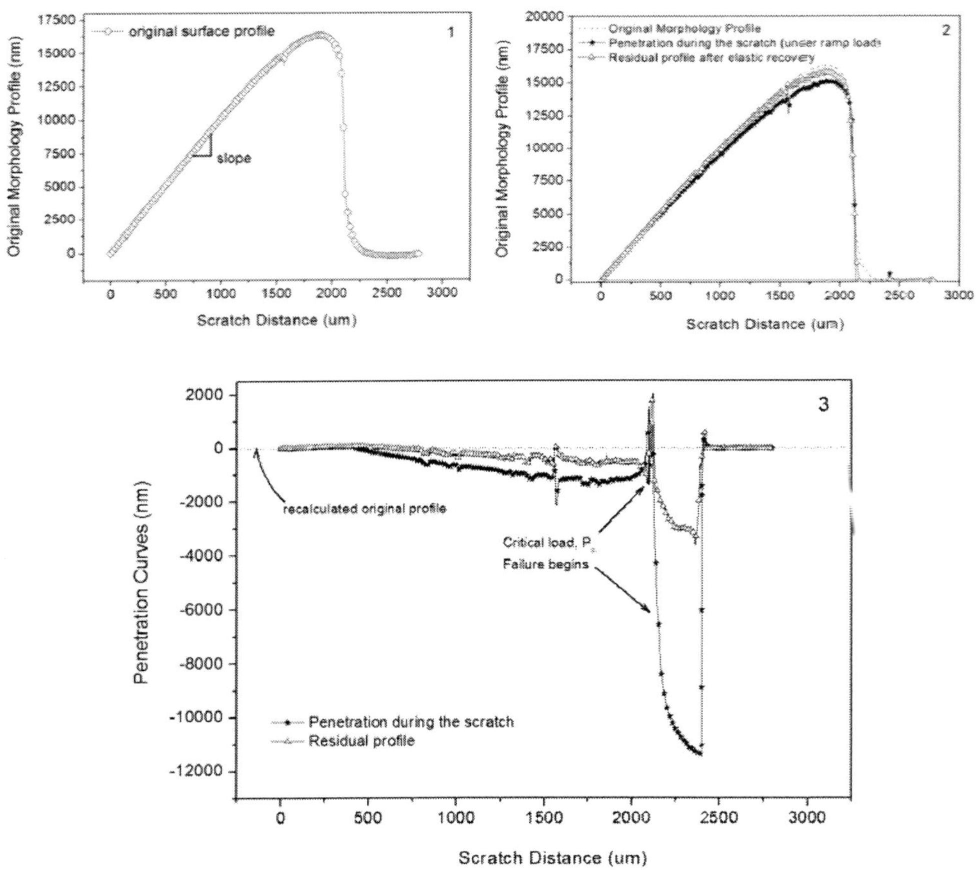

Figure 10. *(1)* Displacement curves on a sintered Y-TZP sample where the original profile corresponds to the vertical indenter displacement along the sample surface before the scratch test is performed. This profile permits to measure the slope of the surface when large displacements are essayed. *(2)* Penetration depth plot reached during the scratch and after the elastic recovery. *(3)* Calculated penetration curves, which have been corrected by means of the slope scanned by the original profile.

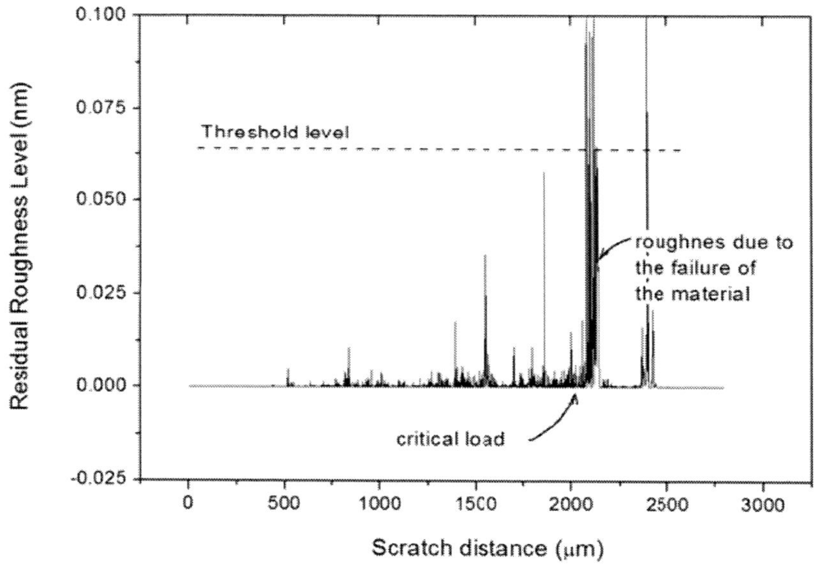

Figure 11. Representation of the residual roughness level along the scratch distance for a sintered Y-TZP sample.

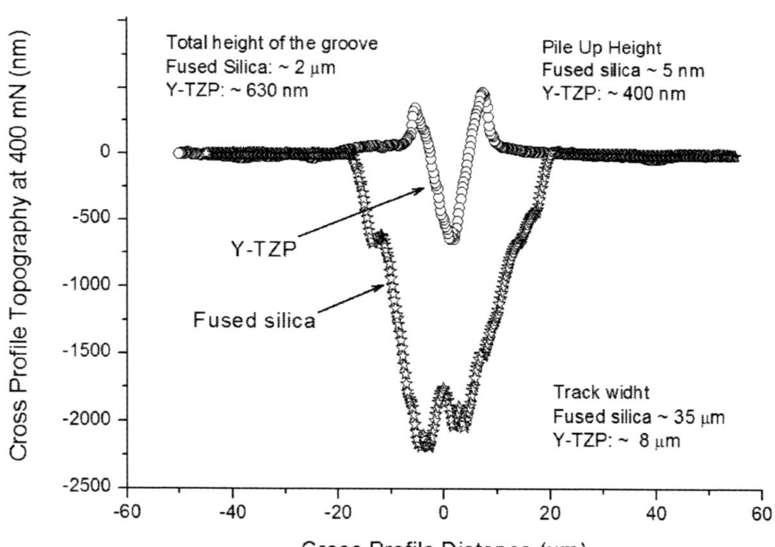

Figure 12. Comparative of the plasticity close to the scratch zone between Fused silica and Y-TZP samples.

The most representative experimental curves are listed below:

- *Original profiling topography. T*he acquisition of the surface profile under a low contact load makes possible to scratch long distances without the influence of the slope of samples. This curve shows the topographic slope of the sample and gives us a qualitative idea of its roughness. All subsequent penetration curves are sensitive to be corrected using this data.

- *Penetration during the scratch profile.* This measurement provides a quantitative value for the penetration underneath the sample surface. If experimentally obtained curves are not slope-profile corrected, the plot is named morphology profile.

After the original morphology is subtracted from the scratch data, the Displacement curve plot along scratch distance is obtained. The next sequence of figures show an example where the slope calculated from the original profile curve is used to correct the penetration curves acquired along the scratch. The tested sample was a sintered Y-TZP specimen with an enormous forced slope to demonstrate how powerful the proposed correction is it. The Figure 10 shows the penetration and residual curves once the slope was subtracted. Once the correction has been performed the location at which failure was provoked, it is easier identified without visualizing the track. The critical load is revealed as a fall of the penetration curves without to visualize the scratched path.

- *Residual roughness level curves. These* curves are used by the nanoindenter software to automatically determine the *Pc* value by means of a threshold signal. Figure11 shows the residual roughness level that corresponds with previous results. Note that roughness level is measured at a constant load, so the graph represents roughness level versus scratch distance.

3.3.4. Quantification of the Plasticity of the Scratch

The plasticity of a material determined by a scratch experiment can be assessed by observing the scratch profiles, as major plastic deformation is usually detected as pile-up material around the wear path. Furthermore, the possibility of measuring the pile-up material in the same scratch experiment makes possible the acquisition of these profiles at several lapsed times, being a very useful procedure to observe viscoelastic relaxations in time - dependent materials or on polymeric composited ceramics. The nanoscratch is a useful technique to analyze the cross profile of the wear track, as shown in Figure 12. In this figure, the cross profiles of the fused silica and Y-ZPT are registered and it is revealed that the tetragonal zirconia ceramic shows the best mechanical performance under scratch features. This fact is due to its intrinsic properties, which are governed by pressure induced phase transformations.

CONCLUSION

The present chapter seeks to present and discuss the most relevant results concerning the analysis of friction, hardness and elastic modulus with AFM-FS and Nanoindentation techniques for different ceramic materials at the nanoscale in order to provide a clear overview of ceramic nanomechanics to researchers entering the field and a useful guide to those seeking for experimental advice. Besides, the theoretical framework necessary to interpret, provide and better understand the results concerning the different nanomechanics disciplines is also reviewed. The works reviewed in this chapter show that a reasonable correspondence between macro- and nanometric results exists, although nanometric measurements are much more sensitive to surface phenomena as water meniscus, electrical charges, superficial defects as porosity or cracking effects and dislocations.

Acknowledgments

The authors acknowledge the Nanometric Techniques Unit of the Scientific and Technological Centers of the University of Barcelona for technical support.

References

[1] D. Dowson. *History of tribology.* London: Longmans; 1979.
[2] G. Amontons. *De la Resistance causée dans les machines. Memoires de l'Academie Royale.* Amsterdam : Chez Gerard Kuiper ; 1706.
[3] D. Dowson D, C. A. Coulom. *J. Lubr. Technol.,* 100, 148 (1785).
[4] L. Euler. *Mem. Acad. Sci. Berl.*, 4,122(1748).
[5] L. Euler. *Mem. Acad. Sci. Berl.*, 18, 256 (1762).
[6] F. P. Bowden, D. Tabor D. Part I. Oxford: Clarendon Press; 1954.
[7] J. F. Archad. *Proc. R. Soc. Lond. A.*, 243, 190 (1957).
[8] M. R. Van Landingham. *J. Res. Natl. Inst. Stand. Technol.*, 108, 249 (2003).
[9] J. Ma. *Mater. Sci. Eng. A*, 457, 63 (2007).
[10] L. Cretegny, A. Saxena. *Acta Mater.,* 49, 3755 (2001).
[11] J. Man, K. Obrtlik, C. Blochwitz, J. Polak. *Acta Mater.,* 50, 3767 (2002).
[12] J. Man, M. Petrenec, K. Obrtlik, J. Polak. *Acta Mater.*, 52 (19), 5551 (2004).
[13] A. Schwab, O. Meibneri, C. Holste. *Philos. Mag. Lett.*, 77, 23 (1998).
[14] M.Risbet, X. Feaugas, C. Guillemer-Neel, M. Clavel. *Scrip. Mater.,* 49, 533 (2003).
[15] L. Cretegny, A. Saxena. *Fatigue Fract. Eng. Mater. Struct.,* 25,305 (2002).
[16] J. J. Roa, G. Oncins, F. T. Dias, X. G. Capdevila, V. N. Vieira, J. Schad, M. Segarra. *Physica C*, 471,544 (2011).
[17] K.Johnson. Cambridge, UK : Cambridge University Press ; 1987.
[18] A. C. Fischer-Cripps. New York : Springer, 2004.
[19] J. D. Kiely, J. E. Houston. *Phys. Rev. B: Condens Matter Mater Phys*, 57, 12588 (1998).
[20] J. Li, L. V. Vliet, T. Zhu T, S. Yip, S. Suresh. *Nature*, 418 (6895), 307 (2002).
[21] J. Knap, M. Ortiz. *Phys. Rev. Lett.*, 90, 226102/1 (2003).
[22] D. Lorenz, A. Zeckzer, U. Hilpert, P. Grau, H. Johansen, H. Leipner. *Phys. Rev. B: Condens. Matter. Phys.*, 67, 172101/1 (2003).
[23] P. F. M. Teran Arce, G. Riera, P. Gorostiza, F. Sanz. *Appl. Phys. Lett.*, 77,839 (2000).
[24] A. Asenjo, M. Jaafar, E. Carrasco, J. M. Rojo. *Phys. Rev. B: Condens Matter. Mater. Phys*, 73, 75431/1 (2006).
[25] T. Filleter, S. Maier, R. Bennewitz. *Phys. Rev. B*, 73, 155433/1 (2006).
[26] J. N. Israelachvili. *Surf. Sci. Rep.*, 14, 109 (1992).

[27] Y. Huang, F. Zhang, K. C. Hwang, W. D. Nix, G. M. Pharr, G. A. Feng. *A model of size effets in nano-indentation*, 54 (8), 1668 (2006).
[28] D. Hull, D. J. Bacon. *Introduction to Dislocations*, Butterworth-Heinermann. 1965. ISBN.: 0 7506 4681 0.
[29] J. W. Christian, S. Mahajan. *Prog. Mat. Sci.*, 39, 1(1995).
[30] M. W. Barsoum. *Prog. Solid St. Chem.* 28, 201 (2000).
[31] Y. Choi, K. J. Van Vliet, J. Li, S. Suresh. *J. Appl : Phys.*, 94, 6050 (2003)
[32] J. N. Israelachvili, P. M. McGuiggan. *J. Mater. Res.,* 5, 2223 (1990).
[33] J. N. Israelachvili. *J. Phys. Condens Matter*, 2, SA89 (1990).
[34] C. M. Mate, G. M. McClelland, R. Erlandsson, S. Chiang. *Phys. Rev. Lett.,* 59, 1942 (1987).
[35] V. L. Popov. *Tribol. Int.,* 34, 277(2001).
[36] A. Socoliuc, E. Gnecco, S. Maier, O. Pfeiffer, A. Baratoff, R. Bennewitz, et al. *Science,* 313 (5784), 207 (2006).
[37] G. Binnig, C. F. Quate, C. Gerber. *Phys. Rev. Lett.,* 56, 930 (1986).
[38] J. Krim, D. H. Solina, R. Chiarello. *Phys. Rev. Lett.*, 66, 181 (1991).
[39] J. Y. Park, D. F. Ogletree, P. A. Thiel, M. Salmeron. *Science*, 313 (5784), 186 (2006).
[40] R. Budakian, S. J. Putterman. *Phys. Rev. Lett*, 85, 1000 (2000).
[41] J. J. Roa, G. Oncins, J. Diaz, X. G. Capdevila, F. Sanz, M. Segarra. *J. Europ. Ceram. Soc.*, 31,429 (2011).
[42] J. J. Roa, G. Oncins, J. Diaz, F. Sanz, M. Segarra. *Recent Patents on Nanotechnology*, 5, 27 (2011).
[43] D. Mazzucco, M. Spector. *J. of Tribology-Trans. of the Asme.*, 128 (1), 113 (2006).
[44] K. L. Johnson. *Wear*, 190 (2), 162 (1995).
[45] H. Hertz. *J. Reine Angew. Math.*, 92, 156 (1881).
[46] K. L. Johnson, K. Kendall, A. D. Robert. Proceedings of the Royal Society of London Series A-Mathematical and Physical Sciences. 324, 301 (1971).
[47] D. Mauguis, M. Barquins. *J. Phys. D: Appl. Phys.,* 16, 1843 (1983).
[48] R. W. Carpick, D. F. Ogletree, M. A. Salmeron. *J. Colloid. Interface Sci.*, 211, 395 (1999).
[49] B. V. Derjaguin, V. M. Müller, Y. P. Toporov. *J. Colloid. Interface Sci.*, 53,314 (1975).
[50] K. L. Jonson. *Langmuir*. 12, 4510 (1996).
[51] D. Maugis. *J. Colloid. Interface Sci.*, 150, 243 (1992).
[52] V. M. Müller, V. S. Yushenko, B. V. Derjaguin. *J. Colloid Interface Sci.*, 77, 91 (1980).
[53] W. C. Oliver, G. M. Pharr. *J. Mater. Res.,* 7, 1564 (1992).
[54] G. M. Pharr, W. C. Oliver, F. R. Brotzen. *J. Mater. Res.,* 7, 613 (1992).
[55] N. E. Kurland, Z. Driea, V. K. Yadavalli. Micron (Oxford, England; 1993) (2011).
[56] H. Hertz, A. Reine. J. Math. 92 (1881) 156 (Translated and reprinted in English in Hertz's Miscellaneous Papers, Macmillan and Co., London (1986) Ch. 5).
[57] E. G. Herbert, G. M. Pharr, W. C. Oliver, B. N. Lucas, J. L. Hay. *Thin Solid Films.* 398-399, 331 (2001).
[58] I. N. Sneddon. *Int. J. Engineering Science*, 3, 47 (1965).
[59] J. S. Field, M. V. Swain. *J. Mater. Res.,* 10,101 (1995).
[60] D. Tabor D. *Hardness of Metals* (Clarendon, Oxford, U. K., 1951).
[61] W. C. Oliver, G. M. Pharr. *J. Mater. Res.* 19, 3 (2004).
[62] M. T. Huber. *Ann. Phys.*, 43, 153 (1904).

[63] A. C. Fischer-Cripps. *Surface and Coatings Technologies*, 200, 4153 (2006).
[64] M. F. Doener, W. D. Nix. *J. Mater. Res.*, 1, 601 (1986).
[65] Y. Gaillard, E. Jiménez-Piqué, F. Soldea, F. Mücklich, M. Anglada. *Acta Mater*, 56, 4206 (2008).
[66] R. B. King. *Int. J. Solids Struct.*, 23,1657 (1987).
[67] M. F. Doener, D. S. Gardner, W. D. Nix. *J. Mater. Res.*, 1,845 (1986).
[68] D.Shuman. *Microscopy and Analysis*, 107, 21 (2005).
[69] A. V. Sumant, D. S. Grierson, J. E. Gerbi, J. Birrell, U. D. Lanke, O. Auciello, J. A. Carlisle, R. W. Carpick. *Adv. Mater.*, 17 (8), 1039 (2005).
[70] R. J. A. Van den Oetelaar, C. F. J. Flipse. *Surf. Sci.*, 384, 1828 (1997).
[71] J. A. Heimberg, K. J. Wahl, I. L. Singer, A. Erdemir. *Appl. Phys. Lett.*, 78, 2449 (2001).
[72] M. Enachescu, R. J. A. Van den Oetelaar, R. W. Carpik, D. F. Ogletree, C. F. J. Flipse, M. Salmeron. *Phys. Rev. Lett.*, 81, 1877 (1998).
[73] C. Corbella, M. Vives, G. Oncins, C. Canal, J. L. Andujar, E. Bertran. *Diam. Relat. Mater.*, 13, 1494 (2004).
[74] [E. Riedo, J. Chevrier, F. Comin, H. Brune. *Surf. Sci.*, 477, 25 (2000).
[75] T. H. Fang, C. I. Weng, J. G. Chang, C. C. Hwang. *Thin Sol. Films*. 396,166 (2001).
[76] G. T. Gao, P. T. Mikulski, G. M. Chateaneuf, J. A. Harrison. *J. Phys. Chem. B,* 107, 11082 (2003).
[77] H. Gang, M. H. Müser, M. O. Robbins. *Science*. 284, 1650 (1999).
[78] H. I. Kim, J. R. Lince, O. L. Eryilmaz, A. Erdemir. *Tribol. Lett.*, 21, 53 (2006).
[79] A. R. Koniced, D. S. Grierson, P. U. P. A. Gilbert, W. G. Sawyer, A. V. Sumant, R. W. Carpick. *Phys. Rev. Lett.*, 100, 235502 (2008).
[80] L. Sun, Q. Wu. *J. Mater. Res.* 14, 631 (1998).
[81] Y. Qi, E. Konca, A. T. Aplas. *Surf. Sci.* 600, 2955 (2006).
[82] B. O'Regan, M. A. Grätzel. *Nautre*. 353, 737 (1991).
[83] S. R. Sousa, P. Moradas-Ferreira, B. Saralago, L. Viseu Melo, M. A. Barbosa. *Lagmuir*. 20, 9745 (2004).
[84] A. K. Jämting, J. M. Bell, M. V. Swain, L. S. Wielunski, R. Clissold. *Thin Sol. Films*.332, 189 (1998).
[85] W. Zang, W. Liu, C. Wang. *Wear*. 253, 377 (2002).
[86] I. Piwonski. *Thin Sol. Films*. 515, 3499 (2006).
[87] F. L. Leite, E. C. Ziemath, J. R. O. N. Oliveira, P. S. P. Herrmann. *Microsc. Microanal.* 11 (supp 3) 2005.
[88] K. L. Johnson K. L. *Contact mechanics and adhesion of viscoelastic solids.* In: Book of abstracts. 216th ACS national meeting. 1998.
[89] K. N. G. Fuller, D. Tabor. *Proc. R. Soc. Lond. A*. 345, 327 (1975).
[90] K. Pöhlmann, K. H. Zlum Gahr. *Mater. Sci. End. Technol.* 31(4), 280 (2003).
[91] Z. H. Zum Garh, R. Blattner, D. H. Hwanf, K. Pöhlmann. *Wear*. 250, 299 (2001).
[92] M. Binggeli, C. M. Mate. *J. Vac. Sci. Technol. B.: Microelectron Nanometer Struct.*, 13(3), 1312 (1995).
[93] T. Aselage, K. Keefer. *J. Mater. Res.* 3, 1279 (1988).
[94] J. J. Roa, G. Oncins, J. Diaz, X. G. Capdevila, F. Sanz, M. Segarra. *J. Eur. Ceram. Soc.* 31, 429 (2011).
[95] F. Sandiumenge, T. Puig, J. Rabier, J. Plain, X. Obradors. *Adv. Mater.*, 12(5), 375 (2000).

[96] J. J. Roa, K. Konstantopoulou, E. Jiménez-Piqué, V. Martín, M. Segarra, J. Y. Pastor. *Ceramics International.* 38, 2035 (2012).
[97] J. J. Roa, G. Oncins, F. T. Dias, V. N. Vieira, J. Schaf, M. Segarra. *Phys. C.*, 471, 544 (2011).
[98] F. Licci, A. Gauzzi, M. Marezio, G. P. Radaelli, R. Masini, C. Chailout-Bougerol. *Phys. Rev. B.: Condens Mater. Phys.*,58(22), 15208 (1998).
[99] J. J. Roa, F. T. Dias, M. Martinez, J. A. Padilla, M. Segarra. *J. Eur. Ceram. Soc.*, 32, 425 (2012).
[100] J. J. Roa, X. G. Capdevila, M. Martinez, F. Espiell, M. Segarra. *Nanotechnology.*18, 385701/1 (2007).
[101] Y. Y. Lim, M. M. Chaudhri. *J. Phys. D : Appl. Phys.*, 34, L70 (2001).
[102] Ya. M. Soifer, A. Verdyan, J. Azoulay, M. Kazakevich, E. Rabkin. *Physica C.* 402, 80 (2004).
[103] R. F. Cook, T. R. Dinger, D. R. Clarke. *Appl. Phys. Lett.*, 51(6), 454 (1987).
[104] A. Verdyan, Y. M. Soifer, J. Azoulay, E. Rabkin, M. Kazakevich. *IEEE Trans. Appl. Supercond.,* 15, 3585 (2000).
[105] J. J. Roa, E. Jiménez-Piqué, T. Puig, X. Obradors, M. Segarra. *Thin Solid Films.*519, 2470 (2011).
[106] T. H. Johansen. *Supercond. Sci. Technol.,* 13(10), R121 (2001).
[107] N. Sakai, F. Murakama, P. Diko, S. Takebashi, S. I. Yoo, M. Murakami. *Adv. Supercond. X.*, 11, 645 (1997).
[108] N. M. Alford, J. D. Birchall, W. J. Clegg, M. A. Harmer, K. Kendall, D. H. Jones. *J. Mater. Sci.* 23(3), 761 (1988).
[109] J. J. Roa, J. Tartaj, M. Morales, X. G. Capdevila, M. Segarra. *Comparative study of mechanical properties of YSZ, GDC and LSGM electrolytes.* Proceedings of 9th European Solid oxide Fuel cell Forum. 2010. Chapter 11, pp. 39-48.
[110] V. Menvie Bekale, G. Sattonnay, C. Legros, A. M. Huntz, S. Poissonet, L. Thomé. *J. Nuclear Mat.*, 384, 70 (2009).
[111] A. Lakki, R. Herzog, M. Weller, H. Schubert, C. Reetz, O. Görke. *J. Eur. Ceram. Soc.,* 20, 285 (2000).
[112] A. Atkinson, A. Selcuk. *Mechanical behavior of ceramic oxygen ion-conducting membranes.* S. S. I. 2000; 134: 59-66.
[113] A. J. A. Winnubst, K. Keizer, A. J. Burggraaf. *J. Mater. Sci.,* 18, 1958 (1983).
[114] J. W. Adams, R. Ruh, K. S. Mazdiyasni. *J. Am. Ceram. Soc.,* 80(4), 903 (1997).
[115] K. Sato, H. Yugami, T. Hashida. *J. Mat. Sci.*, 39, 5765 (2004).
[116] M. Morales, J. J. Roa, X. G. Capdevila, M. Segarra, S. Pinol. *Acta Mater.* 58(7), 2504 (2010).
[117] Y. Wang, K. Duncan, E. D. Wachsman, F. Ebrahimi. *Sol. State Ionics.*, 178, 53 (2007).
[118] B. Bhushan, M. Nosonovsky. *Acta Mater.*, 51, 4331 (2003).
[119] B. Bhushan, M. Nosonovsky. *Nanotechnology.*, 15, 749 (2004).
[120] J. Zhang, F. A. Mosleby, S. L. Rice. *Wear.* 149, 1 (1991).
[121] J. Zhang, F. A. Mosleby, S. L. Rice. *Wear.* 149(1-2), 13 (1991).

In: Recent Advances in Ceramic Materials Research
Editors: Joan J. Roa Rovira and Merce Segarra Rubi ISBN: 978-1-62417-729-3

Chapter 9

RAPID FABRICATION OF BULK CERAMICS AND GLASSES BY HIGH-GRAVITY COMBUSTION SYNTHESIS

Guanghua Liu,[1*] Jiangtao Li[1] and Kexin Chen[2]
[1]Technical Institute of Physics and Chemistry, Chinese Academy of Sciences, Beijing 100190, China
[2]Department of Materials Science and Engineering, Tsinghua University, Beijing 100084, China

1. INTRODUCTION

In contrast to metals and alloys, which are generally produced by casting from melts, polycrystalline ceramics are generally fabricated by sintering from fine powders. For the melt-casting of ceramics, two major technical obstacles are involved. The first one is the difficulty in preparing stable and homogeneous ceramic melts because most ceramic materials have extremely high melting points and some decompose instead of melting at elevated temperatures. Therefore, the melting of ceramic materials is often realized by efficient heating media such as high-frequency induction, lamp mirror furnaces and lasers. Another obstacle in the melt-casting of ceramics is exaggerated grain growth during solidification from hot melts. A large grain size is usually not desirable in ceramic materials because it will impair the mechanical strength. In order to obtain fine microstructure, multiphase eutectic ceramics are more frequently prepared than single-phase ceramics.

The preparation of bulk ceramics by direct solidification can be carried out by unidirectional solidification in a container or pulling of a solid from a melt meniscus [1]. The Bridgman method is a suitable technique for preparing large samples, where the ingot size is limited only by crucibles. By this technique, ceramic melts are contained in molybdenum, tungsten or iridium crucibles, and unidirectional solidification is realized by slowly pulling the crucibles off the hot region. Another approach is the Czochralski method, in which

* Corresponding author: Guanghua Liu, Associate Professor, Technical Institute of Physics and Chemistry, Chinese Academy of Sciences, Tel/Fax: +86-10-82543695, E-mail: liugh02@mails.tsinghua.edu.cn.

crucibles are also required but direct contact between crucibles and growing materials is avoided. In addition, the floating-zone method is often used to prepare eutectic ceramics such as Al_2O_3/YSZ, Al_2O_3/YAG and Al_2O_3/YAG/YSZ [2,3]. This method does not require crucibles, and a relatively small volume of sample is melted by high-frequency induction or lasers. Other methods for solid pulling from a melt meniscus include edge-defined film-growth and micro-pulling down.

Despite the difference in technical details, all the above-mentioned methods require an external heating source to achieve high temperature for melting ceramic materials. In contrast to these methods, a furnace-free technique called high-gravity combustion synthesis has been recently developed to prepare bulk ceramics via melt solidification [4-8]. By this technique, highly-exothermic chemical reactions are utilized to realize high temperature and produce ceramic melts. A high-gravity field is applied to accelerate the separation between ceramic and metallic melts as well as removal of gas bubbles form the melts. After cooling and solidification, bulk ceramics are obtained together with metallic ingots. In high-gravity combustion synthesis, the heat energy for melting ceramic materials is created by chemical reactions, and thus external heating source is not necessary. At the same time, combustion reactions progress very quickly and usually last only tens of seconds. With the above two merits, high-gravity combustion synthesis offers a fast and efficient way to prepare bulk ceramics. Besides ceramic materials, glasses and glass-ceramics can also be produced by high-gravity combustion synthesis [9-11].

This article briefly reviews our recent progress on high-gravity combustion synthesis as well as its application for preparing ceramics and glasses. After a description of experimental procedure, several ceramic and glass materials prepared by high-gravity combustion synthesis are presented as examples. Then, the reaction mechanism in high-gravity combustion synthesis is systematically studied with an emphasis on phase separation and microstructure evolution. Finally, conclusions are drawn followed by a perspective on further development and application of high-gravity combustion synthesis.

2. Experimental Details

Commercial powders of Al, NiO, MgO, Y_2O_3 and ZrO_2 were used as raw materials. The raw materials were mixed in ethanol and homogenized by planetary ball milling for 1 h, and then dried to obtain a uniform mixture of reactant powders. The powder mixture was cold-pressed into a round compact and loaded into a graphite crucible. The crucible was coated by carbon felt and then put into a steel vessel. The vessel was mounted on a Ni-based superalloy rotator, which was placed in a closed reaction chamber. A schematic illustration and the assemblage of the facility for high-gravity combustion synthesis are shown in Figure 1.

The reaction chamber was evacuated to a vacuum of <100 Pa, and then the rotator was started. By centrifugation effect, a high-gravity field was induced. The strength of the high-gravity field was evaluated by a high-gravity factor of g'/g, where g' and g mean the centrifugal acceleration and normal gravitational acceleration, respectively. The strength of the high-gravity field was modulated by controlling the rotation frequency. When a designed high-gravity factor was reached, combustion reaction was triggered by passing an electric current through a tungsten coil closely above the sample. During the reaction, a large amount of heat energy was created and the products were melted. In a high-gravity field, the resultant ceramic and metallic melts were separated because of their density difference. After cooling and solidification, bulk ceramic samples and Ni ingots were obtained in graphite crucibles, as shown in Figure 2.

The prepared ceramic samples were machined and then characterized. The bulk density was measured according to the Archimedes principle. The phase composition was identified by X-ray diffraction (XRD; D8 Focus, Bruker, Germany). The microstructure was examined by scanning electron microscopy (SEM; S-3400, Hitachi, Japan), and chemical analysis was carried out by energy dispersive spectroscopy (EDS; INCA, Oxford Instrument, UK). The hardness was tested by the Vickers indentation method.

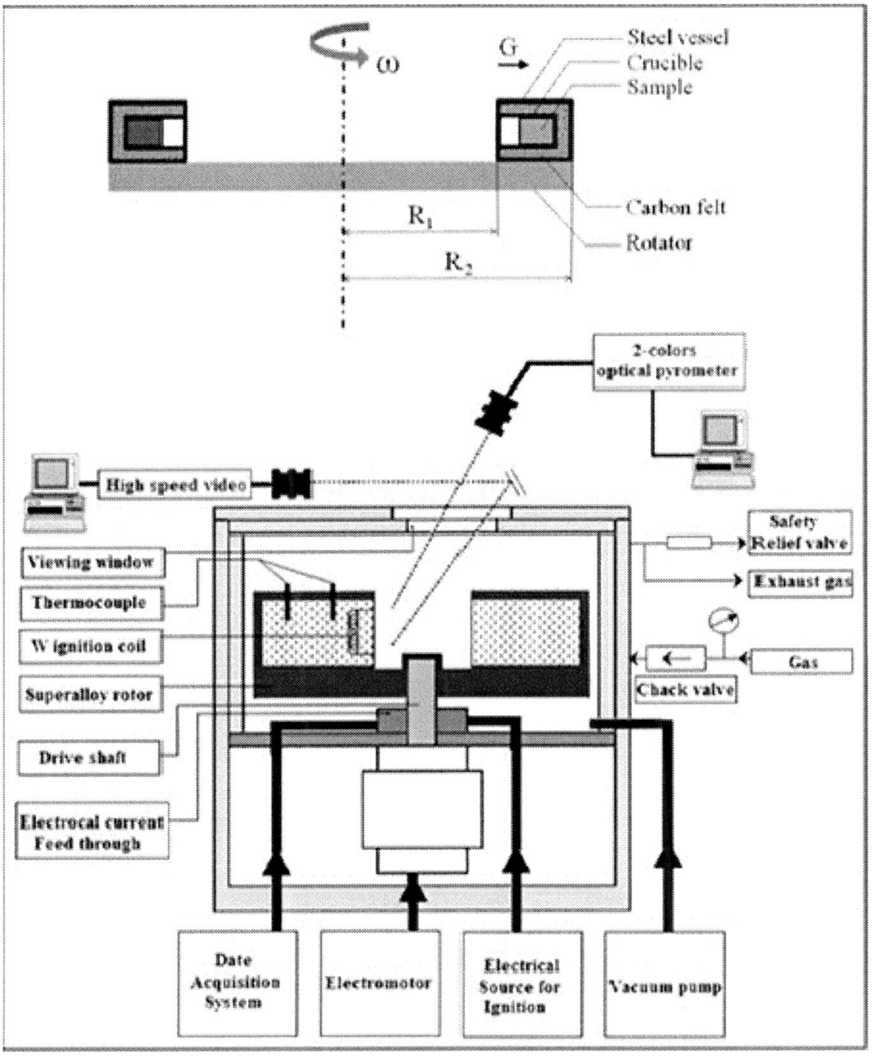

Figure 1. A schematic illustration and the assemblage of the facility for high-gravity combustion synthesis.

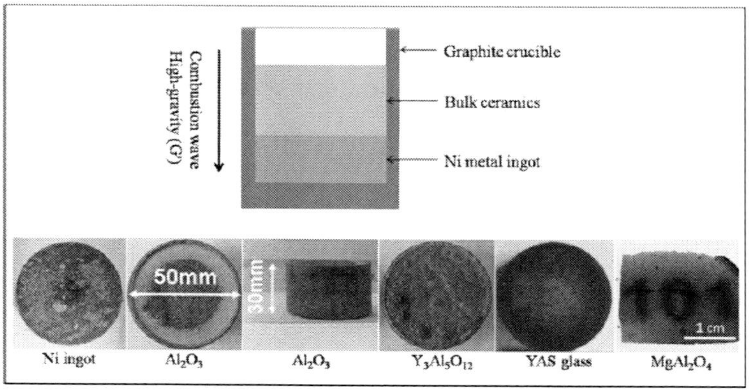

Figure 2. A sketch and photographs of the samples prepared by high-gravity combustion synthesis.

3. Ceramic and Glass Materials Prepared by High-Gravity Combustion Synthesis

High-gravity combustion synthesis is a complex process including combustion reaction, phase separation, and melt solidification. During the combustion reaction, much heat energy is released, which results in an abrupt increase in temperature and make the products melted. Then, phase separation takes place in a high-gravity field, including the separation between ceramic and metallic melts and removal of gas bubbles from the melts. Finally, melt solidification happens during cooling stage and bulk ceramic samples are obtained with Ni ingots.

To prepare bulk ceramics via melt solidification instead of powder sintering, the reaction temperature must exceed the melting point or eutectic point of the ceramic materials. This is satisfied for all the materials involved here, as shown in Table 1, where the adiabatic temperatures are much higher than the melting or eutectic points.

By high-gravity combustion synthesis, various ceramic and glass materials have been prepared, including single-phase ceramics, multiphase ceramic eutectics, glasses, and glass-ceramics, which will be discussed separately as follows.

Table 1. Melting or eutectic point and adiabatic temperature for some ceramic materials

Sample	Nominal chemical composition (mol%)	$T_{m/e}$ (°C)	Calculated phase assemblage (vol%)	T_{ad} (°C)
Al_2O_3	Al_2O_3	2054	Al_2O_3	2884
$MgAl_2O_4$	50 Al_2O_3 + 50 MgO	2135	$MgAl_2O_4$	2476
YAG	62.5 Al_2O_3 + 37.5 Y_2O_3	1950	YAG	2593
AY-1	80 Al_2O_3 + 20 Y_2O_3	1825	41 Al_2O_3 + 59 YAG	2859
AY-2	95 Al_2O_3 + 5 Y_2O_3	1825	84 Al_2O_3 + 16 YAG	2884
AZ	62 Al_2O_3 + 37 ZrO_2 + 1 Y_2O_3	1860	67 Al_2O_3 + 33 ZrO_2	2678
AYZ	65 Al_2O_3 + 16 Y_2O_3+ 19 ZrO_2	1715	37 Al_2O_3 + 49 YAG + 14 ZrO_2	2678

$T_{m/e}$: melting or eutectic point, T_{ad}: adiabatic temperature.

3.1. Single-Phase Ceramics

3.1.1. Al_2O_3

Al_2O_3 ceramics can be prepared by high-gravity combustion synthesis from the aluminothermic reaction of $2Al+3NiO=Al_2O_3+3Ni$. For this reaction, the adiabatic temperature reaches the boiling point of Ni (2884°C). The vaporization of Ni is not desirable because it will increase the porosity of the solidified samples. To decrease the reaction temperature, Al_2O_3 powder is often added as diluent into the thermite. With increasing content of Al_2O_3, the adiabatic temperature for the reaction of $2Al+3NiO+xAl_2O_3=(1+x)Al_2O_3+3Ni$ decreases, as plotted in Figure 3. When too much diluent is added (x=1.6), the adiabatic temperature is below the melting point of Al_2O_3 (2054°C). Considering the fact that the actual reaction temperature is lower than the adiabatic one because of heat dissipation, an appropriate content of Al_2O_3 diluent is expected as $0.2 \leq x \leq 0.6$.

Figure 4 shows the XRD patterns of Al_2O_3 samples prepared under different high-gravity levels from 200g to 800g. No diffraction peaks of Al or NiO are visible, indicating that the aluminothermic reaction has been completed. All the samples are pure α-Al_2O_3, with no appearance of residual Ni.

The density of the Al_2O_3 samples increases with increasing high-gravity factors, as shown in Figure 5. This is reasonable because the phase separation and pore removal can be accelerated in a stronger high-gravity field. Under a high-gravity of 800g, the density of the sample reaches 3.7 g/cm^3 and the relative density is nearly 93%.

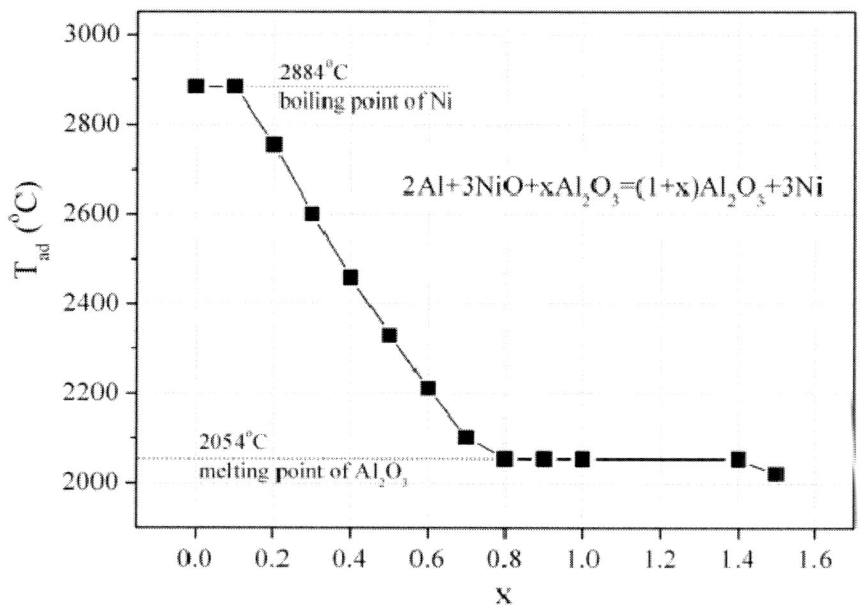

Figure 3. Relationship between the adiabatic temperature and the proportion of diluent for the combustion reaction of $2Al+3NiO=Al_2O_3+3Ni$.

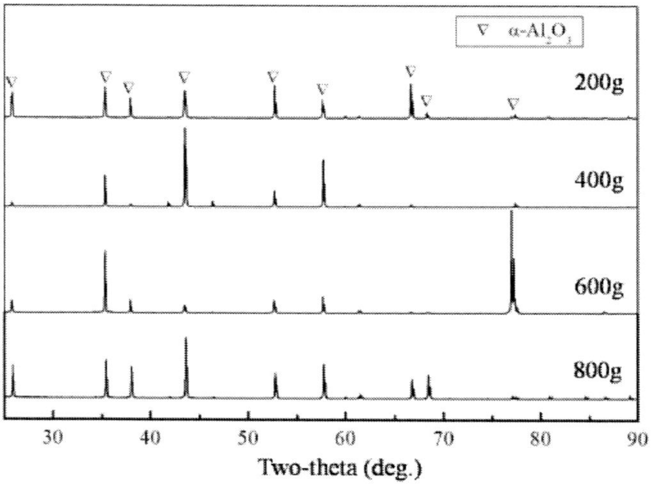

Figure 4. XRD patterns of Al_2O_3 samples prepared under different high-gravity levels.

3.1.2. $MgAl_2O_4$

$MgAl_2O_4$ is a promising transparent ceramic material with the combination of good mechanical and optical properties. However, it is also a hard-to-sinter material and difficult to be consolidated by conventional pressureless sintering. High-gravity combustion synthesis provides an alternative approach to the fabrication of $MgAl_2O_4$ ceramics from the reaction of $2Al+3NiO+MgO=MgAl_2O_4+3Ni$. In order to facilitate the phase separation, some additives such as SiO_2 or Si-Na-Ca-Mg-O glass are used sometimes, as shown in Table 2.

Figure 6 shows the XRD patterns of the products produced under different high-gravity levels. In all the products, $MgAl_2O_4$ was synthesized as the major phase except for minor Ni. The content of Ni decreases with increasing high-gravity factors. In the product prepared in a high-gravity filed of 900g, only the trace of Ni occurred, but in the product obtained under normal gravity (g'=g) much more Ni was found. This reveals that the high-gravity field has an evident effect in improving the phase separation.

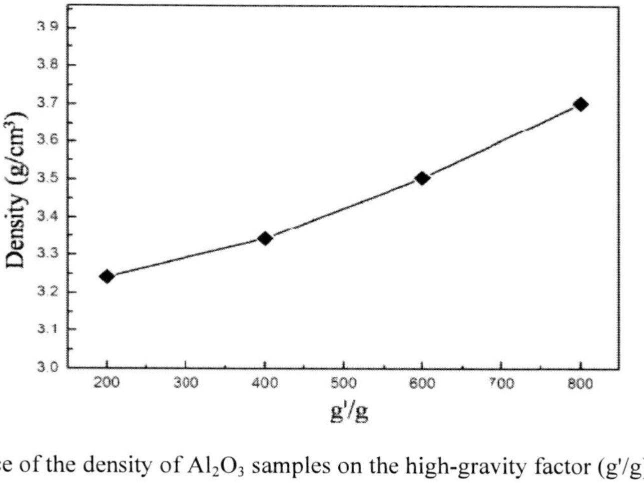

Figure 5. Dependence of the density of Al_2O_3 samples on the high-gravity factor (g'/g).

Table 2. The density and phase assemblage of as-prepared $MgAl_2O_4$ samples

Sample	Additives	g'/g	Density (g/cm^3)	Phase assemblage
1#	no additive	1		$MgAl_2O_4$+Ni
		200		$MgAl_2O_4$+Ni
		900	3.30 (92% TD)	$MgAl_2O_4$+Ni
2#	2% SiO_2	900	3.17	$MgAl_2O_4$+Ni
3#	5% SiO_2	900	3.22	$MgAl_2O_4$
4#	2% Si-Na-Ca-Mg-O glass	900	3.21	$MgAl_2O_4$
5#	5% Si-Na-Ca-Mg-O glass	900	3.24	$MgAl_2O_4$

Note: The Si-Na-Ca-Mg-O glass has a composition of 73%SiO_2-14%Na_2O-9%CaO-4%MgO in weight percent.

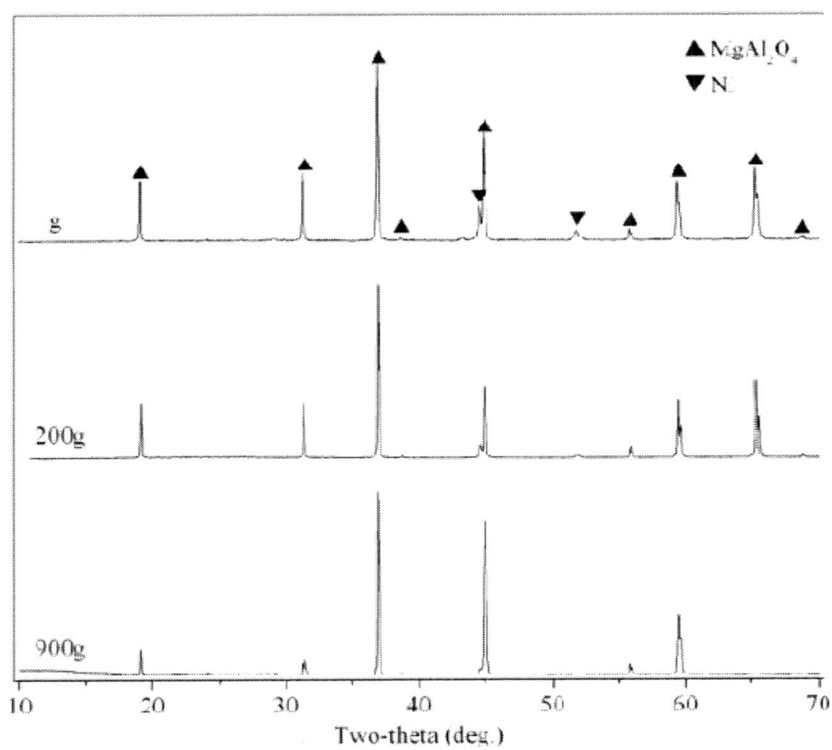

Figure 6. XRD patterns of $MgAl_2O_4$ samples prepared under different high-gravity levels.

The microstructure of the $MgAl_2O_4$ samples also depends on the high-gravity level, as shown in Figure 7. In the sample obtained under normal gravity, large pores were visible with a size of >100 μm. Ni particles of nearly 20 μm were also observed, which is consistent with the XRD result. In the sample prepared under a high-gravity of 200g, both porosity and pore size clearly decreased, and less Ni particles were found. When the high-gravity was further enhanced to 900g, the relative density of the product reached 92% and only small pores about 1 μm existed.

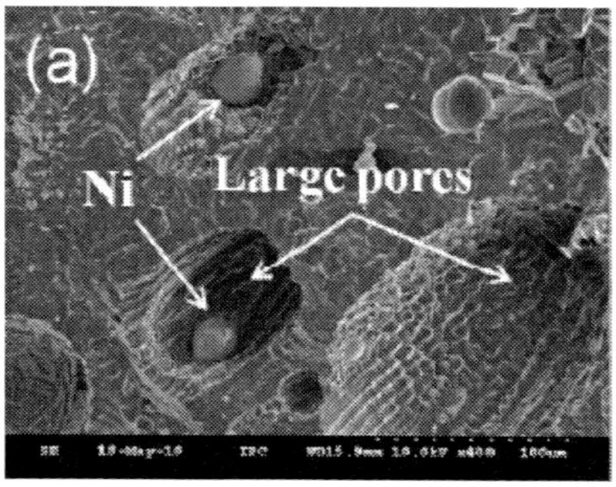

a

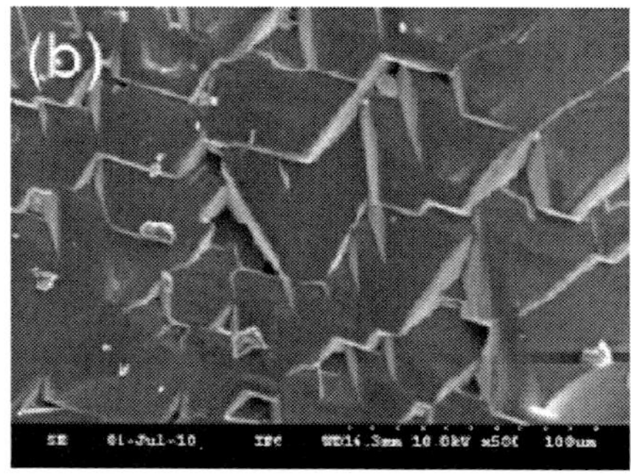

b

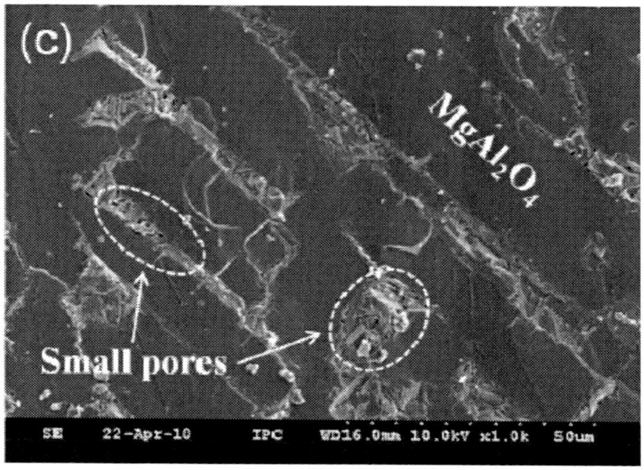

c

Figure 7. SEM images of $MgAl_2O_4$ samples prepared under different high-gravity levels: (a) g'=g; (b) g'=200g; (c) g'=900g.

In addition to the high-gravity field, the additives also affect the phase composition and microstructure of the samples. Compared with 1# sample with no additives, in 2# sample with 2% SiO_2, a higher content of Ni was observed. In 3# sample with 5% SiO_2, however, no residual Ni was found and single-phase $MgAl_2O_4$ was obtained. Such complex influence of SiO_2 can be attributed to its two-fold effect. On one hand, SiO_2 does not take part in the exothermic reaction and acts as an inert component. The addition of SiO_2 will decrease the reaction temperature, which is not desired for phase separation. On the other hand, SiO_2 can react with Al_2O_3 and MgO to form a eutectic melt at a temperature below the melting point of $MgAl_2O_4$. The Si-Al-Mg-O melt has a longer lifetime and lower viscosity than $MgAl_2O_4$, and its formation will favor the phase separation. Single-phase $MgAl_2O_4$ was obtained for the samples with Si-Na-Ca-Mg-O glass additives, as shown in Figure 8.

Compared with the samples with no additives, the samples with SiO_2 or Si-Na-Ca-Mg-O glass additives showed more homogeneous microstructure and lower porosity, as shown in Figure 9. A possible explanation is that, the additives can form liquid phase at a temperature below the melting point of $MgAl_2O_4$. In this case, this liquid will be preserved after $MgAl_2O_4$ has fully solidified, and thus can feed the shrinkage cavities formed during the solidification of $MgAl_2O_4$ and reduce the porosity of final products.

It was noticed that, in the samples with SiO_2 or Si-Na-Ca-Mg-O glass additives, no Si-containing crystalline phase was produced. This result suggests that the additives exist as an amorphous glass phase in the products. This was verified by EDS analysis associated with SEM observation, which revealed that most glass phase was present at the boundary of $MgAl_2O_4$ grains. Figure 10 shows the EDS spectra of $MgAl_2O_4$ grains and the grain-boundary glass, which agrees well with theri nominal compositions.

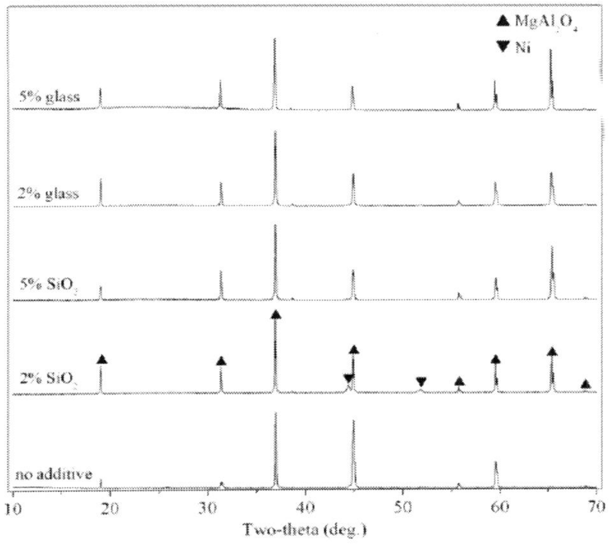

Figure 8. XRD patterns of $MgAl_2O_4$ samples with different compositions (g'=900g).

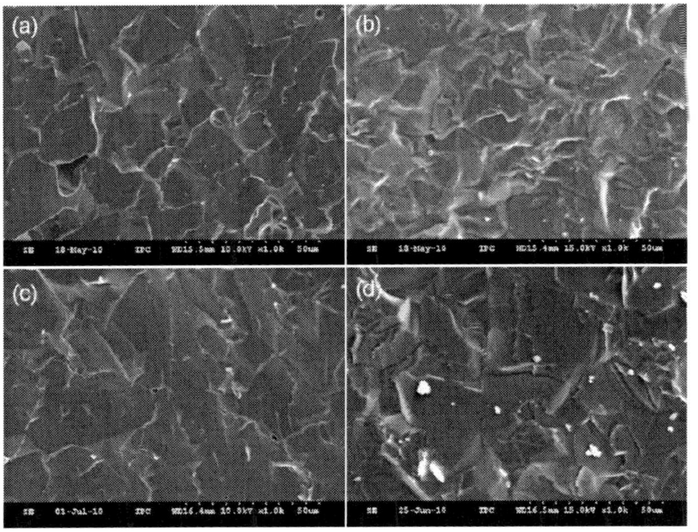

Figure 9. SEM images of $MgAl_2O_4$ samples with different compositions (g'=900g): (a) 2#, 2% SiO_2; (b) 3#, 5% SiO_2; (c) 4#, 2% Si-Na-Ca-Mg-O glass; (d) 5#, 5% Si-Na-Ca-Mg-O glass.

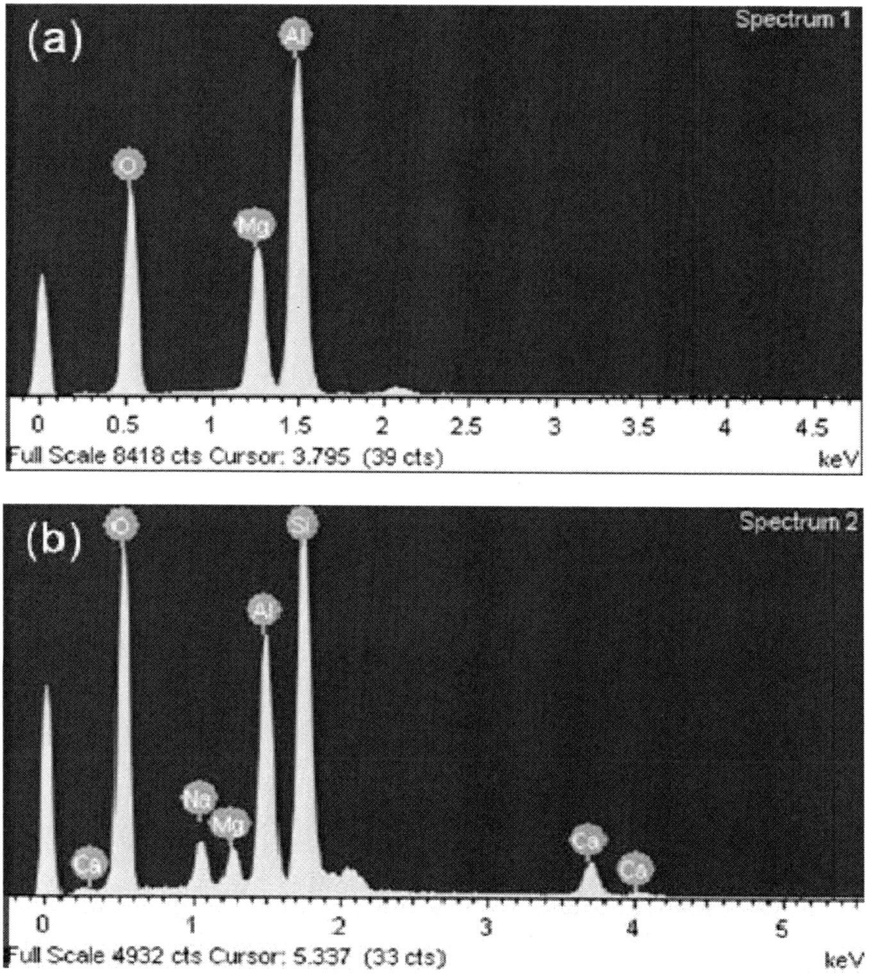

Figure 10. EDS spectra of (a) $MgAl_2O_4$ grains and (b) glass phase at grain boundaries.

3.1.3. $Y_3Al_5O_{12}$

$Y_3Al_5O_{12}$ (YAG) ceramics can be prepared by high-gravity combustion synthesis from the reaction of $10Al+15NiO+3Y_2O_3=2Y_3Al_5O_{12}+15Ni$. Figure 11 shows the XRD patterns of YAG samples prepared under different high-gravity levels from 200g to 800g. No diffraction peaks of Al, NiO, or Al_2O_3 were found and the only product was YAG. This implies that the aluminothermic reaction between Al and NiO was complete and the resultant Al_2O_3 fully reacted with Y_2O_3 to form YAG.

The density of the YAG samples increased with increasing high-gravity factors, approximately showing a linear relationship, as shown in Figure 12. This confirms the positive effect of the high-gravity field in the elimination of pores. Especially, under a high-gravity of 800g, the relative density reached 99% and translucent sample was obtained. A higher relative density with increasing high-gravity factor was also revealed by SEM observation shown in Figure 13.

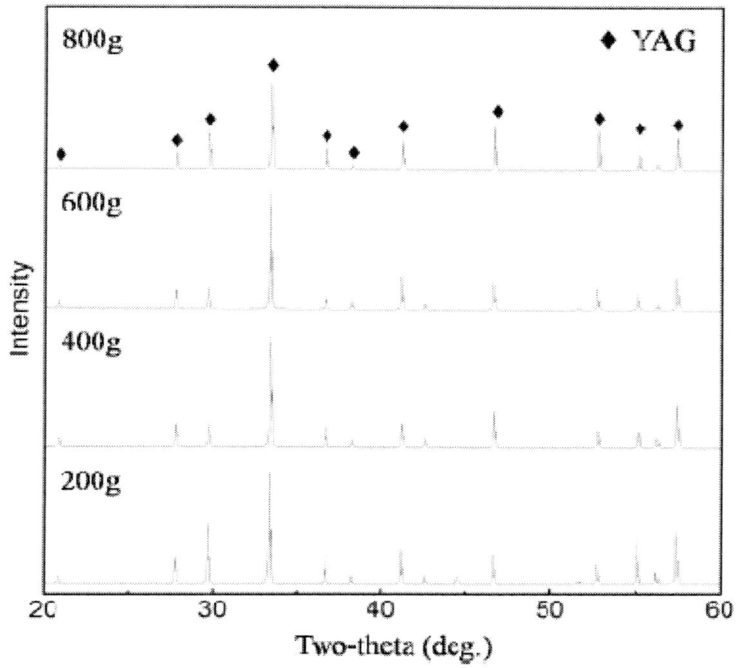

Figure 11. XRD patterns of YAG samples prepared under different high-gravity levels.

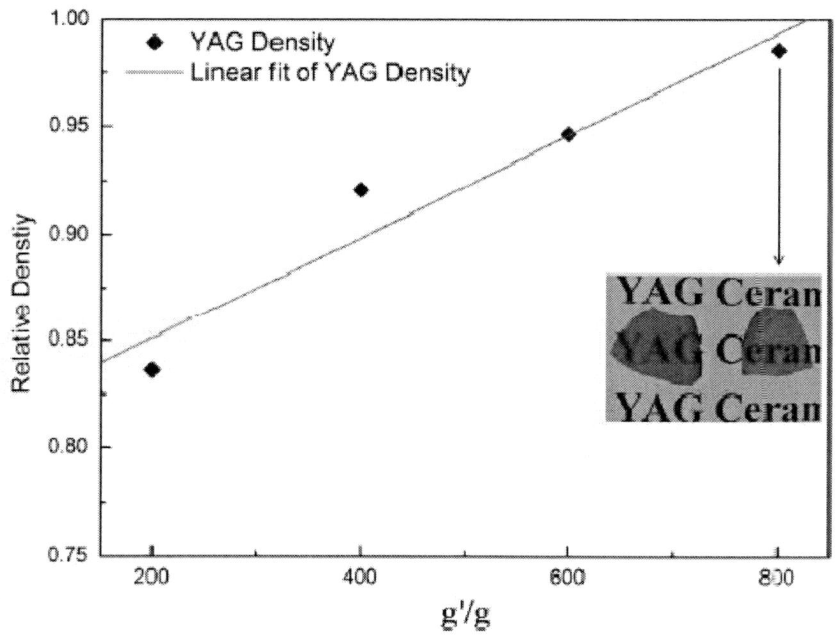

Figure 12. Dependence of the relative density of YAG samples on the high-gravity facto: (g'/g).

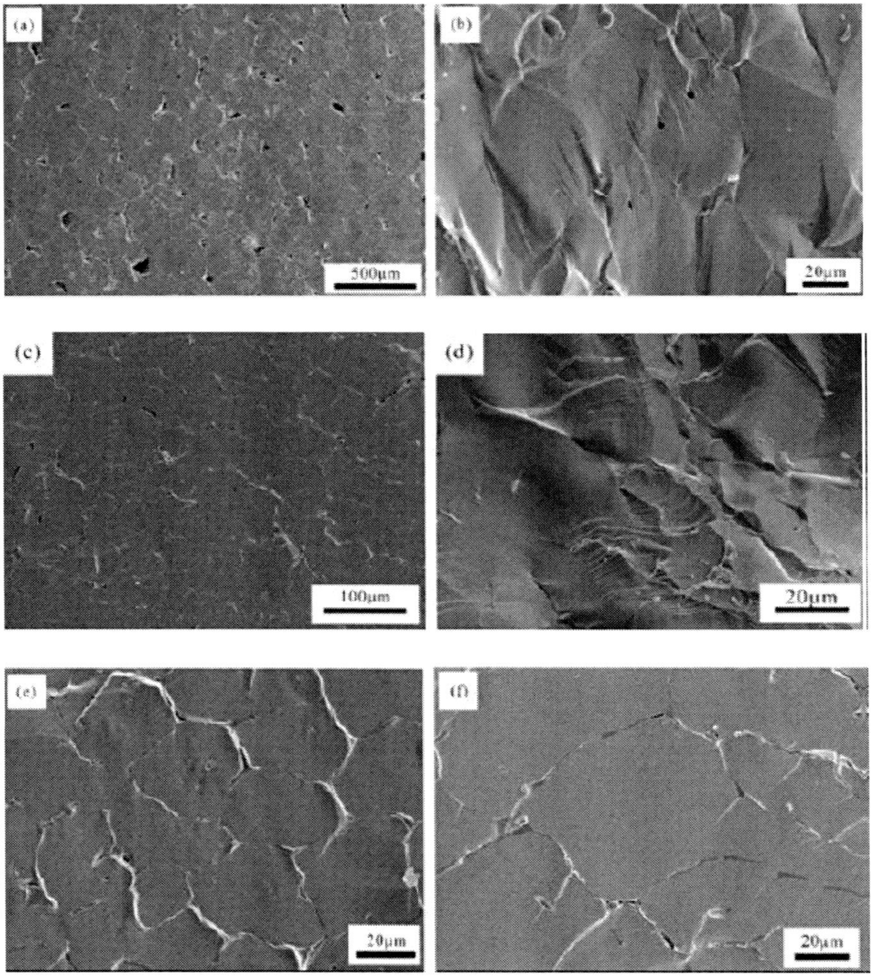

Figure 13. SEM images of YAG samples prepared under different high-gravity levels: (a) g'=400g, surface; (b) g'=400g, internal part; (c) g'=800g, surface; (d)-(f) g'=800g, internal part.

3.2. Multiphase Ceramic Eutectics

Besides single-phase ceramics, multiphase ceramic eutectics can also be produced by high-gravity combustion synthesis. Compared with single-phase ceramics, multiphase ceramic eutectics have two advantages. At first, the eutectic point of two or more phases is usually lower than the melting point of each individual phase. That is to say, eutectic systems have lower melting temperatures than single-phase ceramics. For the same reaction temperature, eutectic melts have a longer lifetime, which is desirable for phase separation and removal of gas bubbles. In this way, the ceramic eutectics usually showed higher relative densities than the single-phase ceramics prepared by high-gravity combustion synthesis. In addition, multiphase ceramic eutectics offer more opportunity for tailoring the microstructure. In single-phase ceramics, only one phase is involved and the microstructure is characterized by grain size and morphology of that phase. In multiphase ceramic eutectics, however, two or more phases co-exist, and the microstructure is affected by more factors, including the proportion, size and morphology of each phase.

3.2.1. Al_2O_3-YAG

The first example of ceramic eutectics is the Al_2O_3-YAG binary system. In this system, two compositions were studied, including a eutectic one (AY-1) and a hypoeutectic one (AY-2), as shown in Table 1. For both the compositions, the adiabatic temperatures are much higher than the eutectic temperature as well as the melting points of Al_2O_3 and YAG.

Figure 14 shows the XRD pattern of the sample AY-1, which is composed of major YAG and minor Al_2O_3. Such phase composition is consistent with that expected from the binary phase diagram shown in Figure 15. By calculation from the nominal chemical composition, the volume fractions of YAG and Al_2O_3 are 59% and 41%, respectively. The sample AY-2 also consists of YAG and Al_2O_3, except that the major phase was Al_2O_3. By calculation, the volume fractions of Al_2O_3 and YAG in AY-2 are 84% and 16%.

In the samples AY-1 and AY-2, different microstructures were obtained, as shown in Figure 16. AY-1 showed a typical fine eutectic structure with interphase spacings on a submicron scale, but in AY-2 large Al_2O_3 crystals existed and eutectic structure was produced only at the boundaries of the large crystals. The different microstructures observed in the two samples can be attributed to the different volume fractions of Al_2O_3 and YAG in them.

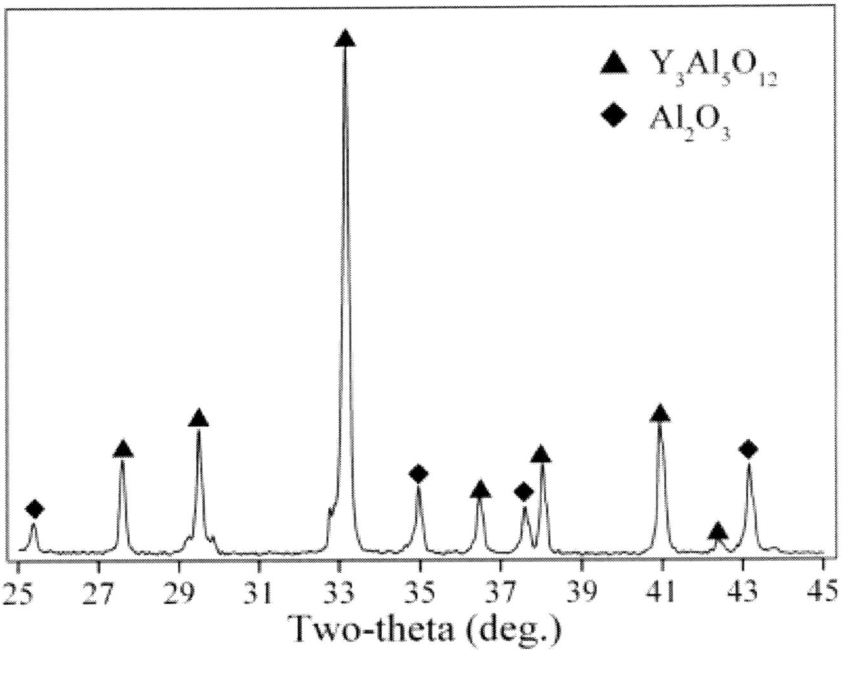

Figure 14. XRD pattern of the Al_2O_3-YAG ceramic eutectics (AY-1).

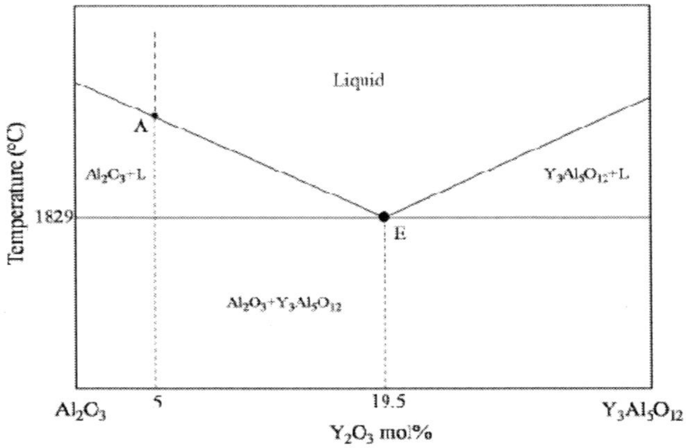

Figure 15. A simplified phase diagram of the Al_2O_3-Y_2O_3 binary system.

According to the binary phase diagram, for AY-1 with a eutectic composition, the eutectic reaction "Liquid→YAG+Al_2O_3" takes place during the whole solidification process, leading to a homogeneous eutectic structure. For AY-2 with a hypoeutectic composition, during solidification primary Al_2O_3 crystals will be precipitated first. The precipitation of Al_2O_3 causes a segregation of Y_2O_3 in the liquid, whose composition gradually changes along the liquidus line (AE line in the phase diagram in Figure 15). When the composition of the liquid reaches the eutectic point, eutectic reaction begins. Consequently, the final structure is composed of primary Al_2O_3 crystals and YAG-Al_2O_3 eutectics at the boundaries of the primary crystals.

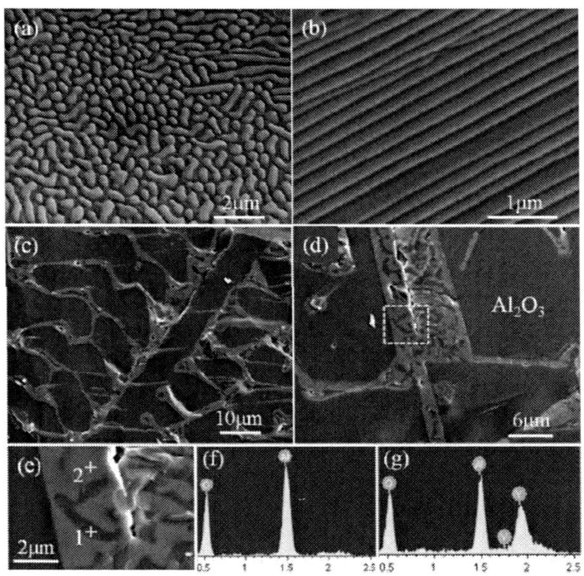

Figure 16. SEM images of Al_2O_3-YAG ceramic eutectics: (a) and (b) uniform and fine eutectic structure observed in AY-1 with a eutectic composition; (c)-(e) mixed structure with large primary Al_2O_3 crystals and fine eutectics observed in AY-2 with a hypoeutectic composition; (f) and (g) EDS spectra for Al_2O_3 and YAG crystallites, as marked with "1" and "2" respectively in (e).

3.2.2. Al_2O_3-ZrO_2

Another example of binary eutectic ceramics is the Al_2O_3-ZrO_2 system. From the Al_2O_3-ZrO_2 phase diagram, the eutectic composition was selected, as shown in Table 1, where 1 mol% Y_2O_3 was added to stabilize the tetragonal ZrO_2. For the eutectic composition, the adiabatic temperature is 2678°C and higher than the eutectic point of 1860°C.

Figure 17 shows the XRD pattern of as-prepared Al_2O_3-ZrO_2 ceramic eutectic sample (AZ). From the pattern, the sample is composed of Al_2O_3 and ZrO_2. In the sample, both polymorphs of ZrO_2 are present, with either tetragonal or monoclinic lattice structure. The occurrence of monoclinic ZrO_2 can be attributed to the relatively low concentration of Y_2O_3. From the nominal chemical composition of the sample AZ, the volume fractions of Al_2O_3 and ZrO_2 are calculated to be 67% and 33%, respectively.

The microstructure of the Al_2O_3-ZrO_2 ceramic eutectics includes two features on different scales, as shown in Figure 18. On a larger scale, the sample consists of colonies with an average size of >10 μm.

In each colony, fine eutectic structure on a smaller scale of < 1 μm is visible. It is noticed that, in the center of each colony the eutectic structure is unaffected, but at the boundaries of the colonies the eutectic structure becomes irregular. At the same time, the interphase spacing in the eutectic structure in the center of a colony is much smaller than that at the boundaries of the colonies.

With a fine eutectic microstructure, the Al_2O_3-ZrO_2 ceramic eutectics showed promising mechanical properties. For example, the hardness of 15.4-15.7 GPa and fracture toughness of 16.5-17.9 MPa·$m^{1/2}$ was reported for Al_2O_3-ZrO_2 eutectics prepared by high-gravity combustion synthesis [12].

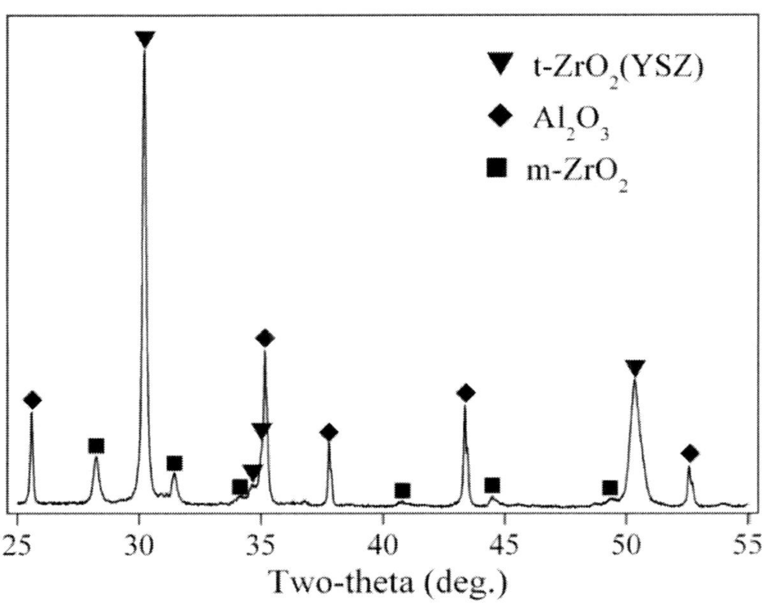

Figure 17. XRD pattern of the Al_2O_3-ZrO_2 ceramic eutectics (AZ).

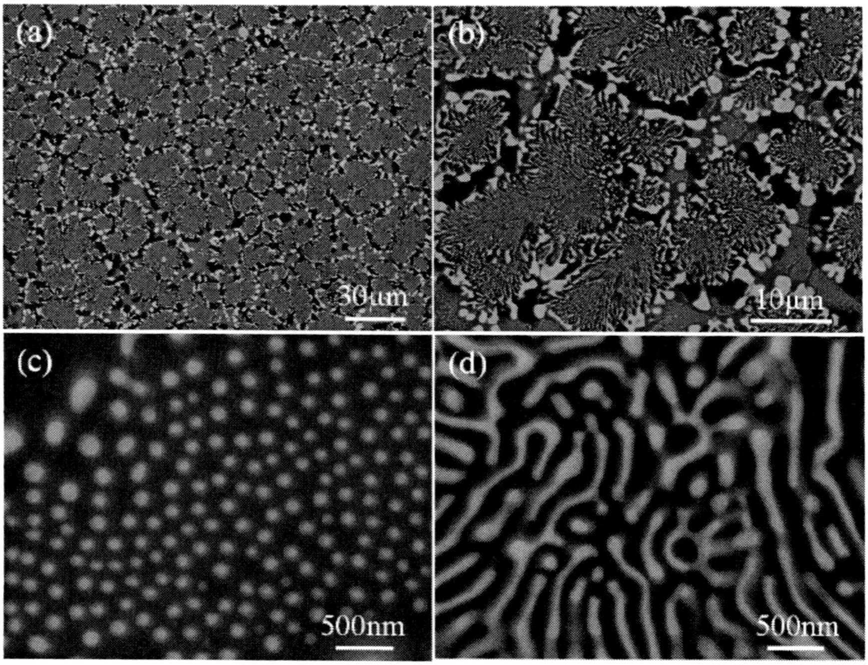

Figure 18. SEM images of Al_2O_3-ZrO_2 ceramic eutectics: (a) and (b) an overview of colonies; (c) and (d) fine eutectic structures in a colony.

3.2.3. Al_2O_3-YAG-ZrO_2

Besides binary eutectic systems, Al_2O_3-YAG-ZrO_2 ternary ceramic eutectics were also prepared by high-gravity combustion synthesis. The eutectic composition (see Table 1) was determined according to the Al_2O_3-Y_2O_3-ZrO_2 phase diagram. For the eutectic composition, the calculated adiabatic temperature is 2678°C and much higher than the eutectic point of 1715°C.

The XRD pattern of as-prepared Al_2O_3-YAG-ZrO_2 eutectic sample (AYZ) is shown in Figure 19. The sample consists of YAG, Al_2O_3 and tetragonal ZrO_2, with no occurrence of monoclinic ZrO_2. By calculation from the nominal chemical composition, in the sample the volume fractions of YAG, Al_2O_3 and ZrO_2 are 49%, 37% and 14%, respectively.

In the Al_2O_3-YAG-ZrO_2 eutectic sample, a typical colony structure was observed, as shown in Figure 20. Most colonies have a size of >30 μm and some are even larger than 100 μm. In the colonies, two microstructure features are visible, including lamellar and three-dimensional interpenetrating structures, as shown in Figure 21. The eutectic structures are composed of YAG and Al_2O_3 crystallites with interphase spacings of <1 μm, and ultrafine ZrO_2 particles are distributed at the interface between YAG and Al_2O_3. At the boundaries of the colonies, the interphase spacings become larger.

The fine eutectic microstructure with three phases intersected offers the Al_2O_3-YAG-ZrO_2 eutectic ceramics good mechanical properties. It was reported that [6], Al_2O_3-YAG-ZrO_2 eutectic ceramics prepared by high-gravity combustion synthesis have a hardness of 17.8 GPa and fracture toughness of 5.5 MPa·m$^{1/2}$.

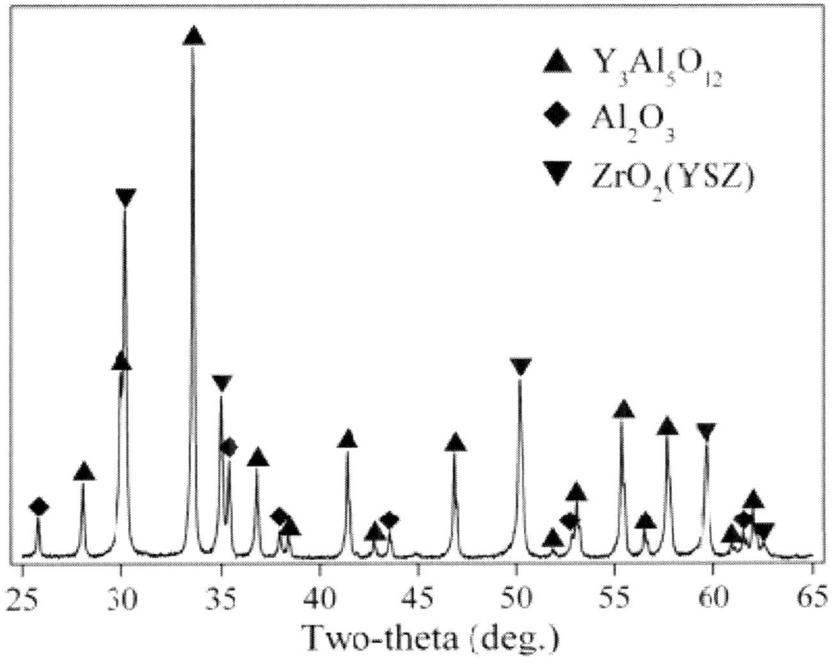

Figure 19. XRD pattern of the Al_2O_3-YAG-ZrO_2 eutectic sample (AYZ).

Figure 20. A general view of the microstructure of Al_2O_3-YAG-ZrO_2 ceramic eutectics.

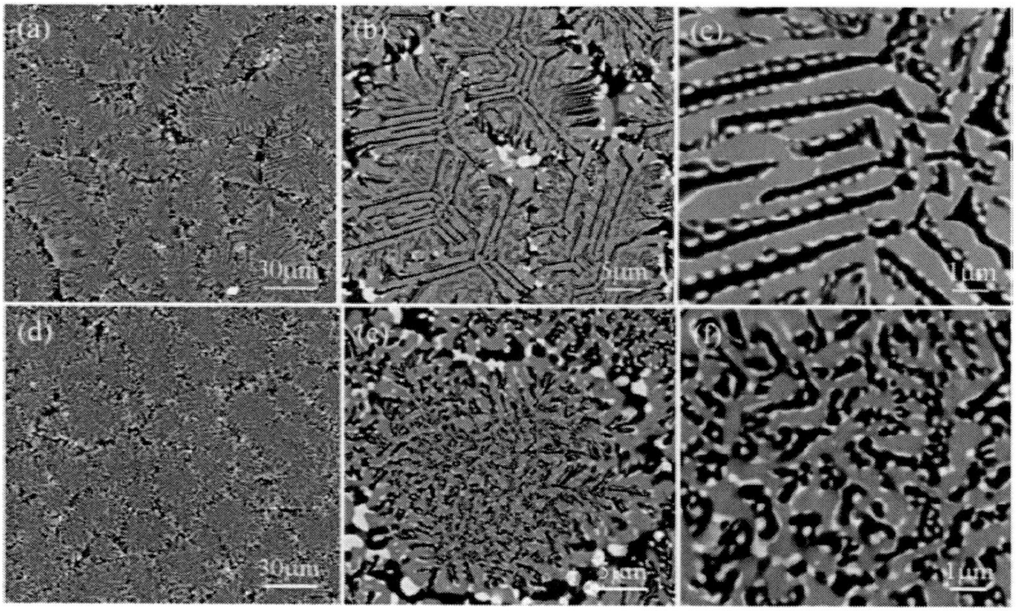

Figure 21. SEM images showing the eutectic structure in Al_2O_3-YAG-ZrO_2 system.

3.3. Glasses

In addition to polycrystalline ceramic materials, non-crystalline glasses can also be produced by high-gravity combustion synthesis. As an example, the preparation of Y_2O_3-Al_2O_3-SiO_2 glasses by high-gravity combustion synthesis has been reported [9]. Y_2O_3-Al_2O_3-SiO_2 glasses are often involved as grain boundary phases in Si_3N_4 and SiAlON ceramics and have a significant effect on the high-temperature properties of these ceramics [13]. At the same time, based on Y_2O_3-Al_2O_3-SiO_2 glasses Ce-YAG glass-ceramics can be prepared as yellow phosphors for white light-emitting diode (LED) [14].

Table 3 summarizes the starting composition, density, and phase assemblage of three Y_2O_3-Al_2O_3-SiO_2 glass samples prepared by high-gravity combustion synthesis. It is clear that the density of the Y_2O_3-Al_2O_3-SiO_2 glasses increases with the decrease of SiO_2 and increase of Y_2O_3 content. XRD analysis reveals that the samples are composed of major glass and minor Ni, as shown in Figure 22. By EDS analysis, the content of Ni is in the range of 0.7-2.8 wt%.

In Table 3, it was also noticed that, a higher content of SiO_2 in the starting composition resulted in more residual Ni in the glasses. In high-gravity combustion synthesis, the heat energy is produced by the aluminothermic reaction of $2Al+3NiO=Al_2O_3+3Ni$. SiO_2 is not involved in the exothermic reaction and thus acts as an inert diluent. An increase in SiO_2 content will decrease the reaction temperature, leading to a shorter lifetime and higher viscosity of the Y_2O_3-Al_2O_3-SiO_2 glass melt. In this case, both the velocity and the time for the settling of Ni droplets in the glass melt are reduced. Consequently, more Ni particles remain in the final glass samples.

Figure 23 shows the SEM images and EDS spectrum of the Y_2O_3-Al_2O_3-SiO_2 glasses. A homogeneous and dense microstructure was observed except for minor small pores and particulate inclusions. From EDS analysis, the actual chemical compositions of the glasses have been determined, showing a good accordance with the nominal compositions, as reported in Table 4. This is reasonable because no crystalline phase is involved in the glass samples.

Table 3. The composition, density, and phase assemblage of Y_2O_3-Al_2O_3-SiO_2 glass samples

Sample	Composition (mol%)	Density (g/cm^3)	Phase assemblage
YAS1	$56.8SiO_2$-$27.0Al_2O_3$-$16.2Y_2O_3$	3.38	Glass + Ni (2.8 wt% by EDS)
YAS2	$52.0SiO_2$-$30.0Al_2O_3$-$18.0Y_2O_3$	3.41	Glass + Ni (1.2 wt% by EDS)
YAS3	$47.2SiO_2$-$33.0Al_2O_3$-$19.8Y_2O_3$	3.52	Glass + Ni (0.7 wt% by EDS)

Note: In all the samples, the molar ratio of Al/Y is fixed to be Al/Y=5/3.

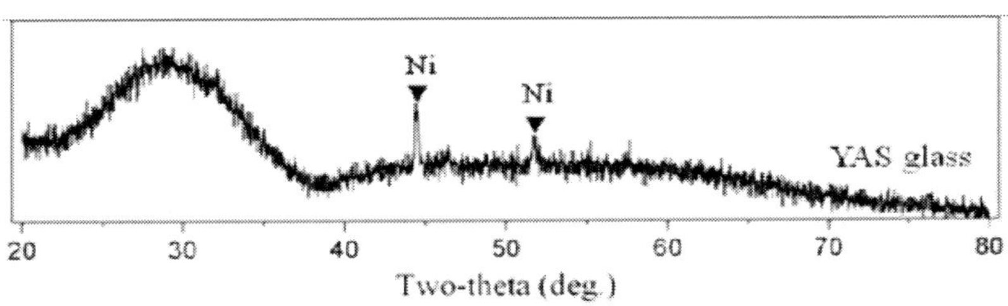

Figure 22. XRD pattern of the Y_2O_3-Al_2O_3-SiO_2 glass sample.

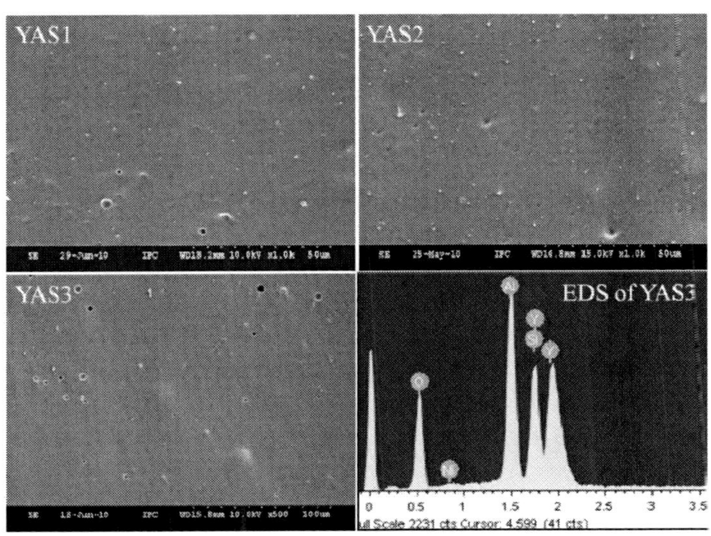

Figure 23. SEM images and EDS spectrum of Y_2O_3-Al_2O_3-SiO_2 glass.

Table 4. A comparison between the nominal and actual compositions of Y_2O_3-Al_2O_3-SiO_2 glasses

Sample	Nominal composition (molar ratio)	Actual composition by EDS (molar ratio)
YAS1	Si : Al : Y = 39.7 : 37.7 : 22.6	Si : Al : Y = 39.9 : 37.5 : 22.6
YAS2	Si : Al : Y = 35.1 : 40.5 : 24.3	Si : Al : Y = 35.6 : 39.8 : 24.6
YAS3	Si : Al : Y = 30.9 : 43.1 : 26.0	Si : Al : Y = 31.2 : 42.2 : 26.5

3.4. Glass-Ceramics

Glass-ceramics are composite materials consisting of non-crystalline glass and crystalline ceramic phases. With special mechanical, thermal, optical, and electrical properties, glass-ceramics are used in a large variety of applications, such as building wall materials, cookware, missile radomes, and dental restoration materials [15]. Glass-ceramics are generally fabricated by conventional glass-forming processes followed by post annealing for crystallization. Glass-ceramics can also be prepared by powder processing, where amorphous glass powders are first sintered and then crystallized. Both the methods require a longtime heat treatment, which is usually fulfilled by an electric furnace.

High-gravity combustion synthesis can provide another approach to the production of glass-ceramics. By this approach, the external heating source is not required and the processing time can be much reduced. Three systems of glass-ceramic materials have been prepared by high-gravity combustion synthesis, as listed in Table 5.

3.4.1. Al_2O_3-SiO_2-ZrO_2 System

In this system, two samples have been prepared and called as ASZ-1 and ASZ-2 (Table 5). The composition of ASZ-1 corresponds to the ternary eutectic point in the Al_2O_3-SiO_2-ZrO_2 phase diagram, and the composition of ASZ-2 is rich in SiO_2.

The XRD pattern of the sample ASZ-1 is shown in Figure 24, where crystalline phases of ZrO_2, mullite, and Al_2O_3 are observed. Such phase assemblage agrees with the prediction from the ternary phase diagram. Besides the crystalline phases, minor glass is also present in the sample, leading to a slightly-arched background in the XRD pattern.

The formation of glass can be attributed to the fast cooling condition in high-gravity combustion synthesis, where a limited time for solidification leads to an incomplete crystallization.

In the sample ASZ-2 with a much higher content of SiO_2, only ZrO_2 and mullite are present as crystalline phases without the occurrence of Al_2O_3. This is reasonable because mullite is expected to be produced by the reaction between Al_2O_3 and SiO_2. In this case, with the increasing of SiO_2 content, more Al_2O_3 will be consumed and converted into mullite. A higher SiO_2 content also results in the formation of more glass.

In both the samples, no complex oxides between ZrO_2 and SiO_2 have been found and the only one crystalline phase containing SiO_2 is mullite. In the sample ASZ-1, the Si/Al molar ratio is 0.26 and smaller than that (0.33) in mullite. That is to say, in the reaction $3Al_2O_3 + 2SiO_2 = Al_6Si_2O_{13}$ (mullite), Al_2O_3 is excessive.

Table 5. The composition, phase assemblage, and properties of the glass-ceramic samples

Sample	Nominal composition (mol %)	Phase assemblage	Density (g/cm^3)	Hardness (GPa)
ASZ-1	$50Al_2O_3$-$26SiO_2$-$24ZrO_2$	ZrO_2+Mullite+ Al_2O_3+Glass	3.32	12.8
ASZ-2	$30Al_2O_3$-$50SiO_2$-$20ZrO_2$	ZrO_2+Mullite+Glass	3 15	12.1
ASY-1	$46Al_2O_3$-$36SiO_2$-$18Y_2O_3$	YAG+Al_2O_3+Glass	3.61	8.5
ASY-2	$35Al_2O_3$-$55SiO_2$-$10Y_2O_3$	Glass+Mullite	3.20	8.2
ASYZ	$30Al_2O_3$-$42SiO_2$-$18Y_2O_3$-$10ZrO_2$	Glass+ZrO_2+YAG+Al_2O_3	3.73	10.1
ASYL	$32Al_2O_3$-$45SiO_2$-$19Y_2O_3$-$4La_2O_3$	Glass+YAG	3.67	11.3
ALZ	$50Al_2O_3$-$25La_2O_3$-$25ZrO_2$	$LaAlO_3$+$LaAl_{11}O_{18}$+ZrO_2 +Glass	3.97	9.2
ASLZ-1	$50Al_2O_3$-$10SiO_2$-$20La_2O_3$-$20ZrO_2$	$LaAlO_3$+$LaAl_{11}O_{18}$+ZrO_2 +Glass	4.35	8.8
ASLZ-2	$35Al_2O_3$-$45SiO_2$-$10La_2O_3$-$10ZrO_2$	Glass+$LaAlO_3$+ ZrO_2+$LaAl_{11}O_{18}$	3.86	8.9

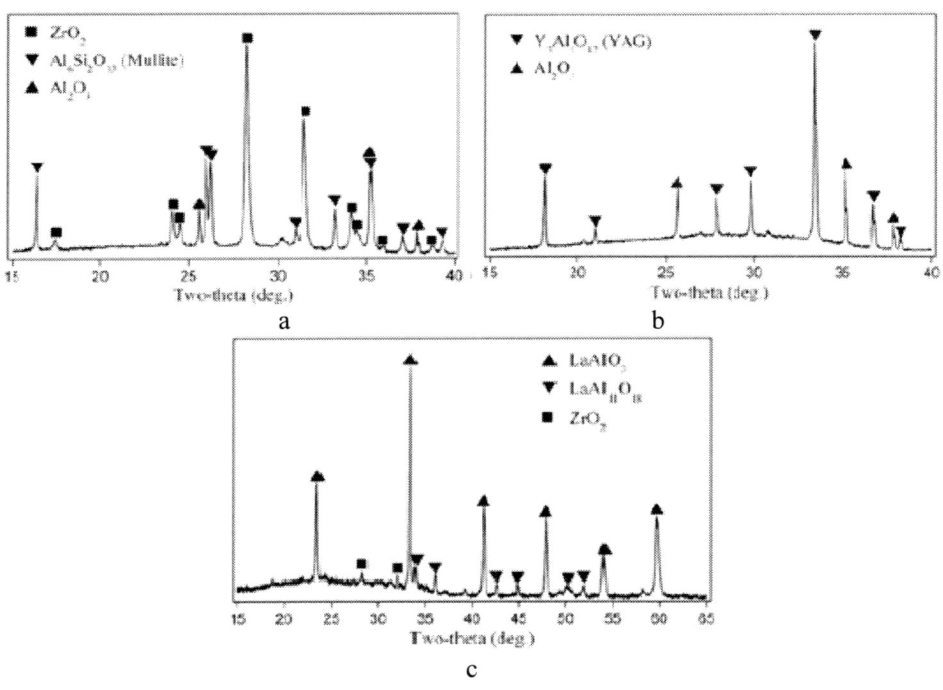

Figure 24. XRD patterns of the glass-ceramics prepared by high-gravity combustion synthesis: (a) ASZ-1; (b) ASY-1; (c) ALZ.

In this case, most SiO_2 will be consumed and some un-reacted Al_2O_3 remains in the product. In the sample ASZ-2, however, the Si/Al molar ratio is 0.83 and much higher than that in mullite. Finally, Al_2O_3 will be almost exhausted and much SiO_2 remains.

Figure 25 shows the SEM images and EDS spectra of the Al_2O_3-SiO_2-ZrO_2 glass-ceramics. In the sample ASZ-1, both primary crystallites and eutectic structure are present. The primary crystallites consist of coarse mullite and Al_2O_3 grains. In the eutectic structure, submicron ZrO_2 fibers are embedded in the mullite/Al_2O_3 matrix. Such eutectic structure was not found in the sample ASZ-2, where fine ZrO_2 particles are distributed at the boundaries of large rod-like mullite grains or in a glass matrix. The different microstructures observed in the two samples are probably associated with their chemical compositions and the content of Al_2O_3 plays a key role. In high-gravity combustion synthesis, the heat energy is produced by the aluminothermic reaction of $2Al+3NiO=Al_2O_3+3Ni$. A higher Al_2O_3 content means a larger proportion of fuels in the starting composition and thus a higher reaction temperature.

When the reaction temperature exceeds the eutectic point of several ceramic phases, eutectics can be produced during solidification. When the temperature is below the eutectic point, no continuous and uniform melt will occur. In this case, a composite structure with simple mixture of several ceramic phases will be produced.

3.4.2. Al_2O_3-SiO_2-Y_2O_3-Based System

In this system, four samples have been prepared, including ASY-1 rich in Al_2O_3, ASY-2 rich in SiO_2, ASYZ with the addition of ZrO_2, and ASYL with the addition of La_2O_3 (Table 5).

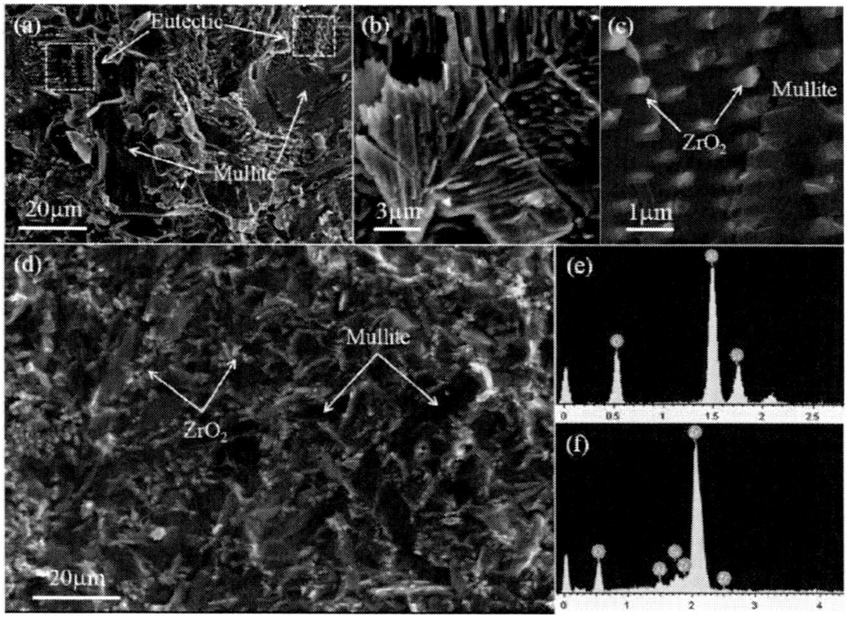

Figure 25. SEM images and EDS spectra of ASZ glass-ceramics: (a)-(c) ASZ-1; (d) ASZ-2; (e) and (f) EDS spectra of mullite and ZrO_2 grains.

For the sample ASY-1, a glass-ceramic composed of glass and crystalline phases of YAG and Al_2O_3 was obtained, as shown in Figure 24 (b). Such phase assemblage is consistent with SEM observation and EDS analysis shown in Figure 26, where polyhedron YAG grains and plate-like Al_2O_3 crystals are embedded in a Al_2O_3-SiO_2-Y_2O_3 glass matrix. From EDS results, the glass shows molar ratios of Si/Al=1.04 and Si/Y=1.22, which are much higher than Si/Al=0.39 and Si/Y=1 in the starting composition. The enrichment of SiO_2 in the glass can be attributed to the precipitation of YAG and Al_2O_3. At the interface between YAG and Al_2O_3 and that between crystalline phases and glass matrix, no cracks or cavities were observed.

The sample ASY-2 is composed of glass and mullite. Two crystal shapes of mullite are visible, including coarse plate-like crystals up to 100 μm and smaller faceted crystals below 10 μm, as shown in Figure 26. In the sample ASYZ, the major crystalline phase is ZrO_2 with minor YAG and Al_2O_3. SEM observation shows that fine ZrO_2 particles are dispersed in an Al_2O_3-SiO_2-Y_2O_3 glass matrix. EDS analysis reveals that no Zr exists in the glass. Thus, it is thought that ZrO_2 will not dissolv into Al_2O_3-SiO_2-Y_2O_3 melt. In the sample ASYL, YAG is the only one crystalline phase and no La-containing crystalline phase has been produced. The YAG grains are homogenously distributed in the glass matrix, and most grains are larger than 10 μm. EDS analysis confirms the presence of La in the glass. It appears that the interaction of La_2O_3 with the Al_2O_3-SiO_2-Y_2O_3 system differs from that of ZrO_2. In contrast to ZrO_2, La_2O_3 is more inclined to dissolve into the Al_2O_3-SiO_2-Y_2O_3 melt.

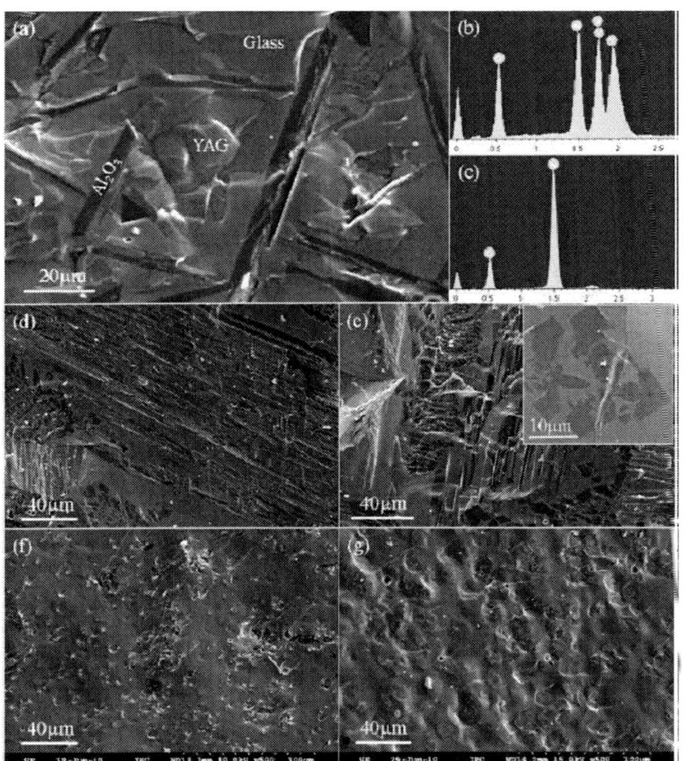

Figure 26. SEM images and EDS spectra of ASY-based glass-ceramics: (a) ASY-1; (b) and (c) EDS spectra of Al_2O_3-SiO_2-Y_2O_3 glass and Al_2O_3grains; (d) and (e) ASY-2; (f) ASYZ; (g) ASYL.

3.4.3. Al_2O_3-La_2O_3-ZrO_2-Based System

In this system, three samples have been prepared, including ALZ, ASLZ-1, and ASLZ-2. Compared with ALZ with a ternary composition of Al_2O_3-La_2O_3-ZrO_2, ASLZ-1 and ASLZ-2 have more complex compositions with the incorporation of SiO_2.

The XRD pattern of the sample ALZ is shown in Figure 24 (c), where $LaAlO_3$ is the predominant phase with minor $LaAl_{11}O_{18}$ and ZrO_2. In the pattern, there is an evident arching behavior of the background line, implying the presence of glass phase. The sample ASLZ-1 has similar phase assemblage to that of ALZ, and the incorporation of 10% SiO_2 brings no significant influence. In the sample ASLZ-2 with 45% SiO_2, however, a high proportion of glass is visible.

The higher proportion of glass in the sample ASLZ-2 than those in ALZ and ASLZ-1 is also revealed by SEM observation (Figure 27). The samples ALZ and ASLZ-1 are mostly composed of faceted $LaAlO_3$ and $LaAl_{11}O_{18}$ grains, and the average grain size is below 10 μm.

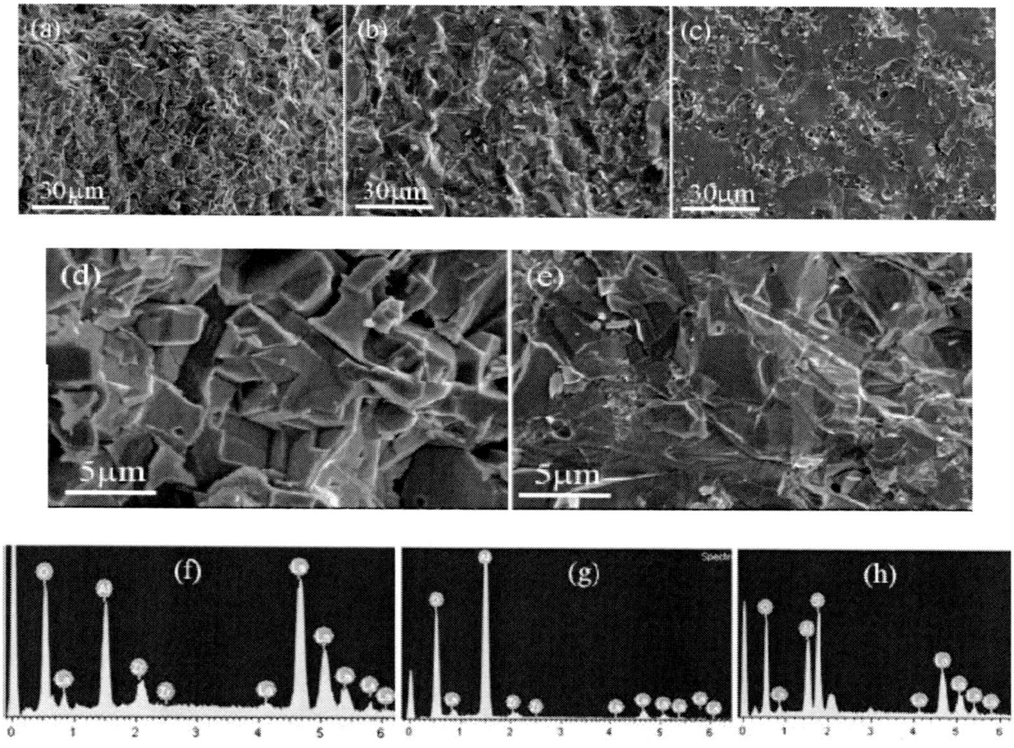

Figure 27. SEM images and EDS spectra of ALZ and ASLZ glass-ceramics: (a) and (d) ALZ; (b) and (e) ASLZ-1; (c) ASLZ-2; (f) and (g) EDS spectra of $LaAlO_3$ and $LaAl_{11}O_{18}$ grains; (h) EDS spectrum of the glass phase in ASLZ-2.

From EDS analysis, the molar ratios in the glass are Si/Al=1.41 and La/Al=0.84, which are much higher than those (Si/Al=0.64 and La/Al=0.29) in the overall composition of the sample. That is to say, compared to the crystalline phases, the glass phase is rich in Si and La.

The enrichment of Si in the glass is reasonable, because no Si-containing crystalline phase is produced.

The hardness of the glass-ceramics prepared by high-gravity combustion synthesis is in the range of 8.2-12.8 GPa. The highest hardness is observed for the sample ASZ-1, which can be attributed to the eutectic microstructure with fine ZrO_2 fibers embedded in a mullite matrix. From the viewpoint of application as a structural component, the ASZ samples with a higher hardness/density ratio are more likely desired than the ALZ-based samples with a lower hardness/density ratio.

4. Phase Separation in High-Gravity Combustion Synthesis

High-gravity combustion synthesis is a complex process involving a series of physical and chemical variations such as combustion reaction, phase transformation (melting and solidification), and phase separation (separation between ceramic and metallic phases and removal of bubbles from melts). In this process, all the variations occur in a short period and thus will not take place in a strict order but overlap with each other. On the other hand, the reaction kinetics strongly depends on the starting compositions of reactants. In this case, it is difficult to give a universal model that can exactly describe the high-gravity combustion synthesis for all reactant systems. Nevertheless, a basic analysis of the synthesis process to outline its essential steps is helpful for technical optimization to promote the quality of products.

The high-gravity combustion synthesis process can be simply divided into three stages: (1) combustion reaction and the formation of melts; (2) phase separation; (3) melt solidification. The first stage is very short because combustion reactions progress very quickly. For example, in combustion synthesis of $Y_3Al_5O_{12}$ from the reaction $10Al+15NiO+3Y_2O_3=2Y_3Al_5O_{12}+15Ni$, the propagation velocity of combustion wave under normal gravity was measured to be 28-100 mm/s (varying with the particle size of reactants). In a high-gravity field, the velocity can be greatly enhanced. It was reported that, for the thermite of (2Al+3FeO) with 18.5% Al_2O_3 dilute, the burning rate in a high-gravity field of g'/g=895 increased by a factor of approximately 6 compared with that under normal gravity [16]. This result should be applicable to other similar aluminothermic reactions. That is to say, for the reaction $10Al+15NiO+3Y_2O_3=2Y_3Al_5O_{12}+15Ni$ in a high-gravity field of g'/g=900, the maximum burning rate can reach 600 mm/s. In this case, for a sample with a height of 50 mm, the combustion reaction will complete in 0.1 s. During such a short period, significant phase separation or melt solidification is not expected. In other words, as the first stage combustion reaction has little overlap with the later two stages.

In high-gravity combustion synthesis, the phase composition and microstructure of the products are closely related to two processes, viz. phase separation and melt solidification. This section deals with phase separation and melt solidification will be discussed in the next section.

Because combustion reactions take place quickly and the reaction temperature is high, the real-time capture of the information on the physical state and chemical composition of samples is difficult.

In this case, the investigation of phase separation has been carried out mostly based on experimental observations of the products.

Figure 28 shows the product by combustion synthesis under normal gravity from the reaction of $2Al+3NiO=Al_2O_3+3Ni$. Hollow Al_2O_3 spheres smaller than 5 mm were obtained instead of bulk ceramic disks. At the surface of the spheres, metallic spots existed, indicating an incomplete separation between the ceramic and metallic phases. In addition, small holes were observed on some spheres, which were probably caused by the escape of gas.

In contrast to the hollow spheres obtained under normal gravity, bulk ceramic samples were prepared by combustion synthesis in a high-gravity field. For most samples, the relative density was higher than 92%, which could be further improved to >99% by the optimization of processing parameters.

This reveals that the high-gravity field has an evident effect to improve phase separation and produce dense bulk ceramics by combustion synthesis.

As examples, several ceramic samples prepared by high-gravity combustion synthesis are shown in Figure 29. Each sample consisted of two distinct parts, where the upper part was porous and the lower part was dense. In the porous part, large pores with a size of nearly 100 μm were present, as shown in Figure 30. Such large pores were not found in the dense part, where only a few pores smaller than 10 μm occurred.

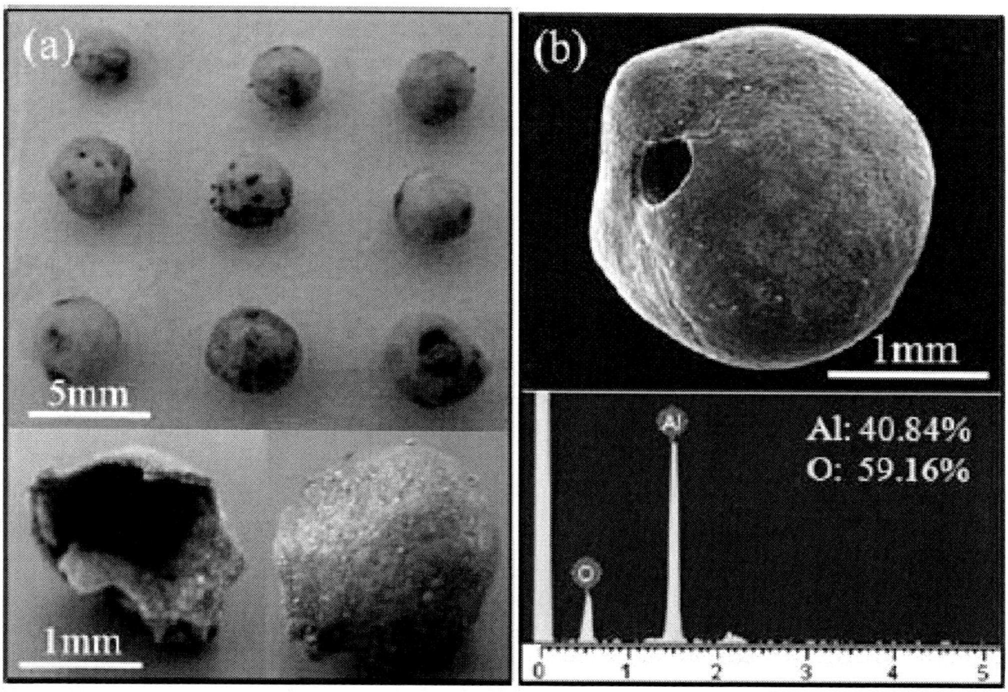

Figure 28. Hollow Al_2O_3 balls prepared by combustion synthesis under normal gravity: (a) photographs; (b) SEM image and EDS spectrum.

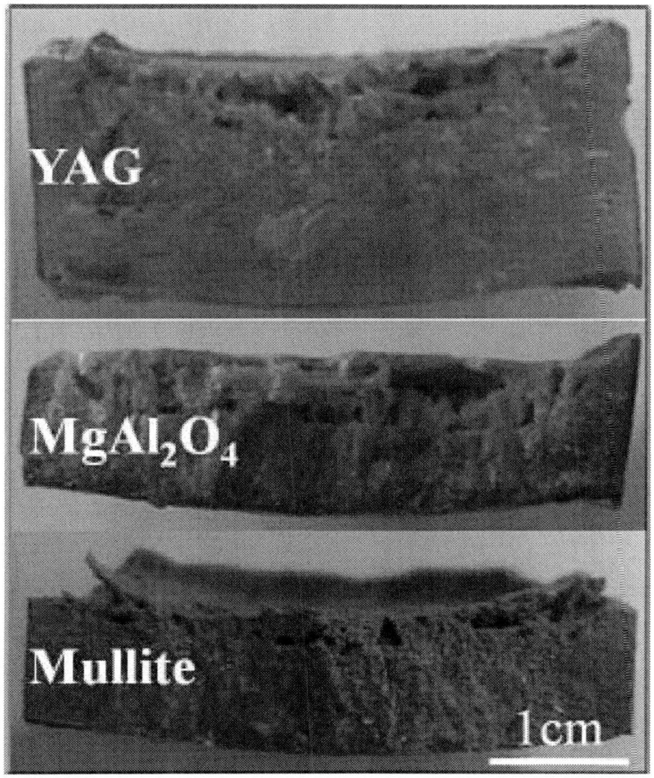

Figure 29. Photographs of several ceramic samples prepared by high-gravity combustion synthesis.

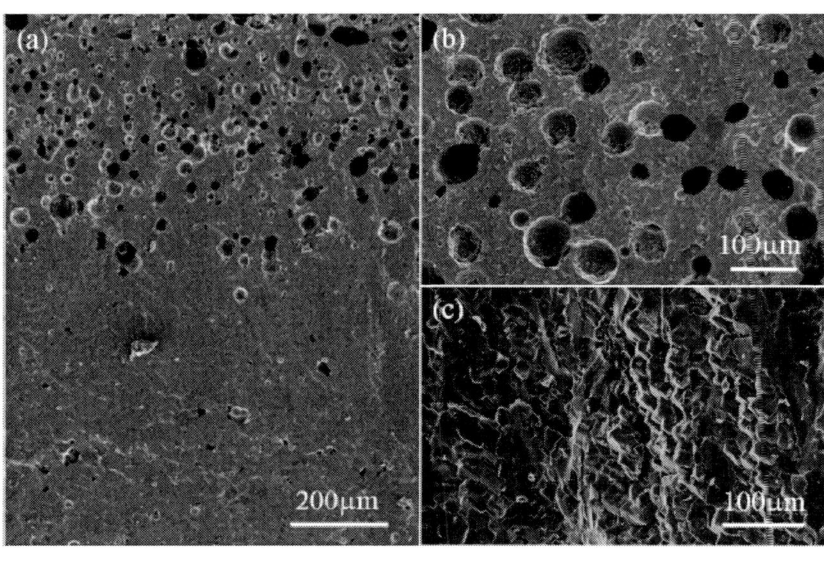

Figure 30. SEM images of bulk Al_2O_3 samples prepared by high-gravity combustion synthesis (g'/g=900): (a) an overview; (b) upper porous part; (c) lower dense part.

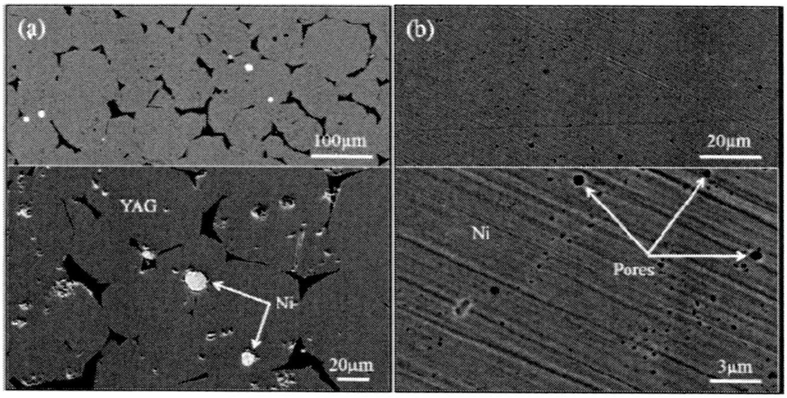

Figure 31. SEM images of (a) YAG ceramics and (b) Ni ingot prepared by high-gravity combustion synthesis (g'/g=900).

Compared with the ceramic products, the metallic ingots had a lower porosity. Figure 31 shows SEM images for YAG ceramics and Ni ingots prepared by high-gravity combustion synthesis. In contrast to the large cavities present in the YAG ceramics, only ultrafine pores below 1 μm were observed in the Ni ingots. It was also noticed that, several Ni particles existed in the YAG ceramics, but no YAG grains were found in the Ni ingots.

From the above experimental results, the phase separation in high-gravity combustion synthesis can be discussed as follows.

In the preparation of bulk ceramic materials by high-gravity combustion synthesis, a complete phase separation is essential to obtain pure and dense products. The phase separation includes both the separation between ceramic and metallic melts and the removal of bubbles from the melts. The separation between ceramic and metallic melts guarantees a high purity of the ceramic product, and the removal of bubbles offers a high density of the ceramic product.

For the separation between ceramic and metallic phases, there are three possible ways: (1) settling of metallic liquid droplets in ceramic melts; (2) rising of ceramic liquid droplets in metallic melts; (3) rising of ceramic solid particles in metallic melts. For high-gravity combustion synthesis of Al_2O_3-based ceramics (such as Al_2O_3 and $Y_3Al_5O_{12}$) and glasses from the reaction of $2Al+3NiO=Al_2O_3+3Ni$, the first way is most operative because of two reasons. At first, in the products, ceramic phases have higher volume fractions (60% for Al_2O_3 and 74% for $Y_3Al_5O_{12}$) than Ni, and thus ceramic phases occur as continuous matrix with Ni particles dispersed as a discrete phase. Secondly, from experimental results, Ni particles were usually observed in the ceramic samples, while few ceramic grains were found in the Ni ingots.

During the settling of Ni droplets in ceramic melts in a high-gravity field, the stable settling velocity can be calculated according to the Stokes law. On a settling Ni droplet, three forces are considered, including high-gravity (G'), buoyant force (F), and viscous resistance (f), as shown in Figure 32 (a). These forces are expressed as

$$G'=4/3\cdot\pi R^3\rho_1 g' \quad (1)$$

$$F=4/3\cdot\pi R^3\rho_2 g' \quad (2)$$

$$f=6\pi R\eta V \quad (3)$$

where ρ_1 and ρ_2 mean the densities of liquid Ni and the ceramic melt, g' means the high-gravity acceleration, R means the radius of the Ni droplet, η means the viscosity of the ceramic melt, and V means the settling velocity of the Ni droplet.

When G'=F+f, the Ni droplet reaches the maximum settling velocity, which is usually called stable settling velocity (V_s). From Equations (1)-(3), the stable settling velocity can be deduced as

$$V_s=2/9\cdot(\rho_1-\rho_2)\, g'R^2/\eta \quad (4)$$

From Equation (4), the stable settling velocity shows a linear dependence on the high-gravity acceleration. With the data in Table 6, the stable settling velocity as well as the time corresponding to a displacement of 1 cm (t_{1cm}) for Ni droplets with a radius of 10 μm under different high-gravity factors (g'/g) have been calculated and plotted in Figure 33.

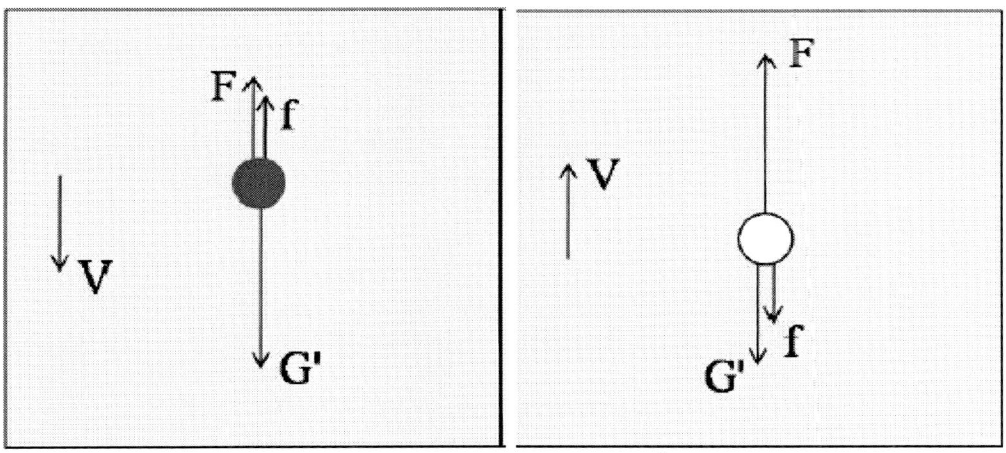

Figure 32. A schematic illustration of the phase separation during high-gravity combustion synthesis: (a) settling of Ni particles in ceramic melts; (b) rising of gas bubbles in ceramic or metallic melts.

Table 6. Properties of ceramic and Ni melts involved in this work

Melts	Density ρ (g/cm^3)	Viscosity η (Pa·s)	Surface energy γ (N/m)	Ref.
Ni	7.9	0.0489	1.778	17
Al_2O_3	3.008	0.04	0.64	18, 19
YAG	3.688	0.046	0.781	20
YAS	3.25	0.08	0.3	21, 22

Note: YAS means Y_2O_3-Al_2O_3-SiO_2 glass.

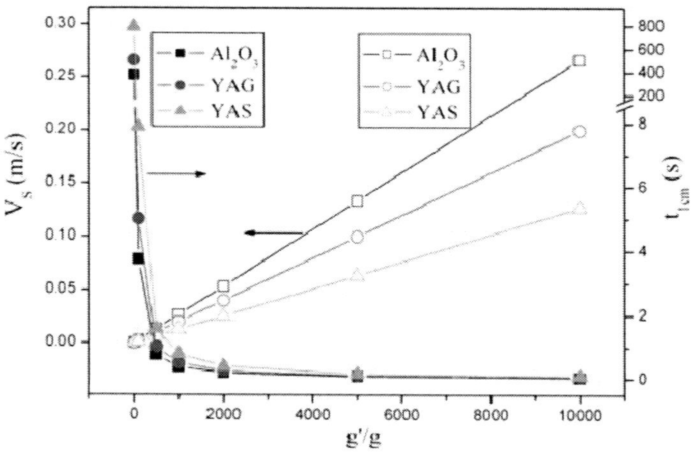

Figure 33. Influence of high-gravity factor on the settling velocity of Ni droplets (R=10 μm) in ceramic melts.

It is evident that a high-gravity field significantly accelerates the phase separation and reduces the time required for complete separation. The settling velocity of Ni particles also depends on the viscosity of ceramic melts and decreases with increasing viscosities.

Similarly, the rising of bubbles in ceramic or metallic melts has also been analyzed, as shown in Figure 32 (b). When F=G'+f, the stable rising velocity is reached as

$$V_s=2/9\cdot(\rho_2-\rho_3)\, g'R^2/\eta \tag{5}$$

where ρ_2 means the density of the ceramic or metallic melt, ρ_3 means the density of gas in the bubbles, and R means the radius of the bubbles. Considering $\rho_3/\rho_2<10^{-3}$, Equation (5) can be simplified as

$$V_s=2/9\cdot\rho_2 g'R^2/\eta \tag{6}$$

From Equation (6), the stable rising velocity and the time corresponding to a displacement of 1 cm for gas bubbles with a radius of 10 μm under different high-gravity factors (g'/g) have been calculated and plotted in Figure 34. The rising velocity is greatly enhanced by increasing the high-gravity factor. The rising velocity also depends on the viscosity of the ceramic or metallic melt, and the bubbles have a higher rising velocity in Ni melt than that in the ceramic melts.

This agrees with the experimental result, where Ni ingots showed a lower porosity than the ceramic samples. Additionally, the size of the bubbles has a strong effect on their rising velocity, and large bubbles rise more quickly than smaller ones, as shown in Figure 35. For this reason, large pores were observed in the upper part of the ceramic samples and small pores in the lower part, as described before. Due to this size effect, it is difficult to remove ultrafine bubbles below 1 μm from ceramic melts, which hinders the fabrication of full-dense ceramics by high-gravity combustion synthesis.

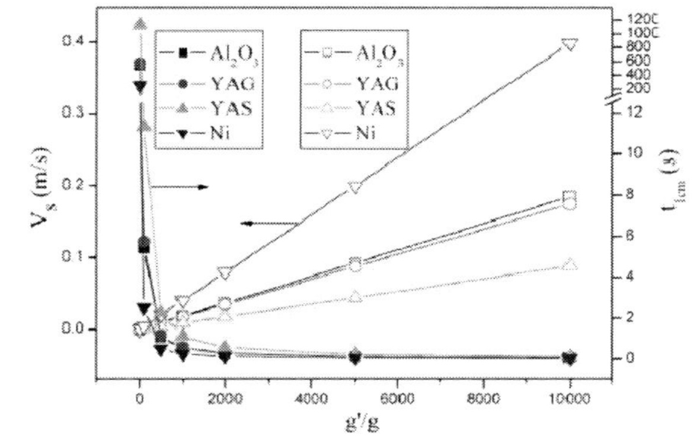

Figure 34. Influence of high-gravity factor on the rising velocity of gas bubbles (R=10 μm) in ceramic or Ni melts.

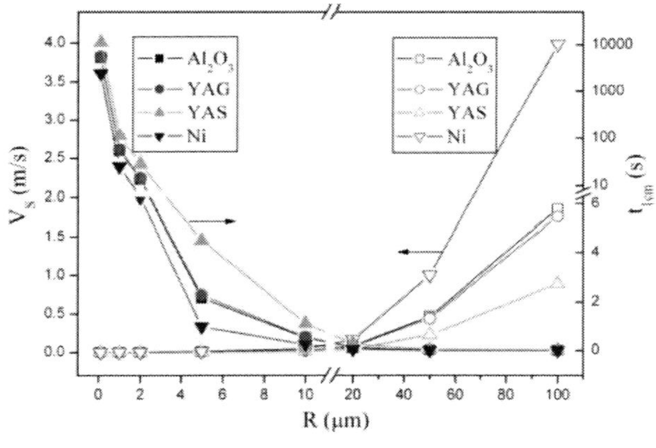

Figure 35. Dependence of rising velocity on the size of gas bubbles in ceramic and Ni melts (g'/g=1000).

Because the rising velocity of the bubbles is related to their size, it is necessary to investigate the critical size of bubbles in ceramic or metallic melts. The Gibbs free energy change for homogeneous nucleation of a bubble in a melt can be expressed as [23]

$$\Delta G=4\pi R^2\gamma-4/3\cdot\pi R^3P \tag{7}$$

where R means the radius of the bubble, γ means the surface energy of the melt, P means the pressure in the bubble. The pressure in the bubble is the sum of gas pressure above the melt, static pressure in the melt, and capillary force

$$P=P_0+\rho_2g'H+2\gamma/R \tag{8}$$

where P_0 means the gas pressure above the melt, H means the depth of the bubble in the melt. Thus, ΔG can be further written as

$$\Delta G=4/3\cdot\pi R^2\gamma-4/3\cdot\pi R^3(\rho_2 g'H+P_0) \quad (9)$$

On the condition of $d(\Delta G)/dR=0$, the critical size (R_c) can be derived as

$$R_c=2/3\cdot\gamma/(\rho_2 g'H+P_0) \quad (10)$$

During high-gravity combustion synthesis in vacuum, when no extra pressure is applied, P_0 is much smaller than the other two terms and can be omitted. In this case, R_c can be simplified as

$$R_c=2/3\cdot\gamma/\rho_2 g'H \quad (11)$$

From Equation (11), the critical size is inversely proportional to the high-gravity factor. This means that a stronger high-gravity field will facilitate the nucleation of bubbles. With the data in Table 6 and provided H=10 mm, the R_c values under different high-gravity factors have been calculated and plotted in Figure 36. It is clear that the critical size of bubbles is much reduced in a high-gravity field compared with that under normal gravity. With a smaller critical size, gas species dissolved in the melt can escape more easily, which is helpful for the reduction of porosity in ceramic bulk products.

According to Equations (4,6,10), the effects of processing parameters on phase separation in high-gravity combustion synthesis can be discussed. At first, the high-gravity factor has an important influence on phase separation. With increasing high-gravity factors, the settling or rising velocity of Ni droplets or bubbles will increase. Secondly, phase separation depends on the viscosity of matrix melt, and a lower viscosity favors phase separation. The viscosity of a melt depends on its chemical composition and temperature. Thirdly, the size of Ni droplets or bubbles affects their settling or rising velocity, and a larger size is desired for phase separation. The critical size for homogeneous nucleation of bubbles in a ceramic melt is determined by the surface energy and density of the melt, the high-gravity factor, and the extra pressure applied on the melt.

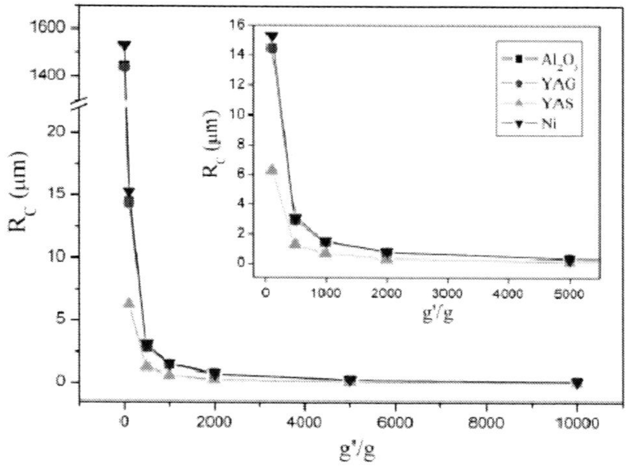

Figure 36. Dependence of critical size of bubbles in ceramic and Ni melts on high-gravity factor.

Finally, the extent of phase separation is directly connected with the lifetime of matrix melt, and a longer lifetime is helpful for phase separation. The lifetime of a melt lies on its formation temperature and the reaction temperature, and can be prolonged by lowering the formation temperature and elevating the reaction temperature.

From the effects of processing parameters, some advices can be raised to improve phase separation in high-gravity combustion synthesis. The first way is to strengthen the high-gravity field by increasing the rotation frequency and radius. A stronger high-gravity field accelerates phase separation and facilitates the escape of gas from melts. The second way is to increase the reaction temperature by choosing more exothermic reactants or by preheating. For most reaction systems, the adiabatic temperature increases with increasing preheating temperature, as shown in Figure 37. A higher reaction temperature prolongs the lifetime of the melt and reduces its viscosity, and thus favors phase separation. The third way is to apply an extra pressure on the melt. This pressure can improve gas removal by reducing the critical size for nucleation of bubbles. At the same time, the pressure can reduce shrinkage cavities during solidification by enhancing the plastic deformation of ceramic samples. The fourth way is to decrease the formation temperature and viscosity of the melt by tailoring its composition, including the selection of low-eutectic points in phase diagrams and the incorporation of appropriate additives. By the decrease of the formation temperature and viscosity of a melt, both the rate and time for phase separation are increased, which is favorable for complete phase separation.

5. MELT SOLIDIFICATION IN HIGH-GRAVITY COMBUSTION SYNTHESIS

In preparing bulk ceramic materials by high-gravity combustion synthesis, the formation of microstructure takes place during the solidification of melts. Therefore, the melt solidification process as well as the microstructure evolution involved there plays an important role. During the solidification process, the microstructure evolution strongly depends on the chemical composition of the melt and the cooling conditions

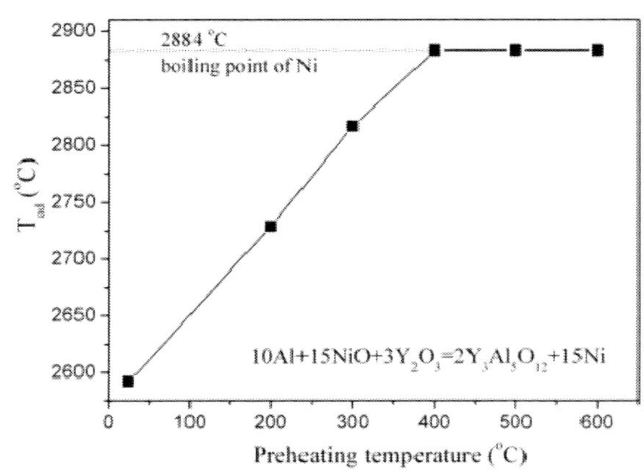

Figure 37. The relation between the adiabatic temperature and preheating temperature for the reaction $10Al+15NiO+3Y_2O_3=2Y_3Al_5O_{12}+15Ni$.

In this section, the microstructure evolution during the solidification will be discussed, where both single-phase ceramics and multiphase ceramic eutectics are involved, with Al_2O_3 and Al_2O_3-ZrO_2 as examples, respectively.

5.1. Single-Phase Ceramics (Al_2O_3)

During the solidification of ceramic melts, the driving force for liquid-solid transformation and growth rate of solid phases are closely related to the under-cooling degree. For this reason, the microstructure evolution during solidification depends on thermal gradient in the melts. By high-gravity combustion synthesis, the cooling and solidification of ceramic melts take place in a graphite crucible. With a good thermal conductivity, the graphite crucible is a major source of heat dissipation. In this case, a thermal gradient occurs in the radial direction in melts, where the temperature decreases from the center to the side. With such a thermal gradient, solidification will take place first at the side and then develop into the center. This has been verified by SEM observation, as shown in Figure 38, where the growth direction of Al_2O_3 crystals is nearly parallel to the radial direction.

Besides crucibles, another source of heat dissipation is the fast gas flow above the surface of melts under the condition of high-speed rotation. In this case, a thermal gradient also exists in the axial direction in the melts. This thermal gradient will cause a directional solidification from the surface to the center in the melts, as shown in Figure 39. Some huge rod-like crystals up to 1 mm growing along the axial direction are visible. The formation of these crystals can be probably attributed to the coupling of combustion reaction with the high-gravity field, because both the propagation of combustion wave and the high-gravity force (G') occurs in the axial direction, where enhanced mass transportation and heat transfer are expected.

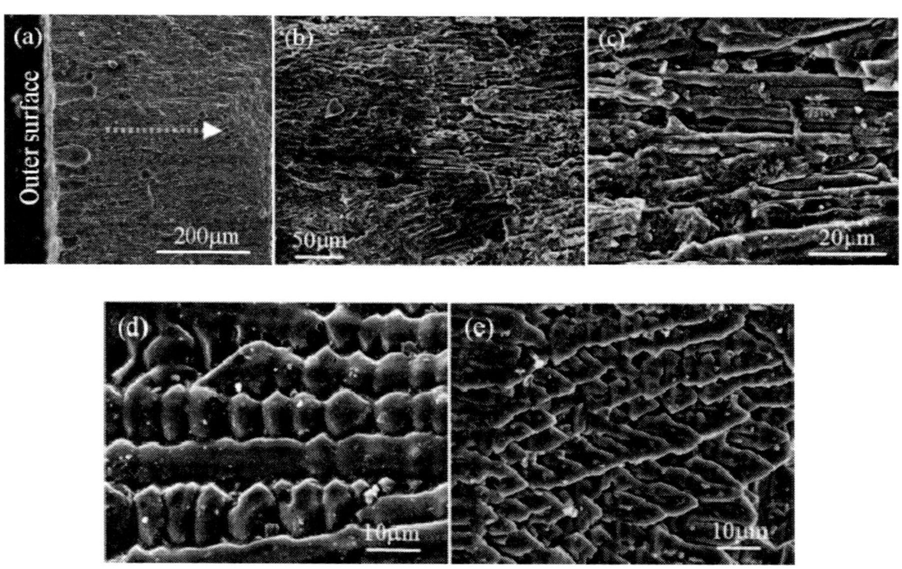

Figure 38. SEM images showing varying microstructures in the radial direction in Al_2O_3 samples.

During solidification, Al_2O_3 showed a strong tendency to develop into faceted crystals, even in large dendrites, as shown in Figure 40. This agrees well with the Jackson's criterion on interface structures and crystal growth modes in solidification of melts. According to the Jackson's criterion [24], the interface structure is identified by an interface roughness parameter defined as $\alpha=A\cdot\Delta S_m/R$, where ΔS_m is the melting entropy, R is the gas constant, and A is a crystallographic parameter equal to the ratio between the number of bonds parallel to the surface and the total number of bonds. For a solidification process, if $\alpha<2$, the liquid/solid interface is atomically rough and non-faceted interface is expected. If $\alpha>2$, the interface is atomically smooth and faceted interface is produced because the crystal growth is limited by nucleation rate. The melting entropies of ceramic compounds investigated here are listed in Table 7. For Al_2O_3, $\Delta S_m/R=5.74$, and A is estimated to be 0.5~1, giving $\alpha>2$. Therefore, atomically-smooth interface structure should be involved during the solidification of Al_2O_3, resulting in faceted crystals.

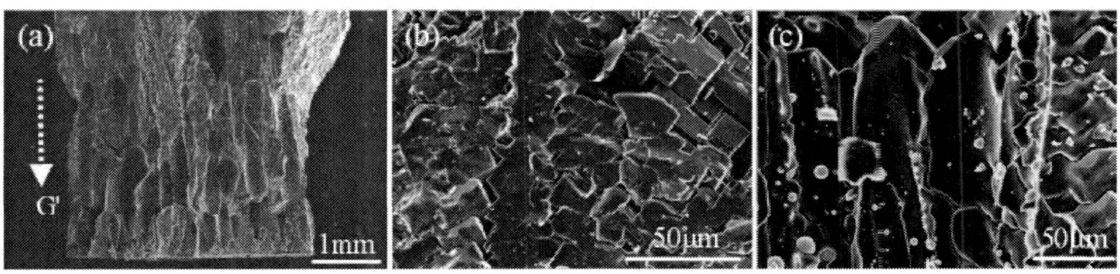

Figure 39. SEM images showing varying microstructures in the axial direction in Al_2O_3 samples.

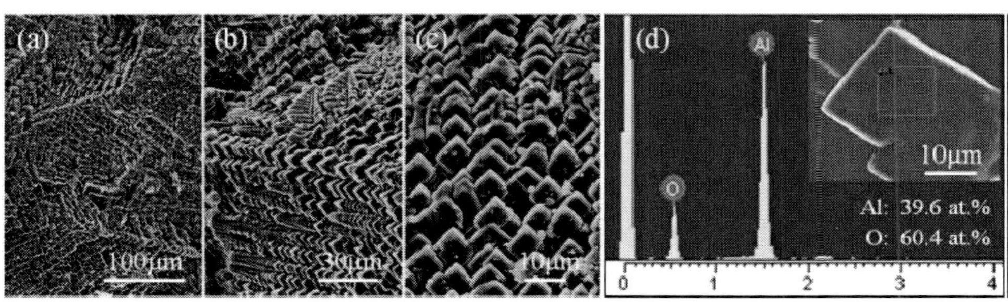

Figure 40. SEM images and EDS results of Al_2O_3 dendrites.

Table 7. Melting entropies of several ceramic materials

Compound	ΔS_m ($J\cdot mol^{-1}\cdot K^{-1}$)	$\Delta S_m/R$
Al_2O_3	47.72	5.74
YAG	122.38	14.72
$MgAl_2O_4$	81.64	9.82
ZrO_2	29.51	3.55

For the solidification of ceramic materials like Al_2O_3 with large melting entropies, the energy barrier for nucleation is high, and the crystal growth is thought to take place by two-dimensional nucleation and growth. With as-existing crystals as nucleation sites, new nuclei form and grow up. By iterative nucleation and growth, large faceted crystals will be produced with a sufficient material supply from the melt at the early stage of solidification. When the solidification is close to the end, however, solid phase is predominant and the melt becomes isolated. In this wise, material supply for solidification is limited and incomplete growth of faceted crystals takes place, resulting in a terraced morphology, as shown in Figure 41.

At the surface of some Al_2O_3 samples, bright layers with a thickness of nearly 0.5 mm were produced. These layers were not entirely smooth but consisted of small flakes with different orientations. The width of the flakes varied in the range of 3~5 mm, and each flake was composed of orderly-arranged faceted Al_2O_3 crystals, as shown in Figure 42. Such strongly-textured structure was also verified by XRD analysis. As shown in Figure 43, an evident difference in relative intensities of the diffraction peaks exists between the XRD patterns of the flakes and that of a powder sample.

The formation of the strongly-textured structure can be attributed to two reasons. One is the intrinsic faceted growth character of Al_2O_3, and the other is the particular solidification condition at the melt surface, which is characterized by fast cooling and free growth. Under high-frequency rotation, the gas flow sweeping the melt surface causes fast cooling and induces rapid solidification as well as crystal growth. On the other hand, crystal growth at the melt surface is not constrained by crucible, and thus the Al_2O_3 crystals can grow freely. It is thought that the strongly-textured structure has been produced by parallel continuous growth.

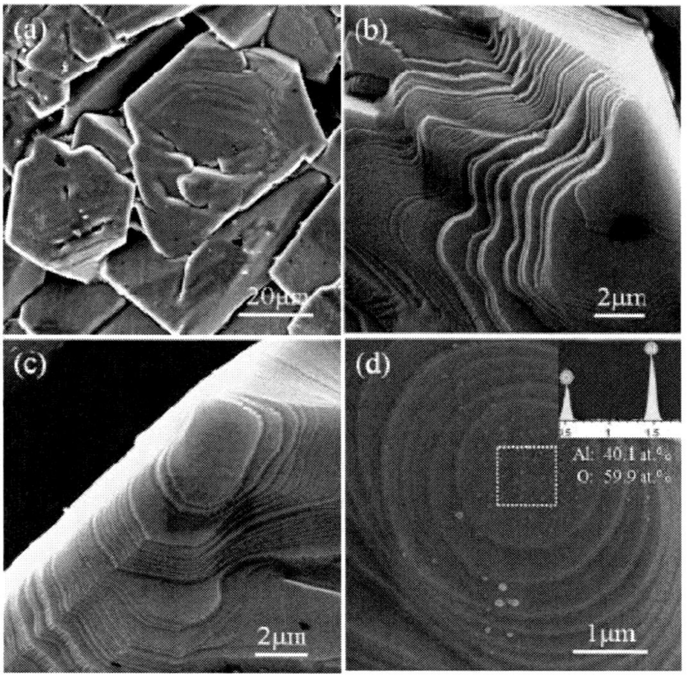

Figure 41. Terraced structures observed in the Al_2O_3 samples.

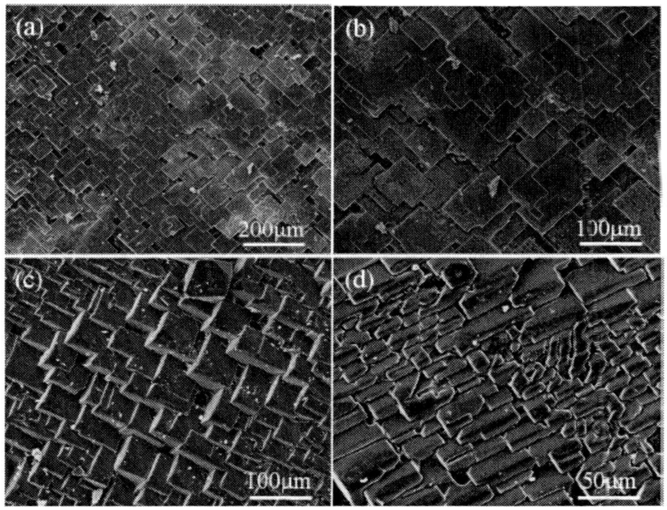

Figure 42. SEM images showing the strongly-textured structure at the surface of Al_2O_3 samples.

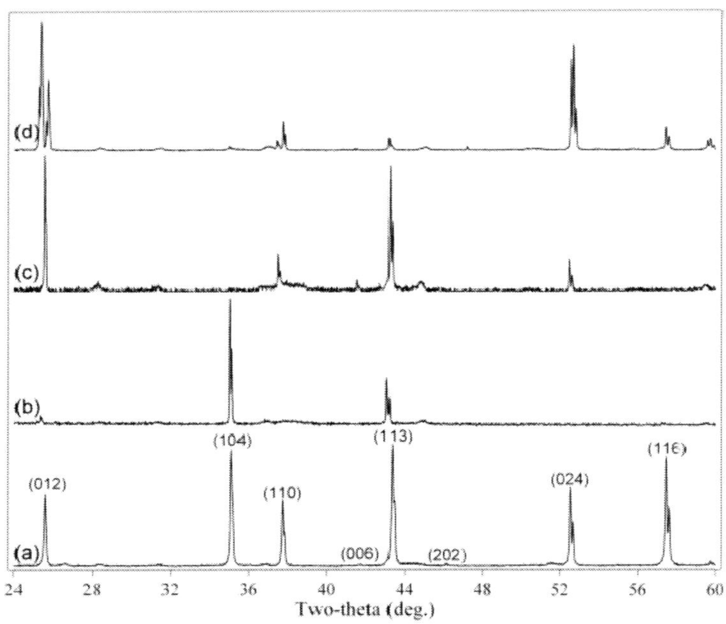

Figure 43. XRD patterns of the Al_2O_3 samples: (a) powder; (b)-(d) flakes in the surface layer.

At first, individual primary nuclei are precipitated in the melt and with different orientations. Then, based on each primary nucleus, more nuclei occur and grow with an orientation identical to the basis. Finally, by continuous nucleation and growth, an array of crystals with the same orientation is produced, which is recognized as a flake on a macroscopic scale. By the assembly of such flakes, a strongly-texture structure is obtained.

In the prepared Al_2O_3 samples, skeletal crystals were also observed, as shown in Figure 44. To some extent, the skeletal crystals can be regarded as intermediate products from incomplete growth of faceted crystals. The skeletal crystals are possibly induced by fast cooling at melt surface. On one hand, fast cooling often results in large under-cooling degree and high growth rate of crystals. On the other hand, fast cooling means a short lifetime of melt, which can lead to the lack of material supply. Without sufficient material supply, the imperfections created during fast crystal growth cannot be timely repaired, leading to the occurrence of skeletal crystals.

In contrast to the surface, in the center part of samples, rather dense microstructure was obtained. In the fracture surface of the center part, typical cleavage planes are visible, as shown in Figure 45. Few secondary particle inclusions were found inside the Al_2O_3 crystals. From EDS analysis for six individual crystals with different size and morphology (two spectra shown in Figures 40 and 41), the atomic ratios of Al and O are 59.5-60.5 % and 39.5-40.5 %, respectively, showing a good consistency with the nominal composition of Al_2O_3.

5.2. Multiphase Ceramic Eutectics (Al_2O_3-Zro_2)

During the solidification of eutectics, the interphase spacing is determined by the balance of two opposite energy contributions. On one hand, transverse diffusion of solutes must occur over greater distance for thicker lamellae or fibers, therefore, favoring smaller interphase spacings.

On the other hand, the interfacial energy associated to interphase boundaries will increase as the width of lamellae or fibers decreases, thus favoring larger interphase spacings. In general, the interphase spacing is related with under-cooling degree and growth rate, which can be expressed as

$$\lambda \Delta T = C_1 \text{ and } V\lambda^2 = C_2,$$

where λ is the interphase spacing, ΔT is the under-cooling degree, V is the growth rate, C_1 and C_2 are constants [25]. From the formulae, a greater under-cooling degree is expected in the center than that at the boundaries of the colonies in the Al_2O_3-ZrO_2 eutectic ceramics.

According to the model proposed by Hunt and Jackson [25], the microstructure of binary eutectics depends on the melting entropies of the components. If the large-volume component has a high melting entropy ($\alpha>2$) and the minor one has a small melting entropy ($\alpha<2$), a "complex regular" microstructure will be produced, which shows many features of the lamellar or fibrous eutectic microstructure.

Figure 44. SEM images of skeletal crystals in as-prepared Al_2O_3 samples.

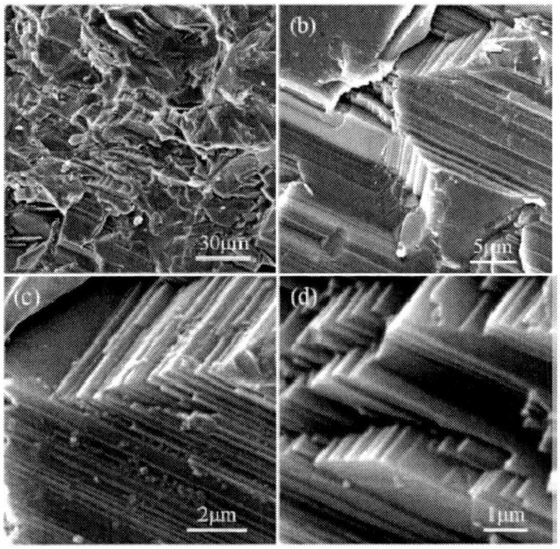

Figure 45. SEM images of fracture surface of Al_2O_3 samples showing typical feature of cleavage planes.

In this work, the Al_2O_3-ZrO_2 eutectic sample possesses a phase composition of 67%Al_2O_3+33%ZrO_2 in volume fraction. The large-volume component of Al_2O_3 has a higher melting entropy of $\Delta S_m/R=5.74$, and the minor ZrO_2 has a lower melting entropy of $\Delta S_m/R=3.55$. In this case, the eutectic sample shows a complex microstructure, where both fibrous and lamellar features are observed, as shown in Figure 18 (c) and (d). In the fibrous structure, fine ZrO_2 fibers with a width of 100-200 nm are embedded in a continuous Al_2O_3 matrix. The lamellar structure is somewhat incomplete, which is more exactly a mixture of lamellar and fibrous features.

The microstructure of binary eutectics is usually connected with the volume fractions of the two components. It has been reported that [26], if the surface energies are isotropic or with no significant anisotropy, the fibrous structure is stable when the volume fraction of the minor component is below 32%, and when the fraction is >32% a lamellar structure will be produced. This prediction has been proved by experimental results on several metal eutectics, where a lamella-to-fiber transition takes place at the volume fraction of ~30%. In the Al_2O_3-ZrO_2 eutectic ceramic sample prepared in this work, the volume fraction of the minor component (ZrO_2) is 33%, which is almost the critical point for the transition between the two structures. As a result, both fibrous and lamellar structures were obtained in the sample.

Another factor affecting the microstructure of eutectics is the growth rate. During the solidification of eutectic ceramics, microstructure evolution is generally related with the thermal gradient, concentration gradient, and growth rate. For given thermal and concentration gradient values, the solidification mechanism depends on growth rate. When the growth rate is below a critical value, planar growth front is stable, and with increasing growth rate shallow cells will occur as a secondary phase. The microstructure transition from coupled to cellular and then to shallow cells with increasing growth rate has been observed in Al_2O_3-ZrO_2 system [2]. In this work, the discrepancy in the growth rates of different parts is probably one of reasons responsible for the diversity of microstructures observed in the Al_2O_3-ZrO_2 eutectic ceramic sample.

In the YAG-Al_2O_3-ZrO_2 ternary eutectic system, YAG and Al_2O_3 have higher melting entropies and ZrO_2 has a lower one. In the sample of YAZ, YAG is the major phase with a volume fraction of 49%, and the minor Al_2O_3 and ZrO_2 have volume fractions of 37% and 14%. According to the model proposed by Hunt and Jackson [25,26], Al_2O_3 and ZrO_2 will form lamellar and fibrous structures, respectively. The experimentally-observed microstructure in the sample YAZ is basically consistent with the prediction, except for the occurrence of the three-dimensional interpenetrating structure.

Conclusion and Perspective

By combining strongly-exothermic combustion reactions and high-gravity field, high-gravity combustion synthesis offers an alternative method to fabricate bulk ceramics via melt solidification instead of conventional powder sintering. In this method, high temperature is fulfilled by exothermic chemical reactions and external heating sources (such as electric furnace) are not required. Thus, the energy consumption for the production of ceramic materials is reduced. Additionally, the processing time for preparing ceramics can be greatly shortened by high-gravity combustion synthesis because the combustion reactions progress very quickly. With such merits, high-gravity combustion synthesis is a fast and efficient way to prepare bulk ceramic materials.

Up to now, high-gravity combustion synthesis has been applied to prepare various ceramic and glass materials, including single-phase ceramics, multiphase ceramic eutectics, glasses, and glass-ceramics. The phase assemblage and microstructure of the products can be manipulated by careful design of starting compositions and optimization of processing parameters, based on a systematic investigation on the phase separation and microstructure evolution.

We believe that high-gravity combustion synthesis can be used to prepare more kinds of materials besides ceramics and glasses. For example, it is appropriate to produce cermets with a metallic matrix reinforced by hard ceramic phases. Another possible application is to prepare refractory alloys and intermetallic compounds. With the extending of the application, further studies on the reaction mechanisms of high-gravity combustion synthesis are required, which can be supported by real-time characterizations.

Acknowledgments

This work is supported by National Natural Science Foundation of China (Grant No. 50932006 and 51002163) and Beijing Natural Science Foundation (Grant No. 2112043).

References

[1] LLorca, J., Orera, V. M., *Prog. Mater. Sci.* 51 (2006) 711.

[2] Pena, J. I., Merino, R. I., Harlan, N. R., Larrea, A., Fuente, G. F., Orera, V. M., *J. Eur. Ceram. Soc.* 22 (2002) 2595.

[3] Mesa, M. C., Oliete, P. B., Orera, V. M., Pastor, J., Martín, A., LLorca, J., *J. Eur. Ceram. Soc.* 31 (2011) 1241.

[4] Zhao, Z., Zhang, L., Zheng, J., Bai, H., Zhang, S., Xu, B., *Scripta. Mater.* 53 (2005) 995.

[5] Pei, J., Li, J., Liu, G., Chen, K., J. *Alloy Comp.* 476 (2009) 854.

[6] Liang, R., Pei, J., Li, J., Chen, K., *J. Am. Ceram. Soc.* 92 (2009) 549.

[7] Liu, G., Li, J., Guo, S., Ning, X., Chen, Y., *J. Alloy Comp.* 509 (2011) 213.

[8] Liu, G., Li, J., Yang, Z., *Mater. Manu. Proc.*, available online, DOI:10.1080/10426914.2011.593250.

[9] Liu, G., Li, J., He, B., *J. Non-Cryst. Solids* 357 (2011) 1764.

[10] He, B., Liu, G., Li, J., Wu, L., Yang, Z., Guo, S., Chen, Y., *Mater. Res. Bull.* 46 (2011) 1035.

[11] Liu, G., Li, J., He, B., *Adv. Appl. Ceram.,* available online, DOI 10.1179/1743676111Y.0000000028.

[12] Mei, L., Mai, P., Li, J., Chen, K., *Mater. Lett.* 64 (2010) 68.

[13] Petzow, G., Herrmann, M., *Structure and Bonding* 102 (2002) 47.

[14] Fujita, S., Tanabe, S., *Jpn. J. Appl. Phys.* 48 (2009) 120210.

[15] Sakamoto, A., Yamamoto, S., *Int. J. Appl. Glass Sci.* 1 (2010) 237.

[16] Serkov, B., Maksimov, E., Merzhanov, A., *Combust Explos Shock Wave* 4 (1968) 349.

[17] Takaimchi, I., Roderick, G., *The physical properties of liquid metals*, Oxford University Press Inc., New York, 1988.

[18] Blomquist, R., Fink, J., Leibowitz, L., *Am. Ceram. Soc. Bull.* 57 (1978) 522.

[19] Paradis, P., Ishikawa, T., *Jpn. J. Appl. Phys.* 44 (2005) 5082.

[20] Fratello, V., Brandle, C., *J. Cryst. Growth* 128 (1993) 1006.

[21] Hyatt, M., Day, D., *J. Am. Ceram. Soc.* 70 (1987) 283.

[22] Saito, N., Kai, K., Furusho, S., Nakashima, K., Mori, K., Shimizu, F., *J. Am. Ceram. Soc.* 86 (2003) 711.

[23] Zhang, H., Li, Y., *Acta Phys. Sinica* 56 (2007) 4864.

[24] Laudise, R. A., Carruthers, J. R., Jackson, K. A., *Ann. Rev. Mater. Sci.* 1 (1971) 253.

[25] Hunt, J. D., Jackson, K. A., *Trans. AIME* 236 (1966) 843.

[26] Jackson, K. A., Hunt, J. D., *Trans. AIME* 236 (1966) 1129.

INDEX

A

absorption spectra, 25, 26
access, 22, 82
acid, 8, 11, 12, 14
acrylate, 19
activation energy, 46, 114
actuators, 32
AD, 195
adaptation, 80, 82, 83
additives, x, 177, 178, 180, 184, 198, 206, 256, 258, 259, 283
adhesion, 217, 219, 220, 222, 223, 225, 231, 232, 233, 242, 248
adhesion force, 217, 223, 225
adjustment, 82
adsorption, 77
aesthetic, viii, 71, 72, 79
AFM, v, x, 100, 101, 102, 103, 110, 111, 215, 216, 217, 221, 222, 223, 226, 230, 231, 232, 233, 235, 245
age, 83
aggregation, 147
aging process, 76, 77
airports, 164
aluminum oxide, 79
amine(s), 2, 3, 11, 12, 14, 16, 19
amino, 11, 13, 19, 20, 24, 26
amino groups, 11, 13, 19, 20, 24, 26
ammonium, 135
amplitude, 33, 93, 134, 182, 221
angiography, 163
anisotropy, 103, 104, 105, 120, 234, 289
annealing, 33, 94, 95, 103, 109, 130, 136, 147, 150, 270
annihilation, 77, 219
apex, 229
aqueous solutions, 11, 135
arginine, 11
Arrhenius equation, 46
asymmetrical polarization hysteresis loops exhibition, vii
atmosphere, viii, 4, 51, 54, 77, 130, 157, 161, 198
atomic force, 110, 240
atomic force microscope, 240
Atomic Force Microscopy-Force Spectroscopy (AFM-FS), x, 215
atoms, 102, 114, 117, 122, 143, 179, 197, 200, 221
attachment, vii, 1, 6, 9, 27
automation, 2
automotive application(s), 210

B

Baars, 28
band gap, 46
bandgap, 132, 161
barium, 128, 154, 156
base, ix, 19, 123
baths, 80
beams, 164
behaviors, 80
Beijing, 251, 290
bending, 54, 59, 61, 63, 222
beneficial effect, 52
bias, vii, 31, 36, 37, 38, 44, 45
binding energy, 133
biocompatibility, viii, 71, 74, 79, 83, 233
biocompatible materials, 72
biological systems, 2
biomaterials, 2
biomedical applications, viii, 71, 73, 236
biopolymers, 11, 14
blends, 52
Boltzmann constant, 89
bonding, 160, 183, 232

bonds, 8, 77, 150, 221, 231, 232, 233, 285
bone, 2, 16, 79, 162
branching, 19
Brazil, 71, 72, 80
breakdown, 206, 207, 210
brittleness, 110
building blocks, 2
bulk materials, 133, 219, 220
bulk optical elements, ix, 127, 129, 151
burn, 33

C

Ca^{2+}, 73, 234
CAD, 72, 79
calcium, 16, 53
calculus, 218
calibration, 230
CAM, 72, 79
candidates, viii, 2, 32, 43, 135, 154, 161, 163, 231
capillary, 2, 7, 8, 184, 233, 281
carbon, 54, 185, 198, 203, 232, 252
carbon film, 232
carboxylic acid(s), 14
casting, 136, 238, 251
catalysis, 5, 11, 25
catalyst, 9, 178
cation, 73, 123, 140, 149, 234
C-C, 231
ceramic materials, vii, viii, ix, x, 2, 18, 71, 74, 88, 89, 114, 117, 122, 127, 128, 129, 135, 140, 147, 163, 165, 216, 217, 218, 228, 245, 251, 252, 254, 268, 270, 278, 283, 285, 286, 290
ceramic membranes, vii, 1, 2, 5, 12, 21, 22, 27
Ceramics, v, vi, 18, 24, 31, 68, 83, 85, 87, 94, 98, 102, 105, 113, 123, 127, 155, 177, 183, 210, 211, 212, 213, 215, 249, 251, 255, 270, 284
cerium, 117, 151
CFI, 212
charge coupled device, 134, 152
chemical, viii, 2, 5, 9, 25, 27, 51, 52, 53, 57, 58, 68, 73, 77, 94, 110, 114, 142, 144, 148, 149, 151, 160, 164, 179, 183, 194, 198, 207, 209, 232, 233, 235, 252, 253, 254, 263, 265, 266, 269, 272, 275, 282, 283, 290
chemical properties, 5, 235
chemical reactions, 27, 252, 290
chemical stability, 73
chemical vapour deposition, 232
China, 198, 251, 290
chiral catalyst, 8
chloroform, 10
chromatography, 2
circulation, 164
clarity, 221
classification, 52
clay minerals, 52, 54, 55, 63
cleaning, 102
cleavage, 90, 91, 288, 289
clusters, 25, 77
CO_2, 114
coatings, x, 154, 215, 219, 220, 222, 230, 232
color, 76, 79, 134, 148, 201
combined effect, 181
combustion, xi, 114, 155, 156, 252, 253, 254, 255, 256, 260, 262, 265, 266, 268, 270, 271, 272, 275, 276, 277, 278, 279, 280, 282, 283, 284, 290
commerce, 163
commercial, 40, 43, 47, 49, 158, 178, 231, 240
communication, 154
community(s), 72, 83, 131
comparative analysis, 94, 103
competition, 88, 98, 146, 156
competitors, 165
compliance, 34, 39, 40, 47, 48
composites, ix, 19, 87, 88, 107, 108, 110, 186, 198, 199, 200, 202, 203, 204, 205, 209, 210
compounds, x, 2, 9, 32, 128, 129, 133, 149, 156, 158, 162, 164, 285
compression, 82, 183, 227
computed tomography, 151
computer, 152, 182
computer-aided design, 182
condensation, 9, 10, 11, 12, 13
conductance, 206, 207
conduction, 130, 141, 148, 151, 156, 160
conductivity, 89, 92, 111, 117, 118, 119, 123, 183, 187, 194, 284
conductors, 181, 183, 234, 235
configuration, 77, 114, 179
conflict, 112
confrontation, 221
Congress, 210
congruence, 95
conjugation, 2
connectivity, 210
consensus, 77, 83
consolidation, 52, 181, 182, 183, 192
constant load, 231, 245
constituents, 129
construction, 2, 72, 88, 163, 226
contamination, 164
contradiction, 221
control group, viii, 71, 82, 83
cooling, 34, 41, 43, 73, 181, 198, 252, 253, 254, 270, 283, 284, 286, 288

coordination, 19, 114, 234
copper, x, 54, 215
correlation, 61, 62, 142, 143
correlation coefficient, 61
corrosion, 77, 183
cost, 154, 158
cracks, 21, 22, 110, 236, 240, 273
creep, 183
critical value, 289
cross-linking reaction, 12
crowns, 79
crust, 72
crystal growth, 144, 145, 285, 286, 288
crystal quality, 164
crystal structure, 73, 114, 149, 150, 200
crystalline, 2, 14, 54, 73, 75, 144, 148, 159, 161, 217, 259, 268, 269, 270, 273, 274
crystallites, 94, 105, 107, 109, 122, 142, 190, 192, 193, 194, 201, 203, 204, 206, 210, 264, 266, 272
crystallization, 270
crystals, x, 2, 14, 74, 98, 110, 119, 129, 135, 140, 144, 147, 148, 149, 155, 157, 165, 180, 215, 263, 264, 273, 284, 285, 286, 287, 288
CSA, 121, 122
CT, ix, 127, 130, 151, 152, 162, 163, 175
culture, 25
cyanide, 128
cycles, 41, 43, 63, 78, 80, 83, 217, 231, 236
cycling, 217
Czech Republic, 127

D

data processing, 95
decay, 129, 132, 133, 134, 140, 141, 143, 144, 146, 148, 150, 151, 152, 153, 154, 155, 157, 158, 161, 164
decay times, 133, 152, 154, 161, 164
decomposition, 55, 73
defects, ix, 24, 25, 46, 68, 72, 74, 75, 77, 87, 88, 89, 99, 102, 103, 105, 109, 110, 119, 122, 130, 140, 142, 143, 144, 145, 147, 151, 152, 160, 164, 165, 195, 197, 219, 234, 246
deformation, 32, 74, 77, 98, 99, 100, 101, 102, 217, 219, 221, 224, 227, 240, 242
degradation, viii, 71, 72, 76, 77, 78, 145, 155
denaturation, 2
dendrites, 285
density functional theory, 156
dentin, 72
dentist, 72
depolarization, viii, 32, 45, 105
deposition, 158, 232
depth, ix, 87, 141, 148, 164, 218, 222, 224, 225, 226, 227, 228, 229, 230, 235, 237, 238, 243, 281
derivatives, 5, 26
destruction, 100, 103
detectable, 152, 201
detection, 25, 128, 131, 134, 141, 145, 148, 158, 162, 163, 164, 201
deviation, 54, 62, 192, 229
DFT, 148
diatoms, 11
dielectric permittivity, viii, 32, 33, 36, 40, 41, 43, 44, 46, 47
dielectrics, 154, 183
diffraction, 33, 34, 40, 41, 88, 192, 196, 197, 207, 208, 255, 260, 286
diffusion, viii, 46, 77, 87, 88, 89, 90, 97, 99, 101, 103, 104, 105, 106, 107, 109, 110, 111, 112, 113, 115, 116, 117, 118, 119, 120, 122, 180, 183, 288
diluent, 255, 268
dipoles, vii, 31, 37, 47
direct measure, 119
directives, 50
dislocation, 98, 99, 183, 195, 197, 217, 218, 219
disorder, 117, 161
disordered systems, 142
dispersion, 8, 106, 136, 233
displacement, 102, 189, 197, 217, 218, 226, 228, 229, 243, 279, 280
dissociation, 77, 232
distilled water, 80
distribution, viii, 11, 16, 22, 23, 25, 51, 52, 53, 57, 60, 68, 76, 78, 80, 99, 102, 112, 114, 119, 140, 155, 163, 187, 200, 201, 203, 204, 209, 234, 235
diversity, 11, 27, 289
DMF, 10
DNA, 25
DOI, 291
domain structure, ix, 88, 117, 119, 123
dopants, 76, 148, 161, 164, 234
doping, 117, 129, 130, 140, 146, 151, 152, 161, 234
draft, 50
drawing, 137
drug delivery, 2, 5
drugs, 163
drying, 6, 21, 52, 53
DSC, 54
dyes, 2

E

economic problem, 144
efficiency criteria, 130
elastic deformation, x, 215, 219, 220, 223

election, 235
electric conductivity, 183
electric current, 181, 252
electric field, 33, 181, 222
electrical conductivity, 114, 198
electrical properties, 49, 234, 270
electrodes, 33
electrolyte, 114, 235
electromagnetic, 206
electron, 7, 12, 18, 54, 94, 99, 103, 110, 112, 114, 123, 128, 130, 131, 140, 141, 142, 144, 145, 146, 147, 148, 149, 151, 154, 160, 164, 168, 185, 186, 197, 200, 201, 203, 222, 240
electron diffraction, 18
electron microscopy, 12, 99, 103, 114
electron paramagnetic resonance, 167, 168
electronic structure, 143, 156
electrons, 112, 130, 141, 160
electrophoresis, 2
e-mail, 1, 31, 51, 71, 87, 127, 177, 215
emission, 98, 130, 131, 132, 133, 134, 141, 142, 144, 145, 146, 147, 148, 149, 151, 152, 154, 155, 156, 157, 161, 163, 164, 165, 232
empirical diagrams, viii, 51, 52, 57
enamel, 72
encapsulation, 2, 13, 19
energy consumption, xi, 2, 290
energy transfer, 131, 133, 141, 148, 149, 151, 165
engineering, x, 178
England, 247
enlargement, 25
entrapment, 112, 113
entropy, 285, 288, 289
environment(s), 2, 77, 78, 82, 141, 232
epitaxial growth, 161
equilibrium, viii, 87, 88, 109, 117, 119, 120
equipment, 50, 82, 226, 230
ESR, 142, 143
ester, 19, 186
etching, 6, 185, 231
ethanol, 33, 198, 252
ethylene, 2, 3, 5, 12, 19
EU, 50
evaporation, 114, 158
evidence, 77, 102, 140, 142, 196
evolution, xi, 236, 241, 252, 283, 284, 289, 290
examinations, 54, 56, 163
excitation, 133, 140, 141, 146, 150, 152, 155, 158, 161, 236
exciton, 131, 133, 154, 156, 157, 159, 161
exclusion, 2
execution, 61, 82
experimental condition, 112, 114, 115, 240
experimental design, 59, 60
exploitation, 128, 164
explosives, 25, 163
external influences, 143
external magnetic fields, 142
extraction, 2, 225
extrusion, 53, 54

F

fabrication, vi, 68, 79, 88, 102, 103, 110, 137, 138, 140, 153, 154, 158, 159, 160, 161, 212, 235, 251, 256, 280
ferroelectrics, 117, 119
fiber(s), 272, 275, 288, 289
fillers, 52
films, 219, 232
filters, 2, 3, 4, 5, 6, 23
filtration, 1, 136
financial, 210
financial support, 210
fine tuning, 149
first generation, 19
fixed rate, 198
flaws, 131
flight, 152, 163
fluctuations, 226
fluorescence, 72
force, x, 74, 80, 181, 183, 184, 194, 195, 215, 216, 217, 219, 221, 222, 224, 226, 228, 232, 233, 234, 236, 241, 278, 281, 284
Ford, 28, 30
formation, vii, ix, 1, 5, 8, 10, 11, 12, 13, 14, 16, 24, 26, 27, 31, 36, 41, 44, 49, 52, 54, 56, 63, 76, 77, 82, 87, 98, 107, 109, 113, 120, 122, 123, 140, 144, 160, 181, 200, 233, 243, 258, 270, 275, 283, 284, 286
formula, 117, 119, 120, 148
fracture resistance, viii, 71, 236
fracture toughness, x, 73, 74, 75, 80, 83, 110, 178, 186, 190, 193, 209, 228, 265, 266
fractures, viii, 71, 73, 162
fragments, ix, 87, 109, 110, 113, 122, 123
France, 69, 215
free energy, 76, 281
frequency distribution, 186, 204
friction, x, 215, 216, 218, 220, 221, 231, 232, 233, 241, 242, 243, 245
fuel cell, xi, 216, 235
functionalization, 8, 12, 19
fusion, 56

G

gadolinium, 114, 117
gallium, 161
gamma rays, 128, 164
gamma-ray detectors, 149
gel, 2, 5, 6, 7, 8, 9, 11, 18, 156
geometry, 79, 220, 228, 229, 230, 239, 241
Germany, 79, 184, 253
glasses, xi, 117, 118, 252, 254, 268, 269, 270, 278, 290
grain boundaries, ix, 87, 88, 89, 102, 106, 109, 110, 112, 119, 122, 123, 129, 135, 140, 150, 158, 260
grain size, viii, ix, 15, 44, 51, 52, 57, 68, 74, 75, 76, 77, 78, 87, 88, 94, 102, 103, 105, 107, 110, 115, 122, 130, 138, 186, 190, 194, 201, 205, 206, 207, 209, 236, 251, 263, 274
grain size distribution, viii, 51, 52, 57, 68
graph, 119, 245
graphite, 181, 182, 183, 185, 187, 189, 194, 198, 203, 232, 252, 253, 284
gravity, xi, 252, 253, 254, 255, 256, 257, 258, 260, 261, 262, 265, 266, 268, 270, 271, 272, 275, 276, 277, 278, 279, 280, 281, 282, 283, 284, 290
Greece, 1
growth, x, 17, 72, 76, 77, 78, 109, 120, 140, 143, 147, 148, 149, 177, 193, 194, 195, 197, 205, 206, 207, 209, 251, 252, 284, 285, 286, 287, 288, 289
growth mechanism, 77
growth rate, 206, 284, 288, 289
guidance, 80

H

halogen, 158
HAP, 16
hardness, x, 78, 96, 110, 177, 178, 186, 215, 218, 220, 222, 225, 227, 228, 229, 230, 232, 233, 235, 236, 237, 238, 245, 253, 265, 266, 275
hazardous substances, 50
heat capacity, 89, 114, 116, 117, 123
heat conductivity, 187
heat release, 232
heat transfer, 284
heating rate, viii, x, 51, 52, 54, 59, 61, 63, 141, 177, 183, 184, 185, 186, 187, 188, 189, 190, 191, 192, 193, 194, 195, 196, 197, 199, 201, 202, 203, 204, 205, 208, 209
height, 114, 140, 146, 147, 161, 275
helium, ix, 89, 90, 98, 111, 112, 117, 119, 120, 122
high strength, 178
high-gravity combustion synthesis, xi, 252, 253, 254, 255, 260, 262, 265, 266, 268, 270, 271, 272, 275, 276, 277, 278, 279, 280, 282, 283, 284, 290
high-power piezoelectric devices, viii, 32, 49
historical overview, 129
history, 128, 130
homeland security, 159, 161, 163, 164
homogeneity, 74, 129
host, ix, 127, 140, 141, 145, 146, 147, 148, 149, 152, 154, 155, 156, 159, 160, 161, 164, 165, 174
hot pressing, 98, 100, 101, 150, 158, 183, 198
HRTEM, 160
human, 74, 78, 82, 163, 164
human activity, 164
human body, 74, 78, 163
humidity, 54, 232, 233
husband, 128
hybrid, vii, 1, 2, 5, 10, 11, 12, 13, 14, 18, 25, 26, 140
hybridization, 232
hydrogen, 135, 161, 232
hydrolysis, 5, 8, 9, 10
hydrophilicity, 231
hydroxyapatite, 14, 16, 17, 18
hydroxyl, 6, 11, 77
hydroxyl groups, 6, 77
hyperbranched polymers, 2
hypothesis, 82
hysteresis, vii, 31, 33, 36, 37, 38, 41, 43, 44, 78
hysteresis loop, vii, 31, 37, 38, 43, 44

I

ideal, 32, 52, 154, 220, 229
identification, 143, 163
image analysis, 204, 205
immersion, 3, 5, 6, 21, 22, 27, 54
immobilization, 24
impact energy, 235
implants, 72, 79, 80
impregnation, 3, 4, 5, 12, 20, 22
improvements, 164
impurities, ix, 87, 113, 123, 146, 178
in vitro, 11, 74, 80, 83
in vivo, 74, 78, 83
incidence, 92
incisors, 80
indentation, 185, 186, 217, 218, 222, 223, 225, 226, 227, 228, 229, 230, 239, 247, 253
indium, 161
induction, 251, 252
industry, 73, 129, 219
inequality, 112, 115
inertia, 2

initiation, 183
insertion, 123
Instron, 80
insulation, 185
integration, vii, 1, 26, 93, 151
interaction effect, 59
interface, 80, 81, 97, 120, 122, 129, 146, 160, 196, 223, 232, 266, 273, 285
intermetallic compounds, 290
interphase, 192, 200, 201, 263, 265, 266, 288
inventions, 144
ionization, 133, 141, 148, 155
ions, vii, 25, 31, 73, 77, 89, 97, 102, 114, 117, 133, 142, 144, 145, 149, 150, 152, 156
iridium, 149, 251
iron, 55, 110, 112, 113, 123
irradiation, 99, 114, 134, 135, 141, 142, 143, 145, 162, 163
isolation, 52, 75, 226
isotope, 140, 163
issues, 160
Italy, 80

J

Japan, 33, 127, 185, 186, 198, 210, 211, 213, 253

K

K^+, 40, 41, 44
KBr, 217
kinetics, ix, xi, 87, 88, 122, 133, 141, 180, 275
Korea, 31

L

laboratory studies, 83
lamella, 289
lamination, 74, 220
landscape, 222
languages, 128
lanthanide, 158
lasers, 129, 251, 252
lattice parameters, 179
laws, 159
lead, vii, 10, 32, 33, 37, 40, 43, 45, 47, 49, 91, 102, 103, 108, 122, 142, 148, 198, 206, 288
leakage, 206
LED, 148, 268
legislation, 50
lifetime, 76, 152, 222, 258, 262, 268, 283, 288
light, 74, 114, 128, 131, 132, 133, 134, 135, 140, 141, 142, 144, 146, 147, 148, 150, 152, 153, 154, 155, 156, 157, 158, 159, 161, 164, 268
light scattering, 135, 158, 159
linear dependence, 279
linear function, 223
liquid chromatography, 2
liquid phase, 34, 35, 41, 44, 49, 179, 180, 181, 184, 192, 194, 195, 207, 258
liquidity, 53
lithium, 98, 195
localization, 156, 160
longevity, viii, 71, 72
low temperatures, 49, 73, 110, 151, 183
LTD, 72, 77
luminescence, ix, 127, 130, 131, 132, 133, 135, 139, 140, 141, 143, 144, 148, 149, 151, 152, 153, 154, 155, 156, 157, 159, 160, 161, 162
luminescence efficiency, 160
Luo, 30
lutetium, 151
lysine, 11

M

macromolecules, 10
magnesium, 79, 178
magnetic field, 89, 142, 143, 183, 233
magnetic moment, 142
magnetic resonance, 142, 165
magnitude, ix, 87, 90, 108, 111, 112, 113, 117, 227
majority, 14, 18, 82, 164, 194, 220
mammography, 162
manipulation, 84
manufacturing, ix, 8, 52, 58, 67, 127, 129, 130, 131, 135, 137, 140, 149, 159, 161, 162, 165
mapping, 200, 201, 202, 203
masking, 74
mass, 77, 112, 154, 184, 193, 194, 216, 284
materials science, 217
matrix, ix, x, 27, 60, 63, 76, 79, 80, 87, 109, 110, 112, 113, 159, 160, 177, 178, 181, 182, 183, 193, 198, 199, 200, 201, 202, 204, 205, 206, 207, 208, 210, 234, 272, 273, 275, 278, 282, 283, 289, 290
matter, viii, 51, 63, 134, 151, 184, 217, 222
measurement(s), 13, 33, 52, 58, 80, 89, 94, 114, 119, 130, 133, 134, 139, 141, 142, 143, 146, 161, 163, 217, 220, 221, 222, 229, 231, 235, 236, 245, 246
mechanical properties, viii, x, xi, 2, 71, 72, 76, 78, 79, 83, 154, 156, 177, 179, 183, 198, 216, 217, 218, 222, 225, 226, 228, 232, 233, 234, 235, 236, 239, 249, 265, 266
mechanical stress, 80

mechanical testing, 218
media, 92, 144, 151, 182, 251
median, 80, 82
medical, ix, 79, 127, 130, 134, 149, 162, 164
medicine, 129
melt(s), 52, 56, 63, 68, 129, 140, 144, 145, 251, 252, 253, 254, 258, 262, 268, 272, 273, 275, 278, 279, 280, 281, 282, 283, 284, 285, 286, 287, 288, 290
melting, 110, 129, 145, 147, 149, 151, 154, 221, 234, 251, 252, 254, 255, 258, 262, 263, 275, 285, 286, 288, 289, 290
melting temperature, 110, 129, 145, 262
membranes, vii, 1, 2, 5, 12, 20, 21, 22, 27, 249
memory, 78
MEMS, 222, 231
mercury, 21
metal complexes, 25
metal fragments, ix, 87, 123
metals, x, 110, 178, 215, 221, 222, 251, 291
meter, 33, 34
methanol, 6, 8
methodology, viii, 51, 52, 67, 68, 130
Mg^{2+}, 73, 161
microfabrication, 231
microhardness, 218, 233, 236, 239
micrometer, 54, 217, 240
microscope, 54, 94, 186, 222, 238, 240, 242
microscopy, 88, 90, 97, 110, 232
microstructure(s), vii, viii, x, xi, 31, 33, 34, 35, 40, 45, 49, 51, 56, 63, 71, 72, 74, 76, 79, 88, 157, 177, 178, 183, 184, 185, 190, 193, 194, 195, 200, 203, 204, 206, 207, 234, 251, 252, 253, 257, 258, 262, 263, 265, 266, 267, 269, 272, 275, 283, 284, 285, 288, 289, 290
microstructure features, 266
microwaves, 142
migration, 45, 46, 130, 147, 149, 206
missions, 163
mixing, 8, 202
modelling, 222
models, 52, 61, 77, 134, 222, 224, 229, 230
modules, 115
modulus, x, 78, 106, 215, 222, 223, 225, 226, 227, 228, 229, 230, 231, 232, 233, 234, 235, 236, 237, 245
moisture, viii, 9, 33, 71, 72, 78
molar ratios, 273, 274
mold, 136, 185, 203
mole, 116
molecular weight, 10
molecules, vii, 1, 10, 13, 14, 133, 221, 232
molybdenum, 251
Morocco, 51, 52
morphology, 14, 16, 155, 180, 186, 190, 192, 193, 195, 200, 203, 243, 245, 253, 286, 288
Moscow, 87, 124, 211
Moses, 167, 170, 172, 174, 175

N

N,N-Dimethylformamide, 10
Na^+, 40, 41
NaCl, 16, 200
nanocomposites, x, 13, 25, 110, 144, 178, 198, 210
nanocrystals, 154, 155, 159, 160
nanoindentation, x, 215, 217, 218, 220, 222, 225, 226, 227, 230, 235, 236, 237, 239
nanomaterials, 11
nanometer, 25, 88, 147, 217, 221, 226, 230
nanometer scale, 230
nanoparticles, x, 2, 8, 11, 12, 13, 14, 16, 25, 79, 178, 195, 198
nanorods, 14
Nanostructures, 11
nanotechnology, 150
Nd, 91, 94, 97, 98, 102, 103, 104, 135, 136, 138, 144
neglect, 98
NEMS, 222, 231
neodymium, 135
Netherlands, 185, 186
neutral, 11
neutrons, 158
nitrogen, 180, 198
noble gases, 221
nonequilibrium, 92
non-equilibrium thermal phonons (NP), viii, 87
normal distribution, 204
nuclear weapons, 163
nucleation, 76, 77, 78, 217, 281, 282, 283, 285, 286, 287
nuclei, 142, 143, 286, 287
nucleus, 17, 287

O

obstacles, 251
OH, 9, 77, 136
olefins, 19
opacity, 74, 78, 106
operations, 120
optical anisotropy, 103, 151, 156
optical microscopy, 218, 240
optical properties, 72, 102, 256
optimization, vii, ix, 1, 41, 88, 151, 153, 159, 161, 275, 276, 290

optimization techniques, vii, 1
oral cavity, 83
ores, 89, 106
organ, 163
organic compounds, vii, 1
oscillation, 98, 144
Ostwald ripening, x, 177, 184, 194, 209
overlap, 134, 275
oxidation, 4, 117, 233
oxygen, vii, 31, 36, 37, 44, 46, 47, 73, 75, 77, 114, 142, 147, 149, 150, 151, 180, 198, 236, 249

P

paints, 219
parallel, 108, 114, 284, 285, 286
parity, 133, 152
passivation, 232
peptide(s), 11
percolation, 206, 209
periodicity, 110, 182
permission, 4, 5, 6, 7, 8, 10, 12, 13, 14, 15, 16, 17, 18, 19, 20, 21, 22, 23, 24, 25, 26, 155, 157, 159
permit, 27, 221, 242
permittivity, 36, 40, 46
PET, 151, 152, 153, 163, 164, 175
pH, 2, 11, 13, 14
phase boundaries, 108
phase diagram, 40, 200, 263, 264, 265, 266, 270, 283
phase transformation, 73, 75, 76, 77, 184, 236, 239, 245, 275
phase transition temperatures, vii, 31, 36, 41
phase transitions, 129, 218
Philadelphia, 175
phonons, viii, 87, 88, 89, 90, 95, 98, 101, 105, 106, 109, 110, 111, 112, 113, 114, 115, 116, 117, 119, 120, 221
phosphate, 11, 12, 14, 15, 16
phosphorescence, 142
photodetector(s), 133
photoelectron spectroscopy, 97
photographs, 145, 254, 276
photoluminescence, 149, 155
photons, 128, 130, 132, 140, 141, 145, 161, 164
physical mechanisms, 165
physical phenomena, 128
physical properties, 88, 189, 291
physics, x, 128, 129, 131, 134, 161
piezoelectric properties, vii, 31, 32, 33, 45, 47, 49
PL spectrum, 152
plaque, 79
plastic deformation, ix, 16, 88, 99, 102, 122, 217, 218, 219, 220, 227, 235, 239, 240, 241, 245, 283
plasticity, viii, 51, 53, 68, 110, 217, 227, 244, 245
platform, 80
playing, 52
Pliocene, 52
point defects, 130, 131, 141, 164
polar, 33
polarization, vii, 31, 33, 37, 43, 89, 120
polonium, 128
polymer(s), vii, 1, 2, 3, 4, 5, 6, 8, 9, 10, 11, 12, 13, 14, 18, 19, 20, 21, 22, 25, 27, 183, 220, 222, 240
polymer solutions, 5
polymerization, 8
population, 219
porosity, 2, 5, 21, 22, 23, 27, 74, 98, 150, 180, 233, 235, 246, 255, 257, 258, 278, 280, 282
positron, 163
precipitation, 12, 17, 25, 180, 264, 273
preparation, x, 1, 2, 14, 18, 19, 25, 128, 129, 131, 135, 137, 138, 144, 156, 165, 177, 251, 268, 278
principles, xi, 156, 162, 163, 216
probability, 90, 92, 93, 163, 186, 203
probe, 97, 221, 222, 225, 233
production technology, 140
profit, 160
proliferation, 161
propagation, viii, 18, 19, 24, 27, 72, 74, 75, 77, 78, 87, 89, 90, 105, 112, 114, 116, 117, 119, 236, 275, 284
proportionality, 158
propylene, 2, 3, 5, 11, 16
prostheses, 72, 74, 78, 79
prosthesis, viii, 71
protons, 77
prototype, 88
purity, 33, 73, 137, 278
PVC, 80

Q

quality control, 217
quantification, 222
quantum dot(s), 25, 159, 164, 174
quartz, 52, 53, 55, 56, 57, 117, 221

R

radiation, 92, 99, 128, 130, 132, 134, 151, 158, 163, 164, 166, 174
radio, 163
radiography, 162
radium, 128

radius, 40, 73, 106, 140, 144, 217, 220, 223, 224, 225, 226, 227, 230, 234, 240, 279, 280, 281, 283
Raman spectra, 154, 239
ramp, 220, 231, 241
rare earth elements, 151
raw materials, 52, 195, 252
RE, 148, 149, 151, 156, 159
reactant(s), 252, 275, 283
reaction mechanism, 252, 290
reaction temperature, 254, 255, 258, 262, 268, 272, 275, 283
reactions, 5, 8, 9, 10, 11, 18, 19, 20, 24, 74, 200, 252, 275, 290
reagents, 7, 20
real time, 162
reality, 221, 225
recombination, 131, 133, 141, 142, 151, 160, 165
recombination processes, 142
reconstruction, 72, 108, 163, 179, 230, 240
recovery, 79, 227, 230, 243
recrystallization, 206
red shift, 148
refractive index, 74
regeneration, 5, 25
regression, 60
regrowth, 206
relaxation, ix, 43, 87, 99, 113, 218, 228
relaxation process(s), 113
remediation, 24
reproduction, 83
requirements, 75, 110, 162, 230
researchers, 11, 72, 78, 110, 217, 245
reserves, 72
resistance, viii, x, 2, 34, 71, 73, 74, 75, 82, 83, 94, 110, 132, 134, 177, 183, 219, 220, 230, 233, 241, 278
resolution, 100, 132, 134, 142, 144, 145, 147, 148, 150, 152, 153, 158, 159, 162, 163, 164, 208, 209, 218, 222, 226, 230
resonator, 161
resources, 68
response, viii, ix, x, 32, 51, 52, 58, 59, 61, 67, 68, 127, 130, 131, 132, 133, 134, 148, 149, 151, 153, 154, 159, 160, 178, 184, 215, 219, 225, 227, 232, 233, 234
response surface methodology (RSM), viii, 51
restoration, 74, 270
restorative material, 74
restrictions, ix
rhodium, 19
RIE, 185
rods, 15
room temperature, viii, 11, 16, 17, 18, 36, 41, 46, 47, 54, 71, 72, 73, 74, 76, 78, 92, 119, 129, 144, 154, 155, 161, 236
roughness, 72, 77, 135, 218, 231, 233, 234, 244, 245, 285
routes, 198
Royal Society, 247
rules, 152
Russia, 87, 124

S

salt concentration, 14, 16
Samsung, 31
saturation, 78
scandium, 151, 157
scanning electron microscopy, 20, 33, 80, 185, 253
scatter, 109
scattering, ix, 88, 89, 90, 94, 98, 99, 102, 103, 105, 106, 108, 109, 110, 111, 112, 113, 115, 117, 119, 120, 122, 123, 129, 130, 134, 135, 136
science, 2, 129
scope, 2
security, 164
sedimentation, 52
sediments, 52
seed, 235
segregation, 140, 264
selected area electron diffraction, 195
SEM micrographs, 13, 21, 22, 56, 190, 191, 192, 203
semiconductor(s), 25, 133, 154, 159, 161, 164, 181, 183
sensing, 222, 225, 226
sensitivity, 53, 133, 162, 230
sensors, 226, 241
September 11, 163
sequencing, 179
shape, vii, 1, 2, 34, 43, 59, 75, 78, 98, 99, 178, 179, 190, 218, 224, 227, 228, 229
shear, viii, x, 32, 33, 47, 49, 115, 215, 221, 225, 227, 232
shear strength, 232
shock, 163, 183
showing, 58, 178, 191, 192, 193, 196, 197, 208, 209, 232, 233, 236, 239, 260, 268, 269, 284, 285, 287, 288, 289
Si_3N_4, x, 177, 178, 179, 180, 183, 184, 185, 186, 190, 191, 192, 193, 194, 195, 197, 198, 199, 200, 202, 203, 204, 205, 206, 207, 209, 232, 258
signals, 104, 114, 115, 117, 118, 119, 151, 164
signs, 232
silanol groups, 19

silica, 2, 6, 7, 8, 10, 11, 12, 13, 14, 15, 18, 19, 20, 24, 25, 78, 79, 159, 180, 229, 231, 240, 241, 242, 244, 245
silicon, x, 154, 178, 179, 180, 181, 193, 210, 215, 222, 231, 233
silver, 25, 33, 234
simulation, 83, 152, 232
simulations, 217, 232
single crystals, ix, x, 88, 89, 90, 98, 99, 101, 103, 117, 119, 120, 122, 123, 127, 129, 131, 135, 144, 145, 148, 149, 150, 153, 158, 160, 162, 215, 235
SiO_2, 8, 19, 53, 72, 79, 98, 154, 160, 179, 195, 256, 257, 258, 259, 268, 269, 270, 272, 273, 274, 279
skeleton, 52
skin, 181
sodium, 16
software, 60, 80, 186, 245
solar cells, 233
sol-gel, vii, 1, 7, 9, 27, 120, 151, 155
solid phase, 194, 284, 286
solid solutions, ix, 87, 89, 103, 112, 113, 114, 115, 117, 122, 123, 148
solid state, 129, 135, 155, 156, 157, 161, 183
solidification, 158, 234, 251, 252, 253, 254, 258, 264, 270, 272, 275, 283, 284, 285, 286, 288, 289, 290
solubility, 4, 5, 180, 184, 195, 209
solution, 3, 4, 5, 6, 7, 8, 11, 14, 22, 25, 26, 33, 89, 107, 117, 135, 152, 156, 180, 184, 198, 199, 200, 217
South Africa, 72
Spain, 215
Spark Plasma Sintering (SPS), v, x, 177, 181
species, 11, 142, 282
specific heat, 112, 113
specific surface, 4, 11, 200
spectroscopy, 88, 100, 102, 113, 140, 141, 143, 158, 165, 168, 222, 233, 236, 253
spin, 133, 140, 142, 143, 152, 168
stability, 2, 79, 132, 134
stabilization, 74, 77, 122, 236
stabilizers, 73, 76, 77
standard deviation, 54, 61, 80, 82
standard error, 15, 16
state(s), ix, 10, 49, 88, 114, 117, 130, 131, 133, 141, 143, 147, 148, 151, 154, 155, 156, 160, 165, 194, 233, 234, 275
steel, 252
stoichiometry, 129, 149, 155, 236
storage, 140, 165, 231, 240
stress, 21, 40, 72, 74, 75, 76, 77, 78, 79, 119, 183, 221, 225, 227
strontium, 156
structural changes, vii, 31, 217
structural defects, 25, 88, 103
structural variation, 158
substitution, 89, 148, 161, 192, 207
substrate(s), 2, 5, 10, 25, 26, 74, 161, 219, 221, 224, 231, 233, 243
sulfate, 16, 53
sulfur, 198
Sun, 29, 211, 248
superconductor, 234
superplasticity, 183
suppression, 144, 145, 147
surface area, 2, 4, 21, 23, 205
surface energy, 281, 282
surface layer, 180, 181, 287
surface modification, 77
surface properties, 233
surface tension, 183
susceptibility, 76, 77
suspensions, 25
swelling, 75
Switzerland, 212
symmetry, vii, 31, 117, 120, 142, 143, 150, 156, 223
synthesis, vii, ix, x, xi, 1, 11, 18, 19, 24, 25, 27, 76, 87, 88, 94, 107, 110, 114, 122, 123, 155, 174, 177, 178, 183, 252, 253, 254, 255, 256, 260, 262, 265, 266, 268, 270, 271, 272, 275, 276, 277, 278, 279, 280, 282, 283, 284, 290
synthetic polymers, 11

T

Taiwan, 177, 185, 210
target, 2, 61, 114
technical support, 246
techniques, vii, x, 1, 2, 27, 78, 128, 134, 140, 143, 162, 183, 215, 218, 234, 235, 236, 239, 245
technological progress, 129
technology(s), ix, x, 1, 72, 88, 90, 103, 127, 128, 129, 131, 137, 140, 144, 150, 159, 163, 165, 177, 181, 235
teeth, 74, 80
TEM, 26, 186, 190, 191, 192, 193, 194, 195, 196, 197, 207, 208, 209, 210
temperature dependence, ix, 33, 36, 41, 47, 49, 87, 89, 90, 92, 94, 101, 103, 110, 111, 112, 113, 115, 117, 119, 122
tension, 74, 75
TEOS, 160
terrorist activities, 164
testing, 80, 81, 217, 225, 227
tetraethoxysilane, 10
tetragonal phase, viii, 47, 71, 73, 75, 76, 77, 236

texture, vii, 8, 234, 287
thermal aging, viii, 71, 80, 82
thermal degradation, 10
thermal energy, 221
thermal expansion, 73, 189, 197
thermal quenching, 155
thermal stability, 2
thermodynamic equilibrium, 194
thermoluminescence, 141, 144, 165
thin films, 219, 220, 230, 235
three-dimensional representation, 64, 65, 66, 67
time resolution, 153
tin, 9
tissue, 2, 74, 162, 163
titania, 2, 8
titanium, x, 20, 74, 198, 199, 200, 201, 202, 203, 204, 205, 206, 207, 209, 210, 215
toluene, 19
tooth, 2, 72
topographic characterization, 240
toxicity, 32, 74
tracks, 220, 240, 241, 242, 243
trafficking, 164
transfection, 25
transformation(s), viii, 51, 52, 72, 73, 4, 75, 76, 77, 78, 79, 82, 128, 130, 161, 179, 180, 195, 218, 236, 239, 284
transistor, 206
transition temperature, vii, viii, 31, 32, 36, 41, 47, 49
transmission, 18, 82, 88, 90, 92, 93, 152, 186
Transmission Electron Microscopy, 213
transparency, 93, 94, 122, 138, 147, 150, 155, 162
transport, viii, 87, 88, 89, 110, 117, 123, 130, 132, 133, 165, 183, 193, 194, 233
transport processes, 88
transportation, 284
treatment, 2, 11, 14, 16, 18, 20, 25, 80, 147, 180, 232, 270
trial, viii, 51, 135
tribology, 217, 219, 246
tumors, 162
tungsten, 32, 122, 251, 252
tunneling, 160, 203, 206
tunneling effect, 203
twinning, ix, 88, 100, 102, 103, 122, 219
twins, 235

U

UK, 246, 253
Ukraine, 72
ultrasound, 32, 94, 97, 120
uniform, 27, 60, 136, 190, 252, 264, 272
United, 72, 158
United States, 72, 158
universal gas constant, 116
uranium, 128
USA, 33, 80
UV, 26, 132, 133, 141, 152, 156, 157

V

vacancies, vii, 31, 36, 46, 113, 123, 149, 198, 234, 236
vacuum, 6, 94, 103, 136, 138, 155, 182, 187, 188, 189, 191, 192, 193, 196, 197, 201, 202, 203, 204, 205, 208, 209, 217, 233, 252, 282
valence, 130, 131, 156, 157, 160
Valencia, 215
validation, 61
vapor, viii, 158
variables, 52, 58, 59, 60, 82, 83
variations, 43, 44, 45, 56, 89, 275
vehicles, 2, 50
velocity, 89, 92, 112, 115, 120, 216, 220, 231, 268, 275, 278, 279, 280, 281, 282
vibration, 221
Vickers hardness, 186, 228
Vickers indentation hardness, 228
viscosity, 4, 5, 180, 207, 258, 268, 279, 280, 282, 283
visualization, 163, 240
volatility, 129
volume component, 288, 289

W

waste, 50
wastewater, 2
water, viii, 2, 5, 8, 11, 12, 13, 51, 52, 54, 56, 59, 61, 63, 71, 72, 76, 77, 78, 80, 82, 83, 136, 232, 233, 246
water absorption, viii, 51, 52, 54, 56, 59, 61, 63
water purification, 2, 12
water vapor, 77
wave number, 93
wave propagation, 93
wave vector, ix, 87, 90, 91, 95
wavelengths, 89, 99
wear, 110, 183, 219, 220, 221, 231, 232, 233, 240, 241, 242, 243, 245
welding, 234
wells, 46
wetting, 184, 194
windows, 219

working conditions, 233
World War I, 128
worldwide, 77

X

X-ray diffraction (XRD), 17, 34, 40, 41, 52, 53, 54, 94, 97, 120, 253
X-ray diffraction data, 120
XRD, 33, 36, 185, 190, 192, 199, 201, 253, 255, 256, 257, 259, 260, 261, 263, 265, 266, 267, 268, 269, 270, 271, 274, 286, 287

Y

yield, 11, 12, 19, 75, 132, 140, 146, 147, 148, 150, 152, 153, 154, 155, 156, 157, 158, 159, 161, 164, 198, 209, 227
yttrium, ix, 73, 79, 89, 113, 135, 136, 151
yttrium-stabilized zirconium (YSZ), ix

Z

zinc, 161
zinc oxide, 161
zirconia, viii, x, 2, 71, 72, 73, 74, 75, 76, 77, 78, 79, 82, 83, 216, 218, 236, 245
zirconia abutments, viii, 71
zirconium, ix, 73
ZnO, 133, 159, 161, 164